Ethernet
Passive Optical
Networks

Ethernet Passive Optical Networks

Glen Kramer

McGraw-Hill

New York Chicago San Francisco Lisbon London Madrid
Mexico City Milan New Delhi San Juan Seoul
Singapore Sydney Toronto

The McGraw·Hill Companies

Cataloging-in-Publication Data is on file with the Library of Congress.

1 2 3 4 5 6 7 8 9 0   DOC/DOC   0 1 0 9 8 7 6 5

ISBN 0-07-144562-5

The sponsoring editor for this book was Stephen S. Chapman, the editing supervisor was David E. Fogarty, and the production supervisor was Pamela A. Pelton. It was set in Century Schoolbook by Digital Publishing Solutions. The art director for the cover was Anthony Landi.

Printed and bound by RR Donnelley.

This book was printed on recycled, acid-free paper containing a minimum of 50% recycled, de-inked fiber.

Figures 7.3, 8.10, 8.12, 8.14-8.18, 8.20, 8.21, and 8.23-8.27 are reprinted with permission from IEEE Standard 802.3ah(TM)-2004,"Information Technology—Telecommunications and Information Exchange between Systems—Local and Metropolitan Area Networks," Copyright 2004, by IEEE. The IEEE disclaims any responsibility or liability resulting from the placement and use in the described manner.

McGraw-Hill books are available at special quantity discounts to use as premiums and sales promotions, or for use in corporate training programs. For more information, please write to the Director of Special Sales, McGraw-Hill Professional, Two Penn Plaza, New York, NY 10121-2298. Or contact your local bookstore.

To Irina and Anton

Contents

Part 2:   EPON Architecture

Part 3:   System-Level Issues

Part 4:   EPON Performance

Preface

Passive optical networks (PONs) have been considered as a solution for the subscriber access network for quite some time, even before the Internet spurred bandwidth demand. One of the first papers describing telephony PON was published by British Telecom researchers in 1988. However, only during the past decade has the optical technology matured enough to make PON-based subscriber access network a reality.

Several alternative architectures for PON-based access networks have been standardized by several standards bodies, one of the main differentiating factors being the choice of the bearer protocol. Currently, standardized specifications exist for ATM-based PON (APON and BPON), gigabit-capable PON utilizing generic framing procedure (GPON), and Ethernet PON (EPON).

Among these PON specifications, EPON is the only access network architecture that traces its ancestry not to public communications, but to private enterprise data networks. EPON is standardized by the IEEE 802.3 work group—a group that rules in the LAN world, but has not ventured much beyond it. The Ethernet in the First Mile project was its first such attempt – a reconnaissance mission into yet unknown territory. EPON took industry by storm, advancing from raw idea to an industry-wide standard in five short years. But will EPON enjoy the same proliferation in access networks that its predecessors have had in corporate LANs? Time will tell.

Motivation for This Book

The technical and economic merits of EPON were recognized by the IEEE Standards Association when, in 2001, it authorized formation of the IEEE Ethernet in the First Mile (EFM) task force and approved EPON architecture as one of the objectives of the task force. The EFM task force completed its charter in June 2004, culminating in ratification of the IEEE 802.3ah standard.

In looking at the completed standard, most task force members realize now that EPON architecture is the most significant departure from traditional Ethernet architecture in the entire 30-year Ethernet history. The architectural decisions made by the task force may not be obvious to readers of the standard unless alternative solutions are reviewed or some background of task force discussions is presented. Of course, the formal and concise text of the standard does not provide either of these. As a result, many task force members suggested that a book expla- ining EPON architecture would be of benefit to telecommunications professionals.

An additional motivation for writing this book was the fact that, unlike the areas of responsibility for other standards, IEEE 802.3 only covers a small portion of a communications system, encompassing only physical and media access control layers. Such issues as security, protection, dynamic bandwidth allocation, quality of service, and utilization, are considered out-of-scope for the standard. This book will investigate several such "left-out" issues and provide a comprehensive analysis of state-of-the-art solutions.

Organization of the Book

This book is divided into four parts.

Part 1 provides an overview of PON-enabling technologies, introduces the history of EPON development, and compares EPON to alternative solutions, such as ATM-based PONs and PONs utilizing the generic framing procedure.

Part 2 provides an overview of a portion of IEEE 802.3ah standard relevant to EPONs.

It is important to emphasize that this book is not intended as a substitution for the standard published by IEEE. Even though, with permission from IEEE, this book reproduces several of the most important state machines, we do not adhere to the paramount level of formality bestowed by the standard. Instead, the book focuses on explaining the ideas behind the EPON specification, where necessary giving specific examples, looking at alternative solutions, or divulging the task force discussions that have led to a particular decision.

The IEEE 802.3ah overview in this book is based on the version of the standard as it was approved by the IEEE Standards Association in June 2004. However, several problems with the specification were identified after the official approval of the standard. Whenever we describe a state machine or a part of the spec with such a mistake, we alert the reader by using the following icon.

> ⚠️
>
> The text in this box points reader's attention to a serious error in the current version of the IEEE 802.3ah standard. It is expected that future revisions of the standard will fix this problem.

The standard is an evolving document. The IEEE 802.3 periodically produces new revisions of the standard by integrating approved amendments and corrigenda.

In Part 3 we investigate several system-level issues which were considered out-of-scope by the IEEE 802.3ah standard. In Chapter 10 we describe an encryption method optimized for EPON. In Chapter 11 we investigate the effectiveness of various path protection schemes in EPON.

Part 4 investigates performance of EPON. It is, to a significant degree, based on author's research conducted at UC Davis Networks Research Lab.

Chapter 12 investigates the various bandwidth overheads associated with EPON, such as optical overhead, framing (encapsulation) overhead, scheduling overhead, and error correction overhead. In Chapter 13 we look at the efficiency of the discovery process in EPON. Then, we turn our attention to various scheduling and bandwidth allocation schemes and their impact on EPON's ability to support applications and services with diverse requirements. In Chapter 14 we consider the simple scheme with a statically allocated bandwidth. When EPONs first emerged, several companies were advocating for and intended to build EPONs based on static allocation of bandwidth.

Chapter 15 introduces the reader to a simple dynamic bandwidth allocation scheme called interleaved polling with adaptive cycle time (IPACT). Comparing the performance of IPACT to a static scheme presented in the previous chapter illustrates the severe penalties imposed by the static resource allocation on bursty network traffic.

EPON is expected to be a truly converged network, supporting voice communications, standard and high-definition video, video-conferencing, real-time and near-real-time transactions, and data traffic. To support this multitude of applications, EPON must guarantee appropriate performance for each such application. In Chapter 16, we examine how EPON can provide differentiated services by combining IPACT with strict (exhaustive) priority scheduling which is a default scheduling algorithm specified in IEEE 802.1D.

Even though providing differentiated services is what most vendors have in mind when claiming that their EPON equipment supports quality of service, this, in many respects, is not sufficient. In

Chapter 17 we take a more formal look at EPON scheduling objectives. We argue that providing service guarantees and fairness, scalability, and isolation requires much more than what priority-based schemes can achieve.

We also find that the above objectives present a conflicting set of requirements. On one hand, the scheduling algorithm should be able to allocate resources to each subscriber individually in order to guarantee SLA and maintain fairness among all the consumers. Hierarchical schedulers cannot be used because they don't guarantee fairness to the consumers located in different groups. On the other hand, sending control messages to each end consumer is not feasible due to excessive bandwidth consumption and so a hierarchical structure is required.

In Chapter 18, we present fair queuing with service envelopes (FQSE) —an algorithm that successfully achieves both conflicting goals: it uses hierarchical control (i.e., each node receives control messages only from its immediate children), yet is maintains fairness among all the consumers located across different groups.

Acknowledgments

I am thankful to Gerry Pesavento and JC Kuo, the founders of Alloptic, Inc and later, the founders of Teknovus, Inc., who were willing to spend scarce resources of a start-up company to create and disseminate new knowledge. A new network architecture conceived by them became the focus of my research. I am grateful to them for giving me the chance to try my ideas in practice and for encouraging me to write this book.

With deep gratitude I thank my adviser Dr. Biswanath Mukherjee. His extensive knowledge, experience, and exceptional ability to find new approaches for seemingly unsolvable problems were pivotal in my research.

I would like to thank my many colleagues at UC Davis Networks Research Lab: Dr. Narendra Singhal (now at Microsoft), Dr. Amitabha Banerjee, Dr. Shun Yao, Dr. Keyao Zhu, Dr. Canhui (Sam) Ou (now at SBC), Dr. Jian Wang (now at Florida International University), and others. I am privileged to be part of this group and benefited immensely from the many interesting discussions and collaborations we had.

Various parts of the draft of this book were reviewed by Eric Lynskey and Swapnil Bhatia, both from Inter-Operability Laboratory at University of New Hampshire. All Eric's comments where dead-on and all were incorporated into the final manuscript. I thank Swapnil for pointing out some inaccuracies in mathematical derivation in Chapter 13 and suggesting a better approach.

I thank the editorial team at McGraw-Hill: editorial director Steve Chapman, editing manager David Fogarty, and copy editor Patti Scott for their trust and patience, and for being the examples of professionalism and perfectionism.

Last, but not least, I am obliged to my family: my wife Irina, my son Anton, my parents Asya and Boris, and my brother Pavel. Without their support, encouragement, and help I would not have been able to complete this work.

Glen Kramer

Ethernet
Passive Optical
Networks

Overview of Access Network Architectures

Chapter

1

Introduction

The past decade has witnessed significant development in the area of optical networking. Such advanced technologies as *dense wavelength division multiplexing* (DWDM), optical amplification, optical path routing (wavelength cross-connect), *wavelength add-drop multiplexer* (WADM), and high-speed switching have found their way into the *wide-area networks* (WANs), resulting in a substantial increase of the telecommunications backbone capacity and greatly improved reliability.

At the same time, enterprise networks almost universally converged on 100 Mbps Fast Ethernet architecture. Some mission-critical *local-area networks* (LANs) even moved to 1000 Mbps rates, courtesy of a new Gigabit Ethernet standard recently adopted by the *Institute of Electrical and Electronics Engineers* (IEEE).

An increasing number of households have more than one computer. Home networks allow multiple computers to share a single printer or a single Internet connection. Most often, a home network is built using a low-cost switch or a hub that can interconnect 4 to 16 devices. Builders of new houses now offer an option of wiring a new house with a *category-5* (CAT-5) cable. Older houses have an option of using existing phone wiring, in-house power lines, or an evermore popular wireless network, based on the IEEE 802.11 standard. Different flavors of this standard can provide up to 11 Mbps bandwidth or up to 54 Mbps bandwidth, with distance being a tradeoff. Whether it is a wireless or wire-line solution, home networks are essentially miniature LANs providing high-speed interconnection for multiple devices.

These advances in the backbone, enterprise, and home networks coupled with the tremendous growth of Internet traffic volume have accentuated the aggravating lag of access network capacity. The "last mile" still remains the bottleneck between high-capacity LANs and the backbone network.

1.1   Existing "Broadband" Solutions

The most widely deployed "broadband" solutions today are *digital subscriber line* (DSL) and *cable modem* (CM) networks. Although they are improvements compared to 56 kbps dial-up lines, they are unable to provide enough bandwidth for emerging services such as *video-on-demand* (VoD), interactive gaming, or two-way video conferencing.

1.1.1   Digital subscriber line

DSL uses the same twisted pair as telephone lines and requires a DSL modem at the customer premises and *digital subscriber line access multiplexer* (DSLAM) in the *central office* (CO). The basic premise of the DSL technology is to divide the spectrum of the line into several regions with the lower 4 kHz being used by *plain old telephone service* (POTS) equipment, while the higher frequencies are being allocated for higher-speed digital communications. There are four basic types of DSL connections.

The *basic* digital subscriber line is designed with *integrated services data network* (ISDN) compatibility in mind. It has 160 kbps symmetric capacity and affords users with either 80 or 144 kbps of bandwidth, depending on whether the voice circuit was supported or not.

The *high-speed digital subscriber line* (HDSL) is made compatible with a T1 rate of 1.544 Mbps. The original specification required two twisted pairs. Later a single-line solution was standardized.

The *asymmetric digital subscriber line* (ADSL) is the most widely deployed flavor of DSL. It uses one POTS line and has an asymmetric line speed. In the downstream direction, the line rate could be in the range of 750 kbps to 1.5 Mbps on the loops of 15,000 ft. On shorter loops, the rate can be as high as 6 Mbps. In the upstream direction, the rate could be in the range of 128 to 750 kbps. The actual rate is chosen by the ADSL modem based on line conditions and anomalies.

Finally, the *very high-speed digital subscriber line* (VDSL) can have a symmetric or an asymmetric line speed. It achieves much higher speed than HDSL or ADSL, but operates over much shorter loops. The rates could range from 13 Mbps for 4500-ft loops to 52 Mbps over 1000-ft loops.

1.1.2 ADSL

Among all the listed variants of DSL, the ADSL is the most common solution. The ADSL technology was developed in the early 1990s at Stanford University by a research group headed by Prof. John Cioffi. This research was funded by Bellcore. Later, Cioffi founded Amati Communications Corp., a company that has built the first ADSL modem.

While the maximum ADSL transmission capacity is 1.5 Mbps, in reality it could go much lower depending on the line conditions. Twisted-pair wires admit a number of impairments, most significant of which are crosstalk, induced noise, bridged taps, and impulse noise. To cope with such impairments, the ADSL employs a multicarrier modulation approach known as *discrete multitone* (DMT). A DMT system transmits data on multiple subcarriers in parallel. DMT adapts to line conditions by varying the bit rate on each subcarrier channel. A good channel may carry as many as 15 bps/baud, while a really noisy channel may carry no data at all.

The asymmetric nature of the ADSL was prompted by observation of user traffic at the time. While the downstream traffic volume was a result of downloading large files and web pages, the upstream traffic primarily consisted of short commands, http requests, and server log-in queries. Consequently, the ADSL adopted a 10:1 ratio of the downstream bandwidth to the upstream bandwidth, with AT&T even advocating for as high as a 100:1 ratio.

It is interesting to note that the highly asymmetric nature of the traffic is a thing of the past. New and emerging applications tend to skew the ratio toward greater symmetry. Such applications as video conferencing or data file repositories (storage-area networks) require symmetric bandwidth in both directions. A big impact on traffic symmetry can be attributed to peer-to-peer applications, such as Napster. It was reported that current ratio of downstream to upstream traffic is approximately 1.4:1 [Ree03].

1.1.3 Community antenna television (CATV) networks

The *community antenna television* (CATV) networks were originally designed to deliver analog broadcast TV signals to subscriber TV sets. Following this objective, the CATV networks adopted a tree topology and allocated most of its spectrum for downstream analog channels. Typically, CATV is built as a *hybrid fiber coax* (HFC) network with fiber running between a video head end or a hub to a curbside optical node, and the final drop to the subscriber being coaxial cable (Fig. 1.1). The

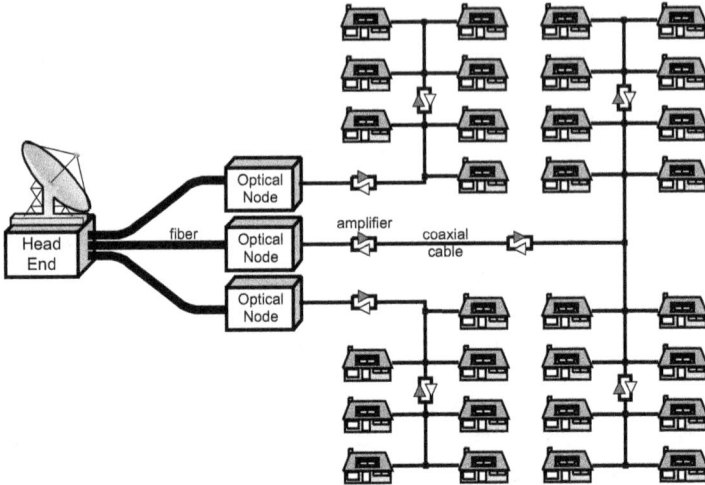

Figure 1.1   Hybrid fiber coax architecture.

coaxial part of the network uses repeaters (amplifiers) and tap couplers to split the signal among many subscribers.

Faced with the competition from telecom operators in providing Internet services, cable television companies responded by integrating data services over their HFC cable networks. This integration required replacing downstream-only amplifiers used for analog video with bidirectional amplifiers enabling an upstream data path. Also a medium access protocol had to be deployed to avoid collisions of upstream data transmitted by multiple subscribers concurrently.

The major limitation of CATV architecture for carrying modern data services is a consequence of the fact that this architecture was originally designed only for broadcast analog services. Out of a total cable spectrum width of about 740 MHz, the 400 MHz band is allocated for downstream analog signals, and the 300 MHz band is allocated for downstream digital signals. Upstream communications are left with about a 40 MHz band or about 36 Mbps of effective data throughput per optical node. This very modest upstream capacity is typically shared among 500 to 2000 subscribers, resulting in frustratingly low speed during peak hours.

1.2   Traffic Growth

Data traffic is increasing at an unprecedented rate. Sustainable traffic growth rate of over 100 percent per year has been observed since 1990. There were periods when a combination of economic and technological factors resulted in even larger growth rates, e.g., the 1000 percent

TABLE 1.1   Annual Growth of DSL Subscribers Worldwide

Year	DSL subscribers
1999	882,000
2000	7,768,900
2001	18,813,700
2002	35,897,700
2003	63,840,000

increase per year in 1995 and 1996 [CO01]. This trend is likely to continue in the future.

There are quantitative and qualitative components to the traffic growth. The quantitative component relates to the increasing number of Internet users. For example, the number of DSL customers increased by 25 million in 2003, according to a recent *DSL Forum* press release [DSL04]. Table 1.1 shows the growth of global DSL subscribers.

The qualitative increase in traffic volume is a consequence of changed behavioral patterns of subscribers. The enhanced users' experience leads them to spend more time online. Market research shows that, after upgrading to a broadband connection, users spend about 35 percent more time online than before [BB01]. Broadband users are also more likely to use bandwidth-intensive applications and services such as video-streaming and peer-to-peer applications. SBC Communications reported that 64 percent of DSL users downloaded video compared to only 36 percent of dial-up users. Another trend contributing to increased Internet usage is telecommuting. More and more subscribers telecommute several days a week, and they perform the same tasks at home as in the office. Accessing large amounts of information on corporate servers, exchanging large documents, and conducting online presentations require the same network performance at home as is available on corporate LANs.

Voice traffic is also growing, but at a much slower rate of approximately 8 percent annually. According to most analysts, the volume of data traffic has already surpassed that of voice traffic. While this event has passed unnoticed by the general population, the network operators realized now that their networks, designed and built to carry voice traffic, are not scalable and not efficient for the data traffic. Data packets of variable lengths incur significant processing and transmission overhead when being fragmented to fit in fixed-size slots or cells. The burstiness of data traffic warrants the bandwidth overprovisioning and, therefore, very low utilization.

Neither DSL nor cable modems can keep up with increased demand. Both technologies are built on top of existing communication infrastructure not optimized for data traffic.

While ADSL provides significantly more bandwidth compared to an analog dial-up modem, it is well shy of being considered "broadband," in that it cannot support emerging voice, data, and video applications. In addition, the physical area that one CO can cover with DSL is limited to distances of less than 18,000 ft (5.5 km), which covers approximately 60 percent of potential subscribers. And even though, to increase DSL coverage, *remote DSLAMs* (R-DSLAMs) may be deployed closer to subscribers, in general, network operators do not provide DSL services to subscribers located more than 12,000 ft from a CO due to increased costs of deployment and maintenance [ANS01].

In cable modem networks, only a few *radio-frequency* (RF) channels are dedicated for data, while the majority of bandwidth is tied up servicing legacy analog video.

Most network operators have come to the realization that a new, data-centric solution is necessary, one which is inexpensive, simple, scalable, and capable of delivering bundled voice, data, and video services to an end-user over a single network. This new architecture will be optimized for *Internet protocol* (IP) data traffic, which is a prevalent communication protocol today.

1.3   Evolution of the "First Mile"

The *first mile*? For telecommunications operators, the access portion of the network always remained the "last mile," unambiguously reflecting its peripheral location in the grand scheme of all telecom things. The last mile was a virtual telecom backyard, unattended, always hidden from views, and not worthy of showing to a houseguest or a traveling Wall Street analyst, for that matter. It is not surprising that the last mile has never received proper attention or sufficient investment.

This dereliction and unsatisfied subscriber demand for new services attracted a new breed of players—companies traditionally involved in data communications. These companies envision a global networking environment with multiple services such as telephony, video, and data, and many different flavors of each service are carried in the digitized format over a single network by a single protocol. For these companies the access network becomes just the first mile in the datacom expansion into the telecom world. Understandably, the networking community has renamed this network segment to the *first mile*, to symbolize its priority and importance.[1]

---

[1] Ethernet in the First Mile Alliance was formed in December 2001 by Alloptic, Cisco Systems, Elastic Networks, Ericsson, Extreme Networks, Finisar, Intel, NTT, and World Wide Packets.

The first mile connects the service provider central offices to businesses and residential subscribers. Also, referred to as the *subscriber access network*, or the *local loop*, it is the network infrastructure at the neighborhood level.

1.3.1 Fiber-to-the-premises

The existing "broadband" solutions are unable to provide enough bandwidth for emerging services such as video-on-demand (VoD), interactive gaming, or two-way video conferencing.

To alleviate bandwidth bottlenecks, optical fibers and thus optical nodes are penetrating deeper into the first mile. This trend is present in both DSL and cable TV worlds. In DSL-based access networks, many remote DSLAMs deployed in the field use fiber-optic links to the central offices. In cable TV networks, optical curbside nodes are deployed close to the subscribers.

The next wave of access network deployments promises to bring fiber all the way to the office or apartment buildings or individual homes. Unlike previous architectures, where fiber is used as a feeder to shorten the lengths of copper and coaxial networks, these new deployments use optical fiber throughout the access network. New optical fiber network architectures are emerging that are capable of supporting *gigabit per second* (Gbps) speeds, at costs comparable to those of DSL and HFC networks.

1.3.2 Next-generation subscriber access network

Optical fiber is capable of delivering bandwidth-intensive, integrated voice, data, and video services at distances beyond 20 km in the subscriber access network. A straightforward way to deploy optical fiber in the local access network is to use a *point-to-point* (P2P) topology, with dedicated fiber runs from the CO to each end-user subscriber (Fig. 1.2a). While this is a simple architecture, in most cases it is cost-prohibitive because it requires significant outside fiber plant deployment as well as connector termination space in the local exchange. Considering N subscribers at an average distance L km from the CO, a P2P design requires $2N$ transceivers and $N \times L$ total fiber length (assuming that a single fiber is used for bidirectional transmission).

To reduce fiber deployment, it is possible to deploy a remote switch (concentrator) close to the neighborhood. This will reduce the fiber consumption to only L km (assuming negligible distance between the switch and customers), but will actually increase the number of transceivers to $2N + 2$, as there is one more link added to the network

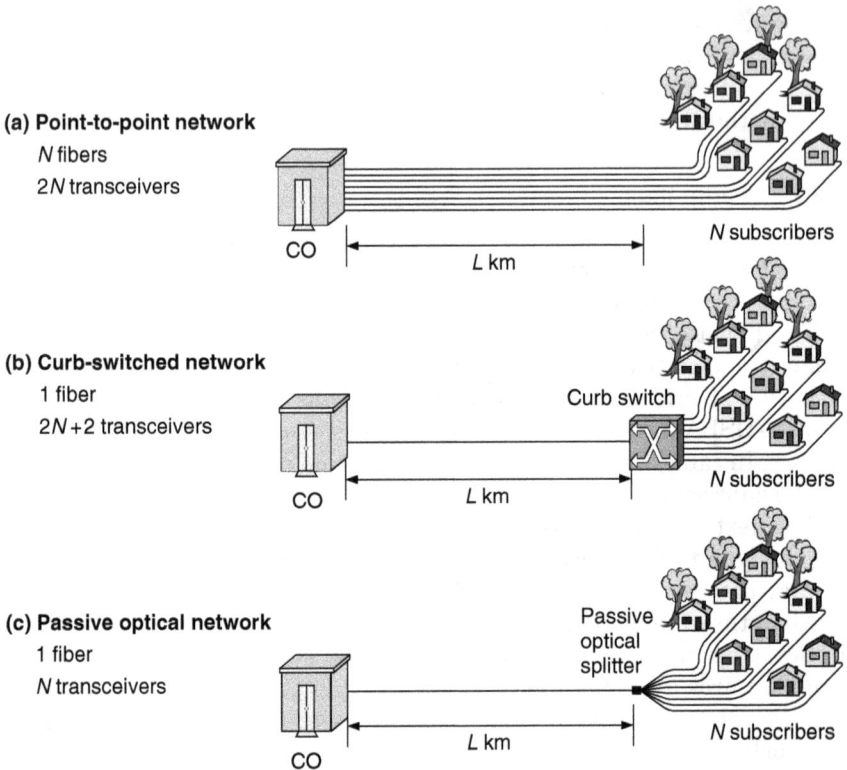

Figure 1.2   Fiber-to-the-home (FTTH) deployment scenarios.

(Fig. 1.2*b*). In addition, a curb-switched network architecture requires electric power as well as backup power at the curb switch. Currently, one of the most significant operational expenditures for *local exchange carriers* (LECs) is that of providing and maintaining electric power in the local loop.

Therefore, it is logical to replace the hardened (environmentally protected) active curbside switch with an inexpensive passive optical splitter. A *passive optical network* (PON) is a technology viewed by many as an attractive solution to the first-mile problem [PK99, Lun99]; a PON minimizes the number of optical transceivers, CO terminations, and fiber deployment.

A PON is a *point-to-multipoint* (P2MP) optical network with no active elements in the signals' path from source to destination. The only interior elements used in PON are passive optical components, such as optical fiber, splices, and splitters. An access network based on a single-fiber PON only requires $N + 1$ transceivers and L km of fiber (Fig 1.2*c*).

1.3.3 PON is the best candidate

PON technology is getting more and more attention by the telecommunication industry as the "first mile" solution. Advantages of using PON for local access networks are numerous:

- PON allows for longer distances between central offices and customer premises. A PON-based local loop can operate at distances of up to 20 km, which considerably exceeds the maximum coverage afforded by DSL.

- PON minimizes fiber deployment in both the local exchange and local loop. Only one strand of fiber is needed in the trunk, and only one port per PON is required in the central office. This allows for a very dense CO equipment and low power consumption.

- PON provides higher bandwidth due to deeper fiber penetration. While the *fiber-to-the-building* (FTTB), *fiber-to-the-home* (FTTH), or even *fiber-to-the-PC* (FTTPC) solutions have the ultimate goal of fiber reaching all the way to customer premises, *fiber-to-the-curb* (FTTC) may be the most economical deployment today.

- As a point-to-multipoint network, PON allows for downstream video broadcasting. Multiple wavelength overlay channels can be added to PON without any modifications to the terminating electronics.

- PON eliminates the necessity of installing multiplexers and demultiplexers in the splitting locations, thus relieving network operators from the gruesome task of maintaining them and providing power to them. Instead of active devices in these locations, PON has passive components that can be buried in the ground at the time of deployment.

- PON allows easy upgrades to higher bit rates or additional wavelengths. Passive splitters and combiners provide complete path transparency.

References

[ANS01]   *Access Network Systems: North America—Optical Access. DLC and PON Technology and Market Report,* Report RHK-RPT-0548, RHK Telecommunication Industry Analysis, San Francisco, June 2001.

[BB01]   *Broadband 2001. A Comprehensive Analysis of Demand, Supply, Economics, and Industry Dynamics in the U.S. Broadband Market,* JP Morgan Securities, Inc., New York, April 2001.

[DSL04]   "2003 Global DSL Subscriber Chart (March 2, 2004)," *DSL Forum*, March 2004. Available at http://www.dslforum.org/ PressRoom/2003_GlobalDSLChart_3.2.2004.pdf.

[CO01]   K. G. Coffman and A. M. Odlyzko, "Internet growth: Is there a "Moore's Law" for data traffic?" *Handbook of Massive Data Sets*, J. Abello, P. M. Pardalos, and M. G. C. Resende, eds., Kluwer, 2001.

[Lun99]   B. Lung, "PON architecture 'futureproofs' FTTH," *Lightwave*, vol. 16, no. 10, pp. 104–107, September 1999.

[PK99]   G. Pesavento and M. Kelsey, "PONs for the broadband local loop," *Lightwave,* vol. 16, no. 10, pp. 68–74, September 1999.

[Ree03]   D. Reed, *Copper Evolution,* Federal Communications Commission, Technological Advisory Council III, Washington, DC, April 2003. Available at http://www.fcc.gov/oet/tac/TAC_- III_04_17_03/Copper_Evolution.ppt.

Overview of PON Enabling Technologies

This chapter presents an overview of major PON building blocks, such as wavelength division multiplexing, single-mode optical fiber, optical splitters/combiners, and burst-mode optical transceivers. Although all these technologies have been known for quite some time, they remained *curiosity technologies*, mostly used in academic research and in occasional optical testbeds. These technologies have matured considerably over the last decade, and in the late 1990s we witnessed an explosive proliferation of optical backbone and *metropolitan-area networks* (MANs). Yet, the access network remained outside the realm of optical networking. The access network aggregates traffic from a relatively small number of subscribers, compared to metro or regional networks. Any capital expenditures in access networks are amortized over a much smaller number of paying customers, making these networks extremely cost-sensitive. The high cost of optical components made it economically unjustifiable to deploy optical technologies in the access networks.

Only during the last several years has the combination of mature technologies, decreased cost of components, and positive experience gained with backbone optical networks made building an optical access network a realistic enterprise. In 1999 Verizon economists proclaimed that deploying fiber in the local loop became cheaper than deploying copper cables [Har00]. In 2003 three major network operators in the United States—Verizon, Bell South, and SBC Communications—announced a *request for proposals* (RFP) for ATM-based PON equipment. Earlier the same year, *Nippon Telegraph and Telephone* (NTT), a Japanese major carrier, announced RFP for Ethernet-based PON.

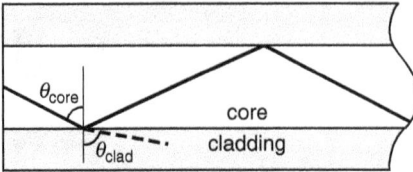

Figure 2.1   Light propagation in fiber.

Below we consider the basic features of the main PON enablers. For a more detailed study of these technologies and components, we recommend [Muk97] and [RS98].

2.1   Optical Fiber

Optical fiber is a truly remarkable communications medium. Fiber is insensitive to, and does not cause itself, any electromagnetic interference. It allows optical signal propagation over extremely large distances with very little distortion. Experiments have been carried out in which optical signals propagated hundreds of kilometers without any amplification. Optical fiber has tremendous information-carrying capacity. A single fiber strand can carry up to 50 THz of bandwidth, which is almost two orders of magnitude higher than the estimated traffic on all U.S. backbone links combined [Odl04].

2.1.1   Light propagation in fiber

A fiber is a very thin filament of glass, which acts as a waveguide. Glass can be characterized by its *refractive index n*, which is the ratio of speed of light in vacuum to the speed of light in glass ($n = c_{\text{vacuum}}/c_{\text{glass}}$).

A fiber consists of two layers of glass, as shown in Fig. 2.1. The inner layer is referred to as the *core,* and the outer layer is called the *cladding.* These two layers of glass are manufactured to have different refractive indices, with the refractive index of the core being larger than the refractive index of the cladding ($n_{\text{core}} > n_{\text{clad}}$). Such fiber is referred to as *step-index* fiber. Fiber can be manufactured with *graded index*, a term referring to a gradual change of the refractive index between core and cladding.

When the light traveling through the core reaches the boundary between the core and the cladding at an angle θ_{core}, it may continue propagating through the cladding at an angle θ_{clad}. The relationship between the two angles is described by a formula known as Snell's law:

$$n_{\text{clad}} \sin \theta_{\text{clad}} = n_{\text{core}} \sin \theta_{\text{core}} \tag{2.1}$$

A phenomenon of *total internal reflection* occurs when $\theta_{\text{clad}} > \pi/2$. According to Eq. (2.1), to achieve the total internal reflection, the angle of incidence θ_{core} should be

$$\theta_{\text{core}} > \sin^{-1}\left(\frac{n_{\text{clad}}}{n_{\text{core}}}\right) \tag{2.2}$$

The minimum value of θ_{core} resulting in total internal reflection is called a *critical angle*. The graded-index fiber typically has a smaller critical angle.

Due to the effects of total internal reflection, light propagates in fiber with very little loss. Signal attenuation in fiber is about 0.45 dB/km for the region from 1270 nm to 1370 nm, and only about 0.2 dB/km for the range of 1430-nm to 1610-nm wavelengths.

In optical fiber, light propagates partially in the core and partially in the cladding. Often, fiber manufacturers, instead of specifying n_{core} and n_{clad}, just specify an *effective refractive index* n_{eff} of the fiber ($n_{\text{clad}} < n_{\text{eff}} < n_{\text{core}}$) and a refractive index difference $\Delta = (n_{\text{core}} - n_{\text{clad}})/n_{\text{clad}}$.

2.1.2   Single-mode fiber versus multimode fiber

Although the total internal reflection may occur at any angle θ_{core}, which is greater than the critical angle, the light may not necessarily propagate for all these angles due to the destructive interference between the incident and reflected light [Muk97]. The angles at which light can propagate are called *modes* of the fiber.

Currently, two types of fiber exist, a *multimode* fiber and a *single-mode* fiber. As the names suggest, the multimode fiber permits light propagation in multiple modes (under multiple angles), whereas the single-mode fiber allows only one mode, called the *fundamental mode*, to propagate. As Fig. 2.2 illustrates, the single-mode fiber has a significantly smaller core diameter compared to the multimode fiber.

The number of modes m supported by the multimode fiber is a function of core diameter d and wavelength λ and is approximately equal to

$$m \approx \frac{\pi^2 d^2}{2\lambda^2}\left(n_{\text{core}}^2 - n_{\text{clad}}^2\right) \tag{2.3}$$

Equation (2.3) is an approximation, which is valid only for cases when the core diameter is much larger than the wavelength of the signal. The step-index fiber becomes a single-mode fiber (SMF) if its core diameter is less than D_{SMF}:

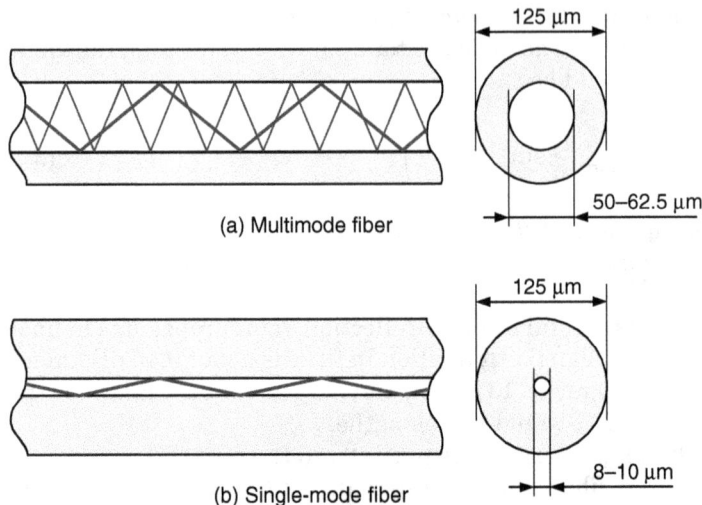

(a) Multimode fiber

(b) Single-mode fiber

Figure 2.2   Comparison of multimode and single-mode fibers.

$$D_{SMF} \approx \frac{2.405 \times \lambda}{\pi n_{eff} \sqrt{2\Delta}} \tag{2.4}$$

Alternatively, if the fiber diameter is given, then Eq. (2.4) allows us to calculate the cutoff wavelength λ_{cutoff}. For any wavelength $\lambda > \lambda_{cutoff}$, the fiber permits propagation of the fundamental mode only; i.e., it becomes a single-mode fiber.

We mentioned above that optical fiber has two usable regions of wavelength: the 1270- to 1370-nm band and the 1430- to 1610-nm band. Single-mode fiber typically places the cutoff wavelength slightly below the lower usable wavelengths, somewhere near 1260 nm.

The standard single-mode fiber typically has $n_{eff} \approx 1.467$ and $\Delta \approx 0.3$ percent. Plugging these values and $\lambda_{cutoff} = 1260$ nm into the formula (2.4), we can calculate the maximum core diameter for fiber to carry a single mode in both usable wavelength bands. We find that for the given parameters the core diameter should not exceed $D_{SMF} \approx 8.5$ μm.

2.1.3 Modal dispersion

The multimode fiber was commercialized first. Its large core diameter facilitated the use of inexpensive large-area light sources and connectors. However, signal transmission in multimode fibers suffers from an impairment called *modal dispersion*. Modal dispersion refers to the widening of a pulse due to multiple modes propagating with unequal velocities. As shown in [RS98], the slowest mode is the one incident to the critical angle, and it takes time $T_{slow} = Ln_{core}^2 / cn_{clad}$ to propagate through a fiber of length L. The fastest mode travels along the axis of the core and takes only $T_{fast} = Ln_{core} / c$. The pulse widening, which is just the difference between T_{slow} and T_{fast} depends on the propagation distance. If the pulse widening exceeds one-half of a bit period, the adjacent bits may begin to interfere. The maximum distance for a given bit rate R can be calculated as follows:

$$L < \frac{c}{2Rn_{eff}\Delta} \tag{2.5}$$

For a graded-index fiber, the maximum distance can be calculated according to the following formula [RS98]:

$$L < \frac{4c}{Rn_{eff}\Delta^2} \tag{2.6}$$

In multimode fiber, the difference between the refractive index of the core and that of the cladding is higher than in single-mode fiber, typically $\Delta = 1.5$ percent. The value of the effective group refractive index $n_{eff} \approx 1.48$. Using the Eqs. (2.5) and (2.6), we can estimate the maximum fiber distance for various bit rates. For example, a 1 Gbps Ethernet link, which has an actual line rate of 1.25 Gbps, could support distances not exceeding 5.5 m. Using graded-index fiber, the maximum distance of a 1 Gbps Ethernet link can be extended to approximately 2.9 km.

Not surprisingly, the multimode fiber is used only for short transmission distances, and mostly the graded-index multimode fiber is used. Single-mode fiber is not affected by modal dispersion and can carry signals over much longer distances. Currently, only the single-mode fiber is considered for applications in access networks. Multimode fiber is used primarily for short transmission distances, such as central office links and intrapremises communications.

2.1.4 Chromatic dispersion

Another cause of pulse widening is the fact that different spectral components of the same signal propagate with different velocities. In other words, the refractive index of glass (silica) is frequency-dependent. This kind of dispersion is called *material dispersion*. Material dispersion depends on the spectral width of the signal. No laser can generate a signal with a single frequency. Even if such a signal could be generated, the process of modulating it introduces additional spectral components and causes its spectral width to increase. Therefore, any information-carrying signal will have nonzero spectral width.

Another component of chromatic dispersion is called *waveguide dispersion*, and it is caused by the light propagating partially in the core and partially in the cladding. The shape of the fiber can determine the proportion of light energy that is split between the core and the cladding. As the refractive index in the core is greater than the refractive index in the cladding, the signal propagating in the cladding will reach the receiver sooner than the signal propagating in the core, thus causing the received pulse to broaden, compared to a transmitted pulse.

The total value of chromatic dispersion is just the sum of the material and waveguide dispersions. For a standard single-mode fiber, the material dispersion component increases monotonically with the wavelength. The waveguide dispersion component monotonically decreases and is always negative. The sum of these two components is equal to zero for $\lambda \approx 1310$ nm. By altering the waveguide dispersion, a *dispersion-shifted* fiber can be manufactured with the zero-dispersion point shifted to the 1550-nm region.

2.2 Optical Splitters/Combiners

A PON employs a passive (not requiring any power) device to split the optical signal from one fiber into several fibers and, reciprocally, to combine optical signals from multiple fibers into one. This device is an optical coupler. In its simplest form, an optical coupler consists of two fibers fused together. Signal power received on any input port is split between both output ports. The power splitting ratio of a splitter can be controlled by the length of the fused region and therefore is a constant parameter.

The $N \times N$ couplers are manufactured by staggering multiple 2×2 couplers (Fig. 2.3) or by using planar waveguide technology.

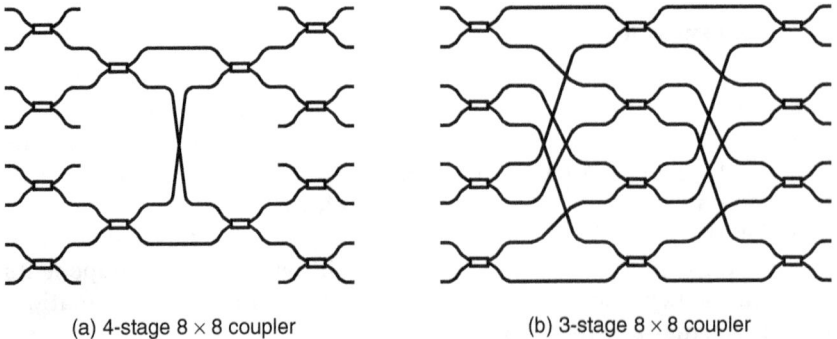

(a) 4-stage 8 × 8 coupler   (b) 3-stage 8 × 8 coupler

Figure 2.3   8 × 8 couplers created from multiple 2 × 2 couplers.

Couplers are characterized by the following parameters:

- *Splitting loss.* Power level at the coupler's output versus power level at its input, measured in decibels. For an ideal 2 × 2 coupler with equal power splitting, this value is 3 dB. Figure 2.3 illustrates two topologies for 8 × 8 couplers based on 2 × 2 couplers. In a four-stage topology (Fig. 2.3a), only one-sixteenth of the input power is delivered to each output. Fig. 2.3b shows a more efficient design, called *multistage interconnection network* [Muk97]. In this arrangement, each output receives one-eighth of the input power.

- *Insertion loss.* Power loss resulting from imperfections of the manufacturing process. Typically, this value ranges from 0.1 to 1 dB.

- *Directivity.* Amount of input power leaked from one input port to another input port. Couplers are highly directional devices with the directivity parameter reaching −40 to −50 dB.

Very often, couplers are manufactured to have only one input or one output. A coupler with only one input is referred to as a *splitter*. A coupler with only one output is called a *combiner*. Sometimes, 2 × 2 couplers are made highly asymmetric (with splitting ratios of 5/95 or 10/90). This kind of coupler is used to branch off a small portion of signal power, e.g., for monitoring purposes. Such devices are called *tap couplers*.

2.3 PON Topologies

Logically, the first mile is a *point-to-multipoint* (P2MP) network, with a CO servicing multiple subscribers. All transmissions in a PON are performed between an *optical line terminal* (OLT) and *optical network units* (ONUs) (Fig. 2.4). The OLT resides in the CO and connects the optical access network to the metropolitan-area network or *wide-area network* (WAN), also known as the backbone or long-haul network. The ONU is located either at the end-user location (FTTH and FTTB) or at the curb, resulting in *fiber-to-the-curb* architecture.

There are several multipoint topologies suitable for the access network, including tree, ring, and bus (Fig. 2.4). Using 1 × 2 optical tap couplers and 1 × N optical splitters, PONs can be flexibly deployed in any of these topologies.

In some critical deployments, the access network may require fast protection switching. This is achieved by providing several alternative, diversely routed paths between the OLT and ONUs. Path redundancy may be added to an entire PON's topology, or to only a part of the PON, say, the trunk of the tree or the branches of the tree. Protection switching is further discussed in Chap. 11.

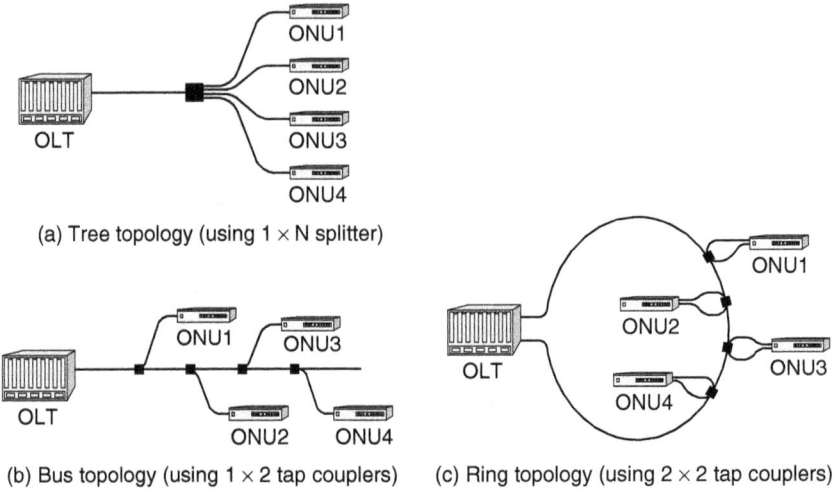

(a) Tree topology (using 1 × N splitter)

(b) Bus topology (using 1 × 2 tap couplers)   (c) Ring topology (using 2 × 2 tap couplers)

Figure 2.4   PON topologies.

2.4   Spectrum Sharing versus Time Sharing

In the downstream direction (from the OLT to ONUs), a PON is a point-to-multipoint network. The OLT typically has the entire downstream bandwidth available to it at all times. In the upstream direction, a PON is a multipoint-to-point network: multiple ONUs transmit all toward one OLT. The directional properties of a passive splitter/combiner are such that an ONU transmission cannot be detected by other ONUs. However, data streams from different ONUs transmitted simultaneously still may collide. Thus, in the upstream direction (from user to network), a PON should employ some channel separation mechanism to avoid data collisions and fairly share the trunk fiber channel capacity and resources.

2.4.1   WDMA PON

One possible way of separating the ONUs' upstream channels is to use a *wavelength division multiple access* (WDMA), in which each ONU operates on a different wavelength. While, from a theoretical perspective, it is a simple solution, it remains cost-prohibitive for an access network. A WDMA solution would require either a tunable receiver or a receiver array at the OLT to receive multiple channels. An even more serious problem for network operators would be wavelength-specific ONU inventory: instead of having just one type of ONU, there would be multiple types of ONUs differing in their laser wavelengths. Each ONU will have to use a laser with narrow and controlled spectral width, and thus will become more expensive. It would also be more problematic for an unqualified user to replace a defective ONU because a unit with the wrong wavelength may interfere with some other ONU in the PON. Using tunable lasers in ONUs may solve the inventory problem, but is too expensive at the current state of technology. For these reasons, a WDMA PON network is not an attractive solution in today's environment.

Several alternative solutions based on WDMA have been proposed, namely, *wavelength routed PON* (WRPON). A WRPON uses an *arrayed waveguide grating* (AWG) instead of wavelength-independent optical splitter/combiner.

In one variation, ONUs use external modulators to modulate the signal received from the OLT and send it back upstream. This solution, however, is not cheap either; it requires additional amplifiers at or close to the ONUs to compensate for signal attenuation after the round-trip propagation, and it requires more expensive optics to limit the reflections, since both downstream and upstream channels use the same wavelength. Also, to allow independent (nonarbitrated) transmission from each of N ONUs, the OLT must have N receivers—one for each ONU.

In another variation, ONUs contain inexpensive *light-emitting diodes* (LEDs) whose wide spectral band is sliced by the AWG on the upstream path. This approach still requires multiple receivers at the OLT. If, however, a single tunable receiver is used at the OLT, then a data stream from only one ONU can be received at a time, which in effect makes it a *time division multiple access* (TDMA) PON.

We refer readers to [RS98] for a detailed overview of these and other approaches.

2.4.2   TDMA PON

In a TDMA PON, simultaneous transmissions from several ONUs will collide when they reach the combiner. To avoid data collisions, each ONU must transmit in its own transmission window (timeslot). One of the major advantages of a TDMA PON is that all ONUs can operate on the same wavelength and be absolutely identical componentwise. The OLT will also need a single receiver. A transceiver in an ONU must operate at the full line rate, even though the bandwidth available to the ONU may be lower. However, this property also allows the TDMA PON to efficiently change the bandwidth allocated to each ONU by changing the assigned timeslot size, or even to employ statistical multiplexing to fully utilize the PON channel capacity.

In a subscriber access network, most of the traffic flows downstream (from network to users) and upstream (from users to the network), but not peer-to-peer (from a user to a user). Thus, it seems reasonable to

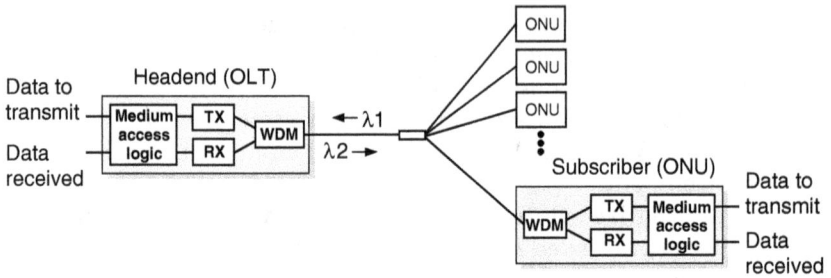

Figure 2.5   PON using a single fiber.

separate the downstream and upstream channels. A simple channel separation can be based on space division multiplexing (SDM), in which separate PONs are provided for downstream and upstream transmissions. To save optical fiber and to reduce the cost of repair and maintenance, a single fiber may be used for bidirectional transmission. In this case, two wavelengths are used: λ_1 for the upstream transmission and λ_2 for the downstream transmission (Fig. 2.5). The channel capacity on each wavelength can be flexibly divided between the ONUs by using time-sharing techniques.

Time-sharing appears to be the preferred method today for optical channel sharing in an access network, as it allows for a single upstream wavelength and a single transceiver in the OLT, resulting in a cost-effective solution.

2.5   Burst-Mode Transceivers

Due to possibly unequal distances between the OLT and the ONUs, optical signal attenuation in the PON may not be the same for each ONU. The power level received at the OLT may be different for each timeslot (called the *near-far problem*). Figure 2.6 depicts power levels of four timeslots received by the OLT from four different ONUs in a TDMA PON. As shown, one ONU's signal strength is lower at the OLT, possibly due to its longer distance. If the receiver in the OLT is adjusted to properly receive a high-power signal from a close ONU, it may mistakenly read 1s as 0s when receiving a weak signal from a distant ONU. In the opposite case, if the receiver is trained on a weak signal, it may read 0s as 1s when receiving a strong signal.

To detect the incoming bit stream properly, the OLT receiver must adjust its 0–1 threshold at the beginning of each received burst, a procedure known as *automatic gain control* (AGC). The mode of operation in which a signal arrives at the receiver in bursts with varying power levels is called *burst-mode reception.*

Tek Run: 50.0MS/s   Sample   **DPO**

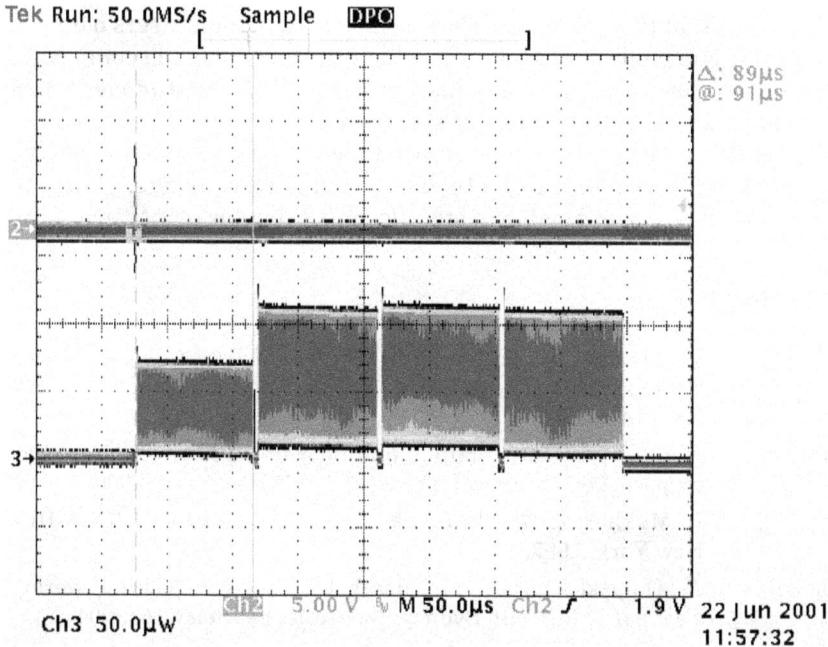

Ch3 50.0μW

△: 89μs
@: 91μs

Ch2   5.00 V   M 50.0μs   Ch2 ∫   1.9 V   22 Jun 2001
11:57:32

Figure 2.6   Illustration of near-far problem in a TDMA PON: a snapshot of power levels received from four ONUs.

Some PON-based access architectures attempt to reduce the necessary dynamic range of the AGC circuitry by forcing the ONUs to adjust their transmitter powers, such that power levels received by the OLT from all the ONUs become nearly equal. This method is not particularly favored by equipment designers, as it makes the ONU hardware more complicated, requires special signaling protocol for feedback from the OLT to each ONU, and most importantly, may degrade the performance of all ONUs to that of a most distant unit.

In addition to performing AGC, burst-mode receivers must be able to acquire phase and frequency lock on an incoming signal. This procedure is called *clock and data recovery* (CDR). The ability to perform AGC and CDR very quickly is paramount for a receiver to operate in burst mode. A burst-mode receiver is necessary only in the OLT. The ONUs receive a continuous bit stream (data or idles) sent by the OLT and do not need to readjust the receiver gain quickly.

In a TDMA PON, it is not enough just to disallow ONUs from sending any data between the assigned timeslots. The problem is that, even in the absence of data, lasers generate *spontaneous emission noise*. Spontaneous emission noise from several ONUs located close to the OLT can easily obscure the signal from a distant ONU (*capture* effect).

To avoid the capture effect, ONUs must shut down their lasers between timeslots. The mode of operation in which the laser is being completely turned off between the transmissions is called *burst-mode transmission*. Because a laser cools down when it is turned off, and warms up when it is turned on, its emitted power may fluctuate at the beginning of a transmission. In burst-mode transmitters, it is important that the laser be able to stabilize quickly after being turned on.

References

[CFL99]   W. Circiora, J. Farmer, and D. Large, *Modern Cable Television Technology: Video, Voice, and Data*, Morgan Kaufmann, San Francisco, 1999.

[Har00]   S. Hardy, "Verizon staffers find fiber-to-the-home cheaper than copper," *Lightwave*, vol. 17, no. 134, p. 1, December 2000.

[Muk97]   B. Mukherjee, *Optical Communication Networks*, McGraw-Hill, New York, 1997.

[Odl04]   A. Odlyzko, "Crisis and mythology in the telecom world," *CommsDaySummit*, Sydney, Australia, February 16, 2004.

[RS98]   R. Ramaswami and K. N. Sivarajan, *Optical Networks, A Practical Perspective*, Morgan Kaufmann, San Francisco, 1998.

Access Network Architectures Based on TDMA PON

TDMA PONs have been considered for the subscriber access network for quite some time, even before the Internet spurred bandwidth demand. One of the first papers describing PON was published by British Telecom researchers in 1988 [SH+88]. This chapter explains the history and various flavors of PON. Several alternative architectures for TDMA PON-based access networks have been standardized by several standards bodies, one of the main differentiating factors being the choice of the bearer protocol. Currently, standardized specifications exist for ATM-based PON, PON utilizing *generic framing procedure* (GFP), and Ethernet PON.

3.1 ATM PON

In 1995, seven network operators formed the *Full Service Access Network* (FSAN) initiative with a goal of creating unified specification for broadband access networks. Current FSAN membership is comprised of the following major network operators: Bell Canada (Canada), BellSouth (USA), Bezeq (Israel), British Telecommunications (UK), Chunghwa (Taiwan), Deutsche Telekom (Germany), France Telecom (France), Korea Telecom (Korea), NTT (Japan), Qwest Communications (USA), SBC (USA), SingTel (Singapore), Telecom Italia (Italy), Telstra (Australia), and Verizon Communications (USA). In addition, almost 30 equipment vendors are members of FSAN.

FSAN members developed a specification for a PON-based optical access network that uses ATM as its layer-2 protocol. Such systems were called *APON*, an abbreviation for ATM PON. The name APON was later replaced with *BPON* for *broadband PON*. The name change was reflective of the system's support of broadband services such as Ethernet access, video distribution, and virtual private line (VPL)/ leased line services.

FSAN is not a standardization body. In 1997 the FSAN group submitted the BPON specification proposals to *International Telecommunications Union—Telecommunication Standardization Sector* (ITU-T) for a formal ratification. Consequently, over the period of several years, ITU-T approved the following series of BPON-related recommendations:

G.983.1   *Broadband Optical Access Systems Based on Passive Optical Networks (PON).* This document was approved in 1998 and specifies the optical physical layer of the APON/BPON system.

G.983.2   *ONT Management and Control Interface Specification for B-PON.* Adopted in 1999, this recommendation defines a common *optical network terminal* (ONT) management control interface.

G.983.3   *A Broadband Optical Access System with Increased Service Capability by Wavelength Allocation.* This document provided specification for wavelength overlay to support additional services such as analog video. This recommendation was ratified in 2001.

G.983.4   *A Broadband Optical Access System with Increased Service Capability Using Dynamic Bandwidth Assignment (DBA).* Also adopted in 2001, this recommendation describes mechanisms necessary to support dynamic bandwidth allocation among multiple ONTs in the same PON.

G.983.5   *A Broadband Optical Access System with Enhanced Survivability.* This document was adopted in 2002. It specifies the protection switching mechanisms for BPON.

G.983.6   *ONT Management and Control Interface Specifications for B-PON System with Protection Features.* This document defines the extensions for control interface necessary for management of protection switching functions at the ONT. This recommendation was adopted in 2002.

G.983.7   *ONT Management and Control Interface Specification for Dynamic Bandwidth Assignment (DBA) B-PON System.* This document defines the extensions for control interface necessary for management of DBA functions at the ONT. This recommendation was adopted in 2001.

G.983.8   *B-PON OMCI Support for IP, ISDN, Video, VLAN Tagging, VC Cross-connections and Other Select Functions.* This recommendation specifies the extensions for control interface necessary for management of various extended services at the ONT. It was approved in 2003.

The original Recommendation G.983.1 specified BPON architecture with symmetric 155 Mbps upstream and downstream bit rates. This specification was amended in 2001 to allow asymmetric 155 Mbps upstream and 622 Mbps downstream transmissions as well as symmetric 622 Mbps transmission.

3.2 Ethernet PON

In January 2001, the IEEE formed a study group called *Ethernet in the First Mile* (EFM). This group was chartered with extending existing Ethernet technology into subscriber access area, focusing on both residential and business access networks. Keeping with the Ethernet tradition, the group set the goal of providing a significant increase in performance while minimizing equipment, operation, and maintenance costs. Ethernet PONs became one of the focus areas of EFM.

Ethernet PON (EPON) is a PON-based network that carries data traffic encapsulated in Ethernet frames as defined in the IEEE 802.3 standard [802.3]. It uses a standard 8b/10b line coding (8 data bits encoded as 10 line bits) and operates at standard Ethernet speed of 1 Gbps. Where possible, EPON utilized the existing 802.3 specification, including the usage of existing 802.3 full-duplex *media access control* (MAC).

3.2.1 Why Ethernet?

In 1995, when the FSAN initiative was started, there were high hopes of the ATM becoming the prevalent technology in the LAN, MAN, and WAN. However, since that time, Ethernet technology has leapfrogged ATM. Ethernet has become a universally accepted standard, with over 320 million ports deployed worldwide, offering staggering economies of scale [Cla00]. High-speed gigabit Ethernet deployment is widely accelerating, and 10-gigabit Ethernet products are becoming available. Ethernet, which is easy to scale and manage, is gaining new ground in MAN and WAN. Given that more than 95 percent of enterprise LANs and home networks use Ethernet, it becomes clear that ATM PON may not be the best choice to interconnect two Ethernet networks.

One of the shortcomings of the ATM is its high overhead for carrying variable-length IP packets, which are the predominant component of the Internet traffic. Below we compare the overheads imposed by Ethernet frame encapsulation and ATM cell encapsulation.[1]

---

[1] It should be mentioned that a 1-Gbps Ethernet link has an actual line rate of 1.25 Gbps. The rate increase is performed by the *physical coding sublayer* (PCS), which preconditions the line signal by encoding each user byte with a 10-bit codeword. The increased rate is only visible at the physical layer; MAC, MAC interface, and MAC clients operate at the stated Ethernet rate of 1 Gbps. Therefore, this encoding generally is not considered an overhead.

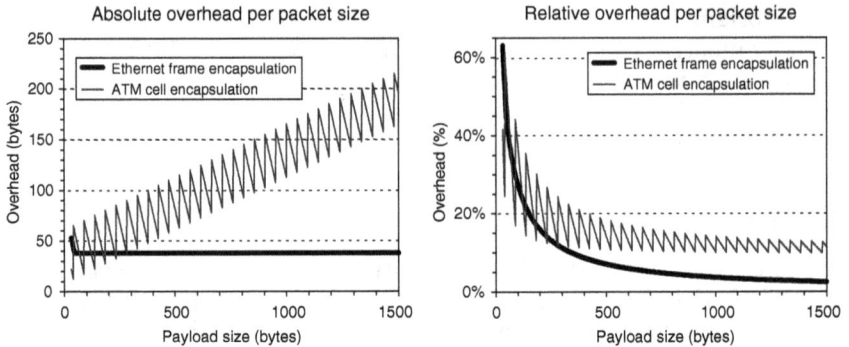

Figure 3.1   Comparison of Ethernet framing overhead and ATM cell tax.

The Ethernet encapsulation (framing) overhead is a result of adding an 8-byte frame preamble, 14-byte Ethernet header, and 4-byte FCS field to a payload comprised of users' data. Additionally, at least 12-byte minimum *interframe gap* (IFG) should be left between two adjacent frames.[2] Thus, the absolute overhead per single frame is constant and equal to 38 bytes[3] (see Fig. 3.1). This encapsulation overhead is not specific to EPON, but is a property common to all Ethernet networks.

In ATM networks, the user's data units, such as IP datagrams, should be broken into multiple cells. The ATM encapsulation overhead (also known as the *cell tax*) is comprised of multiple cell headers, 8-byte *ATM adaptation layer 5* (AAL5) trailer, and variable-size padding. The AAL5 trailer is needed for proper IP datagram reassembly, and the padding is used to fill any remaining portion of the last cell. As seen in Fig. 3.1, the ATM encapsulation overhead depends on the payload size and is considerably higher than the Ethernet overhead.

The average value of the encapsulation overhead depends on the distribution of packet (payload) sizes. These distributions have been reported in the literature; they generally have a trimodal[4] shape with main modes corresponding to 40-byte, 576-byte, and 1500-byte payload sizes.

For a particular IP datagram size distribution obtained in an access network [SG01], the Ethernet frame encapsulation overhead equals 7.42 percent and is significantly lower than the ATM cell encapsulation overhead of 13.22 percent. The improved efficiency is just one of the

---

[2] IFG is specified as a 96-ns time interval, which is equal to 12 byte-transmission times in 1 Gbps (1000BASE-X) Ethernet.

[3] Short payloads are padded to a minimum length of 46 bytes. This padding also contributes to the Ethernet encapsulation overhead and is counted in our calculations.

[4] Term *mode* refers to a frequently encountered packet size.

advantages of using variable-size Ethernet frames to carry variable-size IP packets.

Another of the ATM's shortcomings is the fact that a dropped or corrupted ATM cell will invalidate an entire IP datagram. However, the remaining cells carrying the portions of the same IP datagram will propagate further, thus consuming network resources unnecessarily.

And finally, perhaps most importantly, the ATM did not live up to its promise of becoming an inexpensive technology—vendors are in decline and manufacturing volumes are relatively low. ATM switches and network cards are significantly more expensive than Ethernet switches and network cards [Cla00].

On the other hand, Ethernet looks like a logical choice for an IP data-optimized access network. Newly adopted *quality-of-service* (QoS) techniques have made Ethernet networks capable of supporting voice, data, and video. These techniques include full-duplex transmission mode, prioritization, and *virtual LAN* (VLAN) tagging. Ethernet is an inexpensive technology, which is ubiquitous and interoperable with a variety of legacy equipment. It is not surprising that Ethernet is poised to become the architecture of choice for next-generation subscriber access networks.

3.3 GFP PON

In the presence of ever-growing traffic volume and the emergence of 1 Gbps EPON specification, the FSAN group has realized the need for architecture capable of higher bit rate and improved efficiency for data traffic. However, the physical layer specification adopted for BPON made it quite difficult to achieve upstream bit rates above 622 Mbps. ATM-based PON also is inefficient for IP traffic. To overcome these limitations, in 2001, FSAN undertook a new effort to specify a PON system operating at bit rates exceeding 1 Gbps. The group has directed its attention to the *generic framing procedure* [G.7041] as a means to improve efficiency, while allowing a mix of variable-size frames and ATM cells.

Based on FSAN recommendations, in 2003–2004, ITU-T has approved the new *gigabit-capable PON* (GPON) series of specifications. These specifications are known as ITU-T Recommendations G.984.1, G.984.2, and G.984.3:

G.984.1   *Gigabit-capable Passive Optical Networks (GPON): General Characteristics.* This recommendation describes the general characteristics of a gigabit-capable PON system such as architecture, bit rates, protection, and security.

G.984.2   *Gigabit-capable Passive Optical Networks (GPON): Physical Media Dependent (PMD) Layer Specification.* This recommendation specifies GPON operation at line rates of 1.25 Gbps and 2.5 Gbps in the downstream (central office to customer) direction and 155 Mbps, 622 Mbps, 1.5 Gbps, and 2.5 Gbps in the upstream (customer to central office) direction.

G.984.3   *Gigabit-capable Passive Optical Networks (GPON): Transmission Convergence Layer Specification.* This document covers specifications for the *GPON transmission convergence* (GTC) frame; message; ranging method; operation, administration, and maintenance (OAM) functionality; and security.

3.4   Comparison of BPON/GPON and EPON Approaches

Both BPON and GPON architectures were conceived by the FSAN group, which is driven by major incumbent telecommunications operators. Most of the operators are heavily invested in providing legacy TDM services. Accordingly, both BPON and GPON are optimized for TDM traffic and rely on framing structures with a very strict timing and synchronization requirements.

In BPON, an upstream frame consists of 53 timeslots, where each timeslot is comprised of one ATM cell and 3 bytes of overhead. When two consecutive timeslots are given to different ONUs, these 3 bytes or approximately 154 ns of the overhead should be sufficient to shut down the laser in the first ONU, turn it on in the second ONU, and perform gain adjustment and clock synchronization at the OLT.

Similarly, very tight timing is specified for GPON. For example, in GPON with a 1.244 Gbps line rate, only 16-bit times (less than 13 ns) are allocated for the laser-on and laser-off times. Such short intervals require more expensive, higher-speed laser drivers at the ONU.

A very tight bound of 44-bit times (less than 36 ns) is allotted for the gain control and clock recovery. In many cases, the dynamic range of the signal arrived from different ONUs will require a longer AGC time than the allotted overhead (guard interval). To reduce the range of necessary gain adjustment, BPON and GPON perform a power-leveling operation, in which the OLT instructs individual ONUs to adjust their transmitting power, so that the levels of signals received at the OLT from different ONUs are approximately equal.

The IEEE 802 work group has traditionally focused on enterprise data communication technologies. In EPON, the main emphasis was placed on preserving the architectural model of Ethernet. No explicit framing structure exists in EPON; the Ethernet frames are transmitted in bursts with a standard interframe spacing. The burst sizes and physical layer overhead are large in EPON. For example, the maximum AGC interval is set to 400 ns, which provides enough time to the OLT to adjust gain without ONUs performing the power-leveling operation. As a result, ONUs do not need any protocol and circuitry to adjust the laser power. Also, the laser-on and laser-off times are capped at 512 ns, a significantly higher bound than that of GPON. The relaxed physical overhead values are just a few of many cost-cutting steps taken by EPON.

Another cost-cutting step of EPON is the preservation of the Ethernet framing format, which carries variable-length packets without fragmentation. In contrast, both BPON and GPON break the packets into multiple fragments. BPON uses AAL5, discussed above, to break a packet into cells at the transmitting end and to reassemble multiple-cell payloads into a complete packet at the receiving end. GPON employs the *GPON encapsulation method* (GEM) to enable packet fragmentation. This method uses a complicated algorithm to delineate variable-size GEM segments and reconstruct the packets at the receiving device.

Several operators have deployed BPON systems; however, the foretold mass deployment and corresponding equipment cost reduction have never materialized. At the time of this writing, there are no announced GPON field trials, let alone commercially deployed systems. Given the level of complexity of the GPON or tight specification for various physical-layer parameters, it is very doubtful that the cost of GPON equipment can match that of an EPON.

References

[802.3]   *IEEE Standard for Information Technology—Telecommunications and Information exchange between systems—Local and metropolitan area networks—Specific Requirements.—Part 3: Carrier Sense Multiple Access with Collision Detection (CSMA/CD) Access Method and Physical Layer Specification*, ANSI/IEEE Std. 802.3-2002, 2002 edition. Available at http://standards.ieee .org/getieee802/download/802.3-2002.pdf.

[Cla00]   S. Clavenna, "Metro optical Ethernet," *Lightreading* (www.light-reading.com), November 2000.

[G.7041]   ITU-T Recommendation G.7041, *Generic Framing Procedure (GFP)*, in Series G: Transmission Systems and Media, Digital Systems and Networks, Telecommunication Standardization Sector of ITU, December 2003.

[SG01]   D. Sala and A. Gummalla, "PON functional requirements: Services and performance," presented at IEEE 802.3ah meeting in Portland, OR, July 2001. Available at http://grouper.ieee.org/groups/802/3/efm/public/jul01/presentations/sala_1_0701.pdf.

[SH+88]   J. R. Stern, C. E. Hoppitt, D. B. Payne, M. H. Reeve, and K. A. Oakley, "TPON—A passive optical network for telephony," Fourteenth European Conference on Optical Communication (ECOC'88), vol.1, pp. 203–206, Brighton, UK, September 1988.

Emergence of Ethernet PON

In 2003, the Ethernet protocol celebrated its 30th birthday. All these years, it was adapting and evolving to become a very inexpensive, ubiquitous networking protocol, as we know it today. The idea of passive optical topology also has been around for almost two decades. Yet, strangely enough, the two never met—that is, until the year 1999. What follows is the author's personal account of emerging Ethernet PON architecture. Numerous searches to find any earlier references to EPONs did not bring any results, yet it is entirely possible that the EPON idea materialized somewhere else even earlier that that.

In the summer of 1999, the author joined a small start-up founded by two young entrepreneurs, Gerry Pesavento (CEO) and JC Kuo (CTO). The company name was Alloptic, Inc., reflective of its plans to build an all-optical PON-based digital local loop. The product being built aimed at delivering standard T1/E1 and T3/E3 circuits and employed a sophisticated framing and synchronization scheme. It did not use the ATM as its bearer protocol; rather it could be classified as SONET/SDH PON.

In November 1999, the entire management team (that is, both Gerry and JC) visited the AT&T network planning division. The goal of the visit was to gather AT&T's requirements for user-side and network-side interfaces. The AT&T response was plain and simple: "We want only Ethernet interfaces."

On the flight back to California, JC had a rather disturbing thought: if equipment customers want only Ethernet interfaces, and since Ethernet is known for its relaxed timing requirements and variable-size frames, why would anyone care that inside the PON is this precise clock and framing. Wouldn't a system built specifically for Ethernet be as efficient, but much cheaper?

The following day, Gerry and JC called an all-hands meeting (the company at that time had a total of 12 hands) and announced that the very same day we would abandon the existing, almost-completed design and would start building Ethernet PON.

Alloptic began to actively promote the EPON idea. After raising the first round of funding and hiring additional workforce, the company began to design and build a prototype in the spring of 2000. In September, Alloptic demonstrated the first prototype EPON delivering data, voice, and video. In a 6-month period, the company developed field-programmable-gate-array- (FPGA) based controllers, bidirectional *single-mode fiber* (SMF) burst-mode transceivers, and software. In October, a demonstration was given to Cisco representatives. Cisco liked the idea, especially the fact that it was not an ATM PON.

In November 2000, IEEE announced a call for interest for a new study group, tentatively called Ethernet in the Last Mile, and later renamed as Ethernet in the First Mile. The group was to extend Ethernet into the subscriber access area. EPON technology seemed a very good match for this group, and Pesavento went to the meeting to sell the EPON idea to IEEE 802. Among a dozen presentations given at the call-for-interest meeting, this was the sole presentation mentioning EPON. The EPON idea was accepted very well, and by the following meeting, the list of EPON supporters had grown to include such large equipment vendors as Nokia, Lucent, and Cisco. EPON has been planted in EFM objectives.

4.1 EPON Standardization

Following a very successful call-for-interest meeting, IEEE formed a study group, which had its first meeting in January 2001. EFM quickly became one of the most participated in study groups.

The EFM study group focused on bringing Ethernet to the local subscriber loop, considering the requirements of both residential and business access networks. While at first glance this may appear to be a simple task, in reality the requirements of local exchange carriers are vastly different from those of private enterprise networks for which Ethernet has been designed. To "evolve" Ethernet for local subscriber networks, the EFM study group concentrated on four primary standards areas:

1. Ethernet over copper

2. Ethernet over *point-to-point* (P2P) fiber

3. Ethernet over *point-to-multipoint* (P2MP) fiber (also known as EPON)

4. *Operation, administration, and maintenance* (OAM)

The EFM's emphasis on both copper and fiber specifications, optimized for the first mile and glued together by a common OAM system, was a particularly strong vision, as it allowed a local network operator a choice of Ethernet flavors using a common hardware and management platform. In each of these areas, new physical layer specifications were discussed and adopted to meet the requirements of service providers, while preserving the integrity of Ethernet.

To progress with the project, the study group had to demonstrate that the envisioned architectures satisfy the following five criteria:

- Broad market potential
- Compatibility with 802 architecture, including bridging and *management information bases* (MIBs)
- Distinct identity, i.e., sufficient difference from other IEEE 802 standards
- Technical feasibility
- Economic feasibility

A convincing demonstration that P2MP architecture can meet the above benchmarks has been a major milestone on EPON's road to success. The presentation outlining how and why EPONs should be part of the IEEE 802.3 standard received overwhelming support from study group participants, including representatives from the following companies:[1] Agere, Alloptic, Atrica, Broadcom, Broadlight, Calix, Cisco, Corning, Dominet, E2O, Fiberhood, Fiberintheloop, Infineon, Intel, Lucent, Luminous, Minerva, Nokia, Nortel, OnePath, Optical Solutions, Passave, Pirelli, Quantum Bridge, Redback, Salira, Scientific Atlanta, Vitesse, and Zonu [EFM01]. As a result, the Project Authorization Request (PAR), which included P2MP architecture, was approved by the IEEE-SA Standards Board in September 2001, and consequently, EFM received a status of a task force, with a designation of 802.3ah.

---

[1] The IEEE rules say that group members participate in standard development as individuals, not as corporate representatives. However, usually, a vote from a participant "coincided" with a position taken by the employing company.

Open systems
interconnection (OSI)
reference model

IEEE 802.3
layering diagram

Open systems interconnection (OSI) reference model	IEEE 802.3 layering diagram
Application	Logical link control
	MAC control
Presentation	**Media access control (MAC)**
	Reconciliation sublayer (RS)
Session	Gigabit media independent interface (GMII)
Transport	
Network	**Physical coding sublayer (PCS)**
	Physical medium attachment (PMA)
Data link	**Physical medium dependent (PMD)**
Physical	Medium dependent interface (MDI)
	Medium

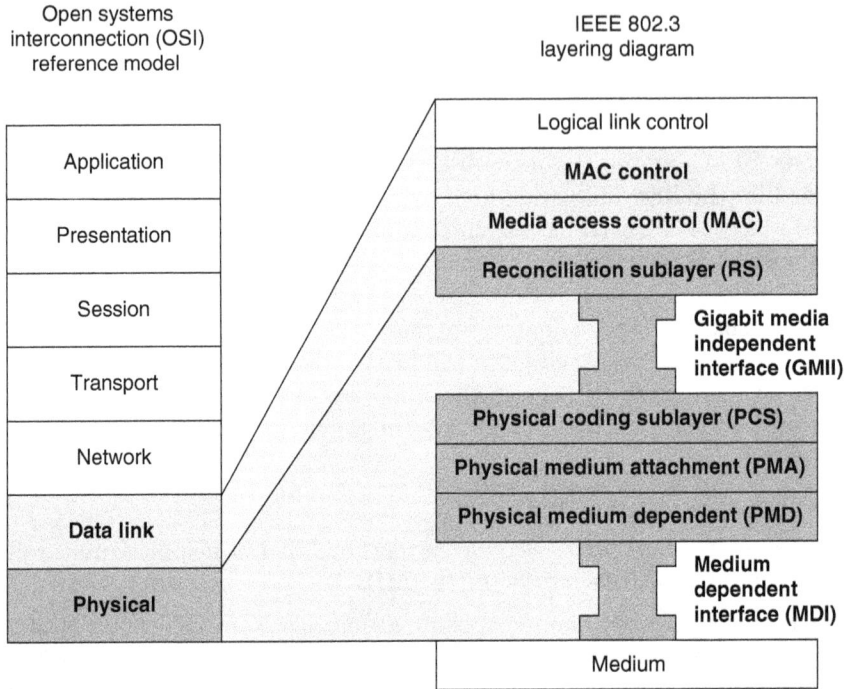

Figure 4.1   Relationship of IEEE 802.3 layering model to Open Systems Interconnection reference model.

4.1.1 Scope of work

The scope of IEEE 802.3 work is confined to two lower layers of the *Open Systems Interconnection* (OSI) reference model [OSI94]: physical layer and data link layer. Each of these layers is further divided into sublayers and interfaces. Figure 4.1 shows the sublayers and interfaces defined for Ethernet devices operating at 1 Gbps data rates.

IEEE 802.3 uses the following subdivision of the physical layer (from lower sublayer to higher):

Medium dependent interface (MDI) specifies the physical medium signals and the mechanical and electrical interface between the transmission medium and physical layer devices.

Physical medium dependent (PMD) sublayer is responsible for interfacing to the transmission medium. The PMD is located just above the MDI.

Physical medium attachment (PMA) sublayer contains the functions for transmission, reception, clock recovery, and phase alignment.

Physical coding sublayer (PCS) contains the functions to encode data bits into code-groups that can be transmitted over the physical medium.

Gigabit media independent interface (GMII) specifies an interface between a gigabit-capable MAC and a gigabit physical layer (PHY). The goal of this interface is to allow multiple Data Terminal Equipment (DTE) devices to be intermixed with a variety of gigabit-speed physical layer implementations.

Reconciliation sublayer (RS) provides mapping for the GMII signals to the media access control service definitions.

The data link layer consists of the following sublayers (from lower to higher):

Media access control sublayer defines a medium independent function responsible for transferring data to and from the pysical layer. In general, the MAC sublayer defines data encapsulation (such as framing, addressing, and error detection) and medium access (such as collision detection, and deferral process).

MAC control sublayer is an optional sublayer performing real-time control and manipulation of MAC sublayer operation. The MAC control structure and specification allow new functions to be added to the standard in the future.

Logical link control (LLC) sublayer defines a medium access independent portion of the data link layer. This sublayer is outside the scope of IEEE 802.3. Correspondingly, MAC and the optional MAC control sublayer are specified in such a way that they are unaware whether LLC is located above them, or any other client, such as a bridge or a repeater.

The point-to-multipoint sub-task force concentrated on the lower layers of an EPON network. The work of defining the EPON architecture was divided into physical medium dependent sublayer specification, P2MP protocol specification, and extensions for reconciliation, physical coding, and physical medium attachment sublayers.

4.1.2 Physical medium dependent sublayer

The EPON PMD sublayer parameters are specified in clause 60 of the IEEE 802.3ah standard. The PMD specification is based on the following set of objectives:

TABLE 4.1   EPON PMD Types

PMD type	1000BASE-PX10-U	1000BASE-PX10-D	1000BASE-PX20-U	1000BASE-PX20-D
Fiber type	SMF	SMF	SMF	SMF
Number of fibers	1	1	1	1
Nominal wavelength, nm	1310	1490	1310	1490
Transmit direction	Upstream (ONU to OLT)	Downstream (OLT to ONU)	Upstream (ONU to OLT)	Downstream (OLT to ONU)
Distance, km	10	10	20	20
Min. channel insertion loss, dB	5.0	5.0	10.0	10.0
Max. channel insertion loss, dB	20.0	19.5	24.0	23.5

1. Support for point-to-multipoint media using optical fiber

2. 1000 Mbps up to 10 km on one single-mode fiber supporting a fiber split ratio of 1:16

3. 1000 Mbps up to 20 km on one single-mode fiber supporting a fiber split ratio of 1:16

4. *Bit error ratio* (BER) better than or equal to 10^{-12} at the PHY service interface

To meet the above objectives, four PMD types were defined in clause 60. These types are summarized and compared in Table 4.1.

The task of selecting PMD timing parameters, such as laser-on and laser-off times and gain control time, has generated debates lasting almost a year. Three competing parties formed in the task force, with none being able to gather 75 percent of the votes required to adopt a technical motion.

The first camp promoted a strict timing similar to BPON and GPON specs (laser-on and laser-off times of 16 ns, gain adjustment time ≤ 50 ns). This group argued that increasing the compatibility with BPON and GPON specifications would result in lower component costs due to economies of scale.

The second party advertised for relaxed parameters (laser-on and laser-off times of 800 ns, gain adjustment time ≤ 400 ns), claiming that this would lead to higher component yields and therefore would lower the costs.

The third group lobbied for negotiable parameters, arguing that devices should be able to exploit faster PMD timing to achieve higher efficiency.

In the end, after a prolonged battle, the task force settled on the following parameters: laser-on time = 512 ns, laser-off time = 512 ns, gain adjustment time $\leq$ 400 ns (negotiable). The winning arguments were that the ONUs, being the mass-deployed device, must be as simple and inexpensive as possible. For this, the PMD components should have high yield and should not mandate implementation of digital interfaces, which otherwise would be mandatory if ONUs were to negotiate laser-on and laser-off times. The OLT device can be more expensive, as only a single device is used per EPON. Therefore, the OLT is allowed to negotiate and adjust its receiver parameters such as AGC time.

4.1.3 Point-to-multipoint protocol

We mentioned previously that, in the upstream direction, a PON should employ some channel arbitration mechanism to share the channel capacity without data collisions.

Almost immediately upon its formation, the study group began technical discussions aimed at selecting a set of baseline technical proposals, including the EPON channel arbitration mechanism. Selecting the baseline proposals was not a consonant process as one might have hoped for. Virtually every interested equipment vendor had an idea of how to "do it right." The reviewed proposals ranged from implementing EPON entirely in the physical layer and using PHY-based messaging, using existing IEEE 802.3 flow control mechanisms (PAUSE MAC control frames), schemes based on DOCSIS, or a unified PHY (similar to what later became GPON) to carry both ATM cells and Ethernet frames.

The study group (and later the task force) reviewed more than 40 presentations related to EPON and had countless conference calls. By November 2001, the opinions began to converge on defining a MAC control-based protocol that would allow the OLT to assign to ONUs the transmission windows. This protocol is currently known as the *multipoint control protocol* (MPCP) and is defined in clause 64 of the IEEE 802.3ah standard.

MPCP uses MAC control messages (similar to the Ethernet PAUSE message) to coordinate multipoint-to-point upstream traffic. There are two modes of operation of MPCP: *autodiscovery* (initialization) and *normal* operation. Normal mode is used to assign transmission opportunities to all discovered ONUs. A detailed description of this mode is given in Sec. 5.3.1. The autodiscovery mode is used to detect newly connected ONUs and learn their parameters such as MAC addresses and round-trip delays. This mode is described in Sec. 5.3.2.

4.1.4 Extensions of the existing clauses

Several existing clauses in IEEE 802.3 require certain extensions in order to be used with P2MP architecture. All these extensions are grouped in a new clause 65.

4.1.4.1 Reconciliation sublayer (RS). The IEEE 802 architecture makes a general assumption that all devices connected to the same media can communicate to one another directly. Relying on this assumption, bridges never forward a frame back to its ingress port. This bridge behavior has led to an interesting problem: A bridge placed in the OLT will see one PON port and will never forward upstream frames back to ONUs. However, due to the directional properties of the splitter, the ONUs are unable to directly communicate with one another. Therefore, it appears that the EPON-based network will have difficulties providing full connectivity among the attached devices. This raises a question of EPON compliance with IEEE 802 architecture, particularly with P802.1D bridging.

To resolve this issue and to ensure seamless integration with other Ethernet networks, devices attached to the EPON medium will use an extended reconciliation sublayer, which will emulate the point-to-point medium. This topology emulation process relies on tagging of Ethernet frames with tags unique for each ONU. These tags are called *logical link IDs* and are placed in the preamble before each frame. Subclause 65.1 of the IEEE 802.3ah standard defines the new format of frame preamble and filtering rules necessary to achieve point-to-point emulation. Chapter 6 provides a detailed overview of the operation of the emulation function.

4.1.4.2 Physical coding sublayer (PCS). It was mentioned in Sec. 2.5 that to avoid spontaneous emission noise from near ONUs obscuring the signal from a distant ONU, the ONUs' lasers should be turned off between their transmissions. To control the laser, the physical coding sublayer is extended (see subclause 65.2 of IEEE 802.3ah) to detect the data being transmitted by higher layers and to turn the laser on and off at the correct times. This data detection function is further discussed in Chap. 7.

An additional PCS extension specifies an optional *forward error correction* (FEC) mechanism, which may increase the optical link budget or the fiber distance. The FEC mechanism uses Reed-Solomon code and adds 16 parity symbols (bytes) for each block of 239 information symbols (bytes). These additional parity data are used at the receiving end of the link to correct errors that may have occurred during the data transmission. The P2MP group has adopted frame-based FEC mechanism, such that each frame is encoded separately and all per-frame parity bytes are added at the end of the frame. This approach would allow devices without FEC capabilities to receive the FEC-encoded frames, albeit with a higher number of bit errors. The FEC mechanisms are further discussed in Chap. 9.

4.1.4.3 Physical medium attachment sublayer. The *physical medium attachment* sublayer is extended to specify a time interval required by the receiver to acquire phase and frequency lock on the incoming data stream. This interval is known as the *clock and data recovery* (CDR) time. The specification requires the PMA sublayer instantiated in an OLT to become synchronized at the bit level within 400 ns and at the code-group level within an additional 32 ns.

4.2 EPON Today: Promise and Challenges

The Ethernet in the First Mile task force completed its charter in June 2004, culminating in ratification of *IEEE 802.3ah (Amendment to—Information Technology—Telecommunications and Information Exchange between Systems—Local and Metropolitan Area Networks—Specific Requirements—Part 3: Carrier Sense Multiple Access with Collision Detection (CSMA/CD) Access Method and Physical Layer Specifications—Media Access Control Parameters, Physical Layers and Management Parameters for Subscriber Access Networks)*. Working materials concerning the P802.3ah standards effort can be found at www.ieee802.org/3/efm.

Recently, subscriber access networks based on EPON became a hot topic in the industry as well as in academic research. Industrial interests stem from the fact that EPON is the first optical technology promising to be cost-effective enough to justify its mass deployment in an access network. The completion of the standard and the expectations that EPON architecture will enjoy the same success and proliferation as its LAN predecessor became a thrusting factor for many telecommunication operators to initiate EPON trials, or to at least study the technology.

Unlike other standards bodies, IEEE 802.3 only specifies a small portion of a communications system (only physical and data link layers). The rest is considered out of scope for IEEE 802.3. The academic research was fueled by a number of interesting challenges brought forward by EPON architecture, but left out by the standard.

One interesting research problem is related to EPON's efficiency and scalability. To support a large number of users, and to exploit multiplexing gains from serving bursty Internet traffic, the EPON scheduler should be able to allocate bandwidth dynamically. Yet, considering very significant propagation delays and the nonfragmentability of Ethernet frames, developing such scheduling algorithm is a nontrivial task. In response to this challenge, the research community has generated many interesting EPON *dynamic bandwidth allocation* (DBA) proposals.

Another set of research problems is related to the fact that EPON is sought for subscriber access—an environment serving independent and noncooperative users. Users pay for service and expect to receive their service regardless of the network state or the activities of the other users. Unlike traditional, enterprise-based Ethernet, the EPON must be able to guarantee *service-level agreements* (SLAs) and enforce traffic shaping and policing per individual user. Providing dynamic bandwidth allocation, while guaranteeing performance parameters such as packet latency, packet loss, and bandwidth, is yet another rich research topic.

Issues of upgradability, encryption, and authentication are also very important for EPON's success in the public access environment.

References

[EFM01]   IEEE 802.3 EFM Study Group, "Ethernet PON (EPON) and the PAR + 5 criteria," IEEE 802 interim meeting, St. Louis, MO, May 2001. Presentation is available at http://www.ieee802.org/3/efm/public/may01/pesavento_1_0501.pdf.

[OSI94]   ISO/IEC 7498-1: 1994, *Information Technology—Open Systems Interconnection—Basic Reference Model: The Basic Model.*

EPON Architecture

5

EPON Overview

The IEEE 802.3 standard defines two basic modes of operation for an Ethernet network. In one configuration, it can be deployed over a shared medium using the *carrier-sense multiple access with collision detection* (CSMA/CD) protocol. In the other configuration, stations may be connected through a switch using full-duplex point-to-point links. Correspondingly, Ethernet MAC can operate in one of two modes: CSMA/CD mode or full-duplex mode.

Properties of the EPON medium are such that it cannot be considered either a shared medium or a point-to-point network; rather, it is a combination of both. It has a connectivity of a shared medium in the downstream direction, and it behaves as a point-to-point medium in the upstream direction.

5.1 Downstream Transmission

In the downstream direction, Ethernet packets transmitted by the OLT pass through a $1 \times N$ passive splitter or cascade of splitters and reach each ONU. The value of N is typically between 4 and 64 (limited by the available optical power budget). This behavior is similar to a shared-medium network. Because Ethernet is broadcasting by nature, in the downstream direction (from network to user) it fits perfectly with the Ethernet PON architecture: Packets are broadcast by the OLT and selectively extracted by their destination ONU (Fig. 5.1).

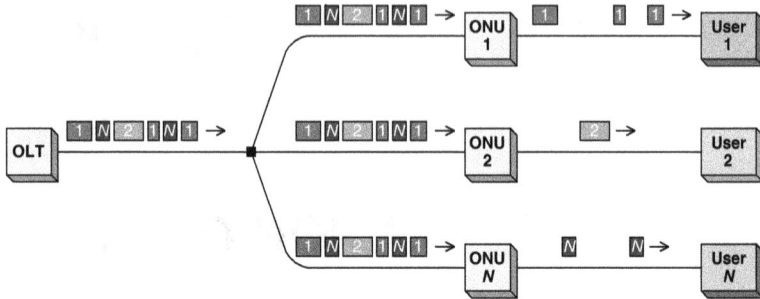

Figure 5.1   Downstream transmission in EPON.

5.2   Upstream Transmission

In the upstream direction (from users to network), due to the directional properties of a passive optical combiner, data packets from any ONU will reach only the OLT, and not other ONUs. In this sense, in the upstream direction, the behavior of EPON is similar to that of a point-to-point architecture. However, unlike a true point-to-point network, in EPON, all ONUs belong to a single collision domain—data packets from different ONUs transmitted simultaneously still may collide. Therefore, in the upstream direction, EPON needs to employ some arbitration mechanism to avoid data collisions and fairly share the channel capacity among ONUs.

5.2.1   Contention-based versus guaranteed media access

A contention-based media access mechanism (something similar to a CSMA/CD) is difficult to implement in EPON because ONUs cannot detect a collision due to the directional properties of optical splitter/combiner. An OLT could detect a collision and inform ONUs by sending a jam signal; however, propagation delays in PON, which can exceed 20 km in length, can greatly reduce the efficiency of such a scheme. Contention-based schemes also have a drawback of providing a nondeterministic service; i.e., node throughput, channel utilization, and medium access delay can only be described as statistical averages. There is no guarantee of a node getting access to the medium in any small interval of time. The nondeterministic access is only a minor nuisance in CSMA/CD-based enterprise networks where links are short, typically overprovisioned, and traffic predominantly consists of delay-tolerant data. Subscriber access networks, however, in addition to data, must support voice and video services and thus must provide certain guarantees on timely delivery of these traffic types.

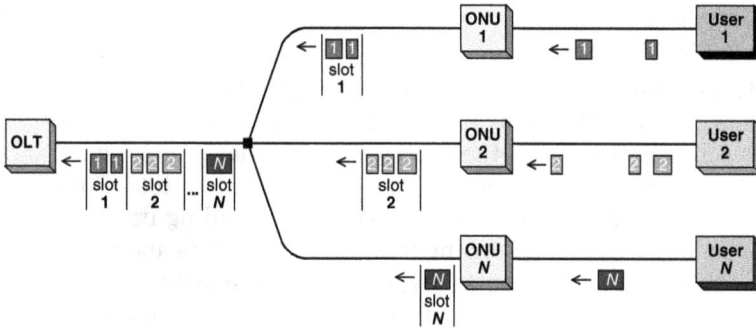

Figure 5.2   Upstream transmission in EPON.

To introduce determinism in the delivery of packets, different noncontention schemes have been proposed. All such schemes grant ONUs an exclusive access to the media for a limited interval of time, commonly referred to as *transmission window* or *timeslot*. Figure 5.2 illustrates an upstream timeshared data flow in an EPON.

All ONUs are synchronized to a common time reference, and each ONU is allocated a timeslot. Each timeslot is capable of carrying several Ethernet packets. An ONU should buffer frames received from a subscriber until its timeslot arrives. When the timeslot arrives, the ONU "bursts" all stored frames at full channel speed which corresponds to a standard Ethernet rate of 1000 Mbps. If there are no frames in the buffer to fill the entire timeslot, 10-bit idle characters are transmitted as specified for full-duplex Ethernet MAC.

The performance of an EPON depends on a particular capacity allocation scheme. The possible timeslot allocation schemes range from static allocation (fixed TDMA) to dynamic adjustment of the slot size based on instantaneous queue load in every ONU (statistical multiplexing scheme). Choosing the best allocation scheme, however, is not a trivial task.

Fixed TDMA schemes are easier to implement. In a simplest form, each ONU would be programmed to start and stop transmission at the predetermined repeating intervals. However, as will be discussed in Chap. 14, fixed TDMA schemes suffer from low efficiency in the presence of bursty data or variable-size packets.

If all users belonged to the same administrative domain, say a corporate or campus network, the full statistical multiplexing would make sense—network administrators would like to get the most out of the available bandwidth, regardless of how much of it each particular user gets. However, subscriber access networks are not private LANs, and the objective is to ensure service-level agreement compliance for each individual user.

5.2.2 Centralized versus distributed arbitration

Noncontention (guaranteed) schemes require channel arbitration. This arbitration can be either centralized or distributed. In a distributed arbitration scheme, the ONUs themselves decide when to send data and for how long. These schemes are somewhat similar to a token-passing approach. In such a scheme, every ONU, before sending its data, will send a special message announcing how many bytes it is about to send. The ONU that is scheduled next (say, in round-robin fashion) will monitor the transmission of the previous ONU and will time its transmission such that the transmission arrives at the OLT right after the transmission from the previous ONU. Thus, there will be no collision, and no bandwidth will be wasted. However, this scheme has a major limitation: it requires connectivity (communicability) between ONUs. This imposes some constraints on the PON topology; namely, the network should be deployed as a ring or as a broadcasting star. This requirement is not desirable as (1) it may require more fiber to be deployed or (2) fiber plant with different topology might be already predeployed. In general, a preferred algorithm should support any point-to-multipoint PON topology.

In an optical access network, we can count only on the connectivity from the OLT to every ONU (downstream traffic) and from every ONU to the OLT (upstream traffic). Therefore, the OLT remains the only device that can arbitrate time-division access to the shared channel.

The challenge in implementing a centralized (OLT-based) dynamic arbitration scheme is the fact that the OLT does not know how many bytes of data each ONU has buffered. The burstiness of data traffic precludes a queue occupancy prediction with any reasonable accuracy. If the OLT is to make an accurate timeslot assignment, it should know the state of a given ONU exactly. One solution may be to use a polling scheme based on grant and request messages. Requests are sent from an ONU to report changes in an ONU's state, e.g., the amount of buffered data. The OLT processes all requests and allocates different transmission windows (timeslots) to ONUs. Slot assignment information is delivered to ONUs using grant messages.

The advantage of having centralized intelligence for the slot allocation algorithm is that the OLT knows the state of the entire network and can switch to another allocation scheme based on that information; the ONUs don't need to monitor the network state or negotiate and acknowledge new parameters. This will make ONUs simpler and cheaper and the entire network more robust. In the end, the IEEE 802.3ah task force has settled on a noncontention centralized model for upstream channel access.

Given that the bandwidth allocation algorithms may depend on many parameters, such as deployment environment, supported services, and mix of SLA plans, the IEEE 802.3ah task force decided that it would be too presumptuous to select a specific *dynamic bandwidth allocation* (DBA) algorithm. Instead, the group has declared the DBA to be out of scope for the standard and has left the choice to equipment vendors.

While the algorithm's decision-making process is left open, to ensure device interoperability, the message exchange protocol needed to be specified. To support dynamic capacity allocation, the IEEE 802.3ah task force has developed the multi-point control protocol. The MPCP is not concerned with a specific DBA algorithm; rather it is a supporting mechanism that facilitates implementation of various bandwidth allocation schemes in EPON.

5.3   Multi-Point Control Protocol

One of the most important conditions EPON has to comply with, in order to be part of the IEEE 802.3 standard, is the use of the existing Ethernet MAC (either CSMA/CD or full-duplex). Should EPON adopt different medium access logic, it most likely would become a new standard, separate from IEEE 802.3 Ethernet. Notwithstanding that transmission arbitration is a MAC function, the IEEE 802.3ah task force had to find a protocol which will achieve the same without any modifications to the MAC sublayer. In was decided to implement MPCP as a new function of the MAC control sublayer.

The scope of MAC control is to provide real-time control and manipulation of MAC sublayer operation. The MAC control sublayer resides between the MAC sublayer and MAC client (see Fig. 4.1). Before MPCP was developed by the IEEE 802.3ah task force, the only function of the MAC control sublayer was flow control—an operation allowing a station to inhibit transmission from its peer for a predetermined interval of time. To achieve this, the flow control protocol uses a PAUSE MAC control message.

Transmission arbitration in EPON required a method exactly opposite to flow control—an operation allowing a station to enable transmission from its peer for a predetermined interval of time. To avoid collisions, the OLT would allow only one ONU to transmit at any given time.

An important difference between MPCP and flow control is their default state, i.e., the state to which a link will eventually converge after control messages have stopped being issued. In flow control, the default state allows communication to be carried over a link; this communication may be explicitly paused by a control message. On the contrary, in MPCP, the default state inhibits the communication. Only when the control message arrives, the transmission will be enabled for a limited time. This behavior necessitated the following MPCP modes of operation:

- *Bandwidth assignment* mode. To sustain communication between OLT and ONUs, the MPCP should provide periodic granting for each ONU.

- *Autodiscovery* mode. To discover newly activated ONUs, the MPCP should initiate the discovery procedure periodically.

While the MAC control sublayer is optional for other configurations, in EPON it is mandatory, because EPON cannot operate without MPCP.

5.3.1   Bandwidth assignment

The bandwidth assignment mechanism relies on grant and request messages, or GATE and REPORT, in IEEE 802.3ah terminology. Both GATE and REPORT messages are MAC control frames, which are identified by a predefined type value of $88\text{-}08_{16}$.

A GATE message is sent from the OLT to an individual ONU and is used to assign a transmission timeslot to this ONU. A timeslot is identified by a pair of values {*startTime, length*}. The values for *startTime* and *length* are decided upon by a *DBA agent* or *scheduler*, located in MAC control client, a sublayer outside the scope of IEEE 802.3ah (see Fig. 5.3). The values of *startTime* and *length* are passed to the gating process at the OLT. The gating process, formally specified in the standard, forms a GATE message and transmits it to the ONU. In the ONU, the received GATE message is parsed and demultiplexed to the ONU's gating process, which is responsible for allowing the transmission to begin within the timeslot assigned by the received message. Additionally, an indication of the received GATE message is passed to the DBA agent at the ONU to allow it to perform any necessary DBA-specific functions, e.g., select the order of frames to be sent out. Indeed, in some scheduling algorithms, such as those based on packet deadlines, the order of frames may depend on the time when the timeslot starts or on the size of the timeslot.

A REPORT message is a feedback mechanism used by an ONU to convey its local conditions (such as buffer occupancy) to the OLT to help the OLT make intelligent allocation decisions. Such information as number of egress queues and their status is not available to the MPCP, and so the REPORT message, similarly to the GATE, is initiated by the DBA agent (see Fig. 5.3). It is then passed to the reporting process at

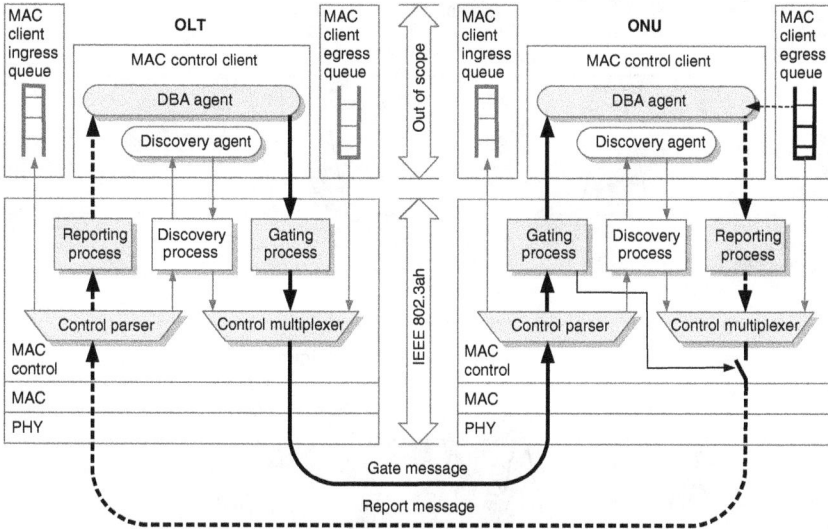

Figure 5.3 Processes and agents involved in bandwidth assignment.

the ONU, which forms and transmits the REPORT frame. REPORT frames can be sent only in previously assigned timeslots. At the OLT, the received REPORT frame is parsed and demultiplexed to the OLT's reporting process, which, in turn, passes it to the DBA agent. The DBA agent may use this information to make timeslot allocations for the next round.

5.3.1.1 Pipelined timeslot assignment. An interesting question is how the OLT can make sure that timeslots assigned to different ONUs do not overlap. In a *sequential* timeslot assignment mode, the OLT assigns a timeslot to ONU i only after the data from ONU $i - 1$ have been received (see Fig. 5.4a). As simple as the timeslot assignment may be, this scheme is very inefficient, because after a GATE has been sent, the channel would remain idle for the entire round-trip time. This idle time is often called *walk-time*. In EPON, the distance between the OLT and an ONU can reach 20 km, so the walk-time could be as high as 200 μs.

To eliminate the walk-time overhead, MPCP allows a *pipelined* timeslot assignment. In this mode, the OLT may send a GATE to ONU i before data from ONU $i - 1$ have arrived (Fig. 5.4b). The pipelined mode requires the OLT to know the round-trip time to each ONU. Having this knowledge, the OLT is able to calculate future time when all pending transmissions will complete and the upstream channel will become idle, and to schedule the following timeslot to start at that time. The measurement of the round-trip time for a newly connected ONU is one of the main tasks of the autodiscovery procedure.

(a) Sequential timeslot assignment

(b) Pipelined timeslot assignment

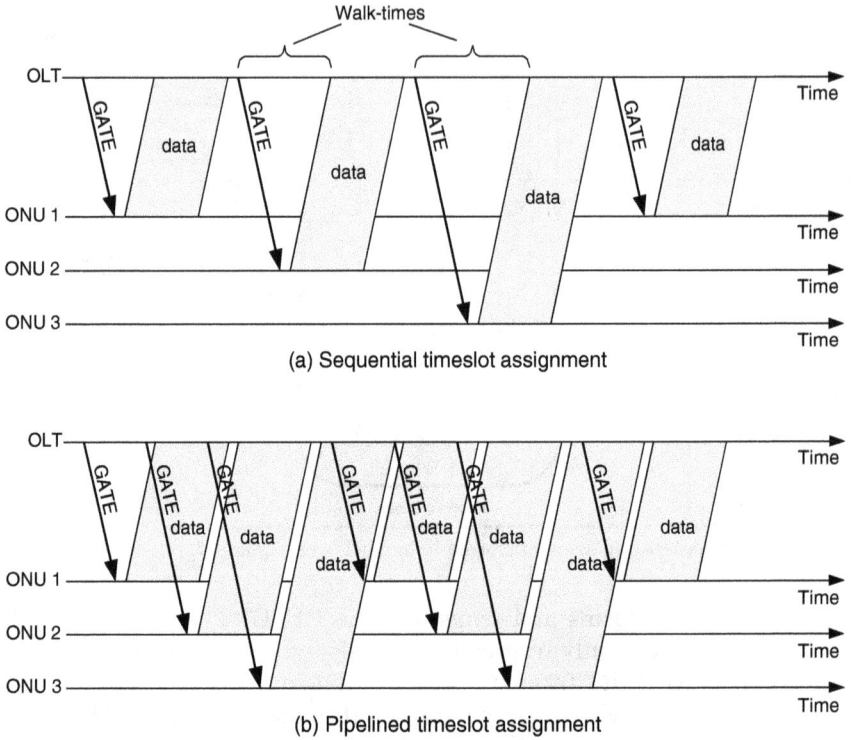

Figure 5.4   Sequential and pipelined timeslot assignments.

5.3.1.2 Decoupled downstream and upstream timing. In several schemes considered by the EFM task force, timeslots were represented in the GATE messages only by *length* instead of by a {*startTime, length*} pair. Proponents of these schemes argued that the GATE message arrival time can explicitly serve as the timeslot start time. If the DBA agent in the OLT desires to receive data from ONU k at time t, the GATE message should be sent to this ONU exactly at time $t - RTT_k$, where RTT_k is the round-trip time to ONU k (including any message parsing and processing delays). This idea of "just-in-time" GATE transmission hit a snag, because Ethernet frames cannot be preempted or fragmented. Very conceivably, a GATE message could be blocked behind a long data frame that started its transmission a moment before the GATE message was to be transmitted. In addition, GATE messages could be blocked behind other GATE messages. A blocked GATE message will be transmitted with some delay, and will cause a corresponding delay in the upstream transmission from an ONU, degrading the upstream channel utilization. A more serious problem may arise if a GATE message to ONU k is scheduled for transmission

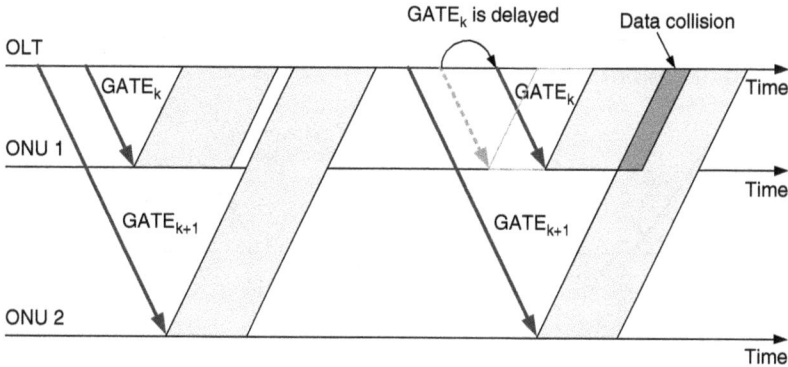

Figure 5.5 In "just-in-time" granting schemes, data collisions are possible due to delayed GATE message.

after the GATE message to ONU $k + 1$. As illustrated on Fig. 5.5, if GATE $k + 1$ was transmitted on time but GATE k was delayed, the transmission from ONU k will be delayed and some data frames may collide with transmission from ONU $k + 1$.

To resolve the difficulties associated with just-in-time granting, it was deemed very desirable to decouple GATE transmission timing from the upstream transmission timing. Such decoupling is achieved by explicitly specifying the timeslot start time in each GATE message (Fig. 5.6). Rather than using the GATE message arrival as the base time, the ONU would start transmission when its local clock became equal to the *startTime* value conveyed in the GATE message. The delay experienced by the GATE message itself will not affect the upstream transmission timing, as long as GATE message arrives before the intended timeslot start time. Since there may be a significant time lag between the GATE message arrival and the timeslot start time, this scheme requires clocks in the OLT and ONUs to be well synchronized.

5.3.1.3 MPCP clock synchronization. To allow decoupling of GATE transmission time from the timeslot start time, the OLT and each ONU should maintain a local clock, called the *MPCP clock*. The MPCP clock is a 32-bit counter which counts time in units of *time quanta* (TQ). The TQ is defined to be a 16-ns interval, or the time required to transmit 2 bytes of data at 1 Gbps line rate. Correspondingly, the timeslot start times and lengths in GATE messages, as well as queue lengths in REPORT messages, are expressed in TQ.

To synchronize ONU's MPCP clock to the OLT's clock, each MPCP message defines a field called *timestamp*. The OLT's control multiplexer, shown in Fig. 5.3, writes the value of the MPCP clock into the timestamp field of an outgoing GATE message. When a GATE message

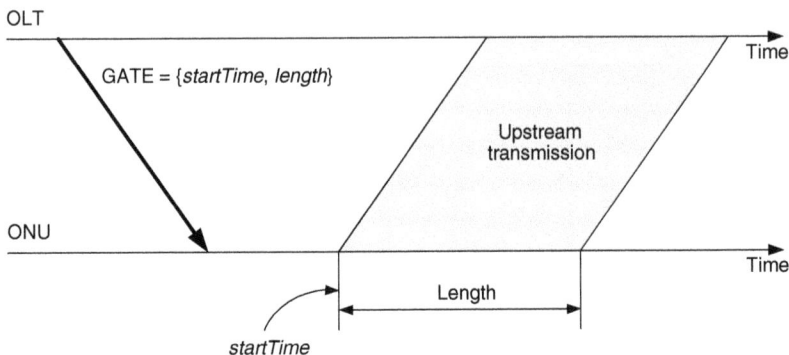

Figure 5.6   Decoupled GATE arrival time and timeslot start time.

arrives to an ONU, the control parser sets its local MPCP clock to the value received in the timestamp field.

This clock synchronization scheme is based on the assumption that frame propagation delay between the control multiplexer at the transmitting device and the control parser at the receiving device is nearly constant. In other words, frames cannot be blocked or delayed in the MAC and PHY sublayers.

5.3.1.4 Loop timing. In traditional Ethernet, clocks were allowed to deviate from the nominal frequency by 100 *parts per million* (ppm). Such clock tolerance was a great asset of Ethernet specification, allowing very inexpensive devices to be built. But in EPON, it becomes a handicap. Consider a situation where the time interval between the GATE message arrival and the start of timeslot is large, say 20 ms. ONU will synchronize the MPCP clock to the received timestamp when the GATE message arrives. If the MPCP clocks in the OLT and ONU are free-running, the OLT clock runs at frequency $f + 100$ ppm, and the ONU clock runs at $f - 100$ ppm, then the ONU will initiate its upstream transmission late by 4 µs, which equals the clock drift during the 20 ms interval since the last synchronization.

To remedy this situation, MPCP mandates loop timing for the ONU, which means that the ONU's MPCP clock should track the receive clock, recovered from the data transmitted by the OLT. The OLT's clock is still allowed to be ±100 ppm from the nominal frequency. Since the OLT constantly transmits data or idle characters, ONUs are able to recover the clock and remain synchronized at all times.

5.3.2 Auto discovery

Recall that in the default state the MPCP does not allow transmission from an ONU. An ONU cannot transmit any data (cannot even turn its

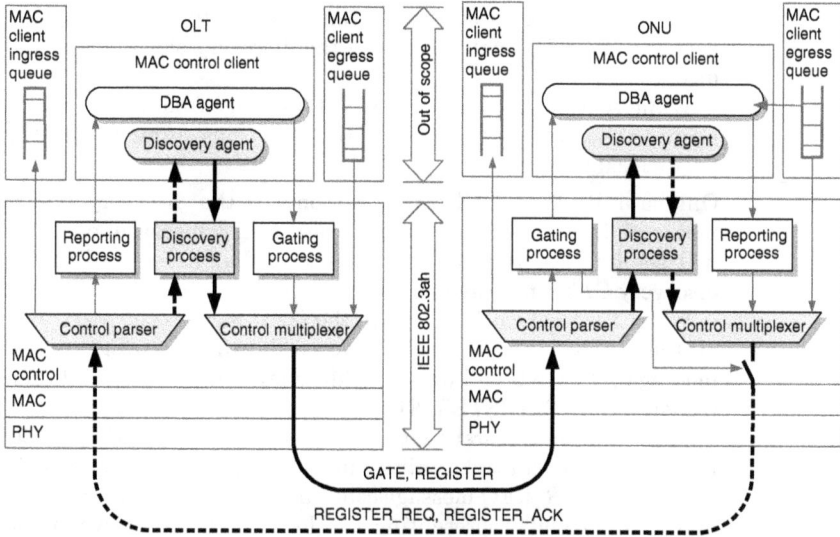

Figure 5.7 Processes and agents involved in autodiscovery.

laser on) unless it is granted by the OLT. Thus, after boot-up, an ONU would silently wait for a grant from the OLT. This grant, however, will never arrive because the OLT doesn't know and cannot know that a new ONU has joined, since the ONU has to remain silent. To resolve this sort of chicken-and-egg situation, MPCP defines an autodiscovery mode.

The autodiscovery mechanism is used to detect newly connected ONUs and learn the round-trip delays and MAC addresses of these ONUs. Both the OLT and ONUs implement the discovery process, which is driven by the *discovery agent* (Fig 5.7).

Autodiscovery employs four MPCP messages: GATE, REGISTER_REQ, REGISTER, and REGISTER_ACK. These messages are carried in MAC control frames, which are distinguished by a predefined type value of 88-08$_{16}$. At a high level, the autodiscovery is a four-step procedure, and it works as follows.

Step 1.   The discovery agent at the OLT decides to initiate a discovery round and allocates a *discovery window*—an interval of time when no previously initialized ONUs are allowed to transmit. It is assumed that the discovery agent may freely communicate with the DBA agent and that both agents will agree on the discovery window size and its start time. The DBA agent ensures that no active ONUs are scheduled to transmit during the discovery window.

The discovery agent instructs the discovery process to send a special GATE message, called *discovery GATE*, advertising the start time of the *discovery slot* and its *length*. Section 5.3.2.1 explains the relationship between the discovery slot size and the discovery window size.

While relaying the discovery GATE message from the discovery agent to the MAC sublayer, the MPCP will timestamp it with the OLT's local time.

Step 2.   Only uninitialized ONUs will respond to the discovery GATE message. Upon receiving the discovery GATE message, an ONU will set its local time to the timestamp that it received in the discovery GATE message.

When the local clock located in the ONU reaches the start time of the discovery slot (also delivered in the discovery GATE message), the ONU will wait an additional *random delay* and then transmit the REGISTER_REQ message. The random delay is applied to avoid persistent collisions when REGISTER_REQ messages from multiple uninitialized ONUs consistently collide. The REGISTER_REQ message contains the ONU's source address and a timestamp representing the local ONU's time when the REGISTER_REQ message was sent.

When the OLT receives the REGISTER_REQ from an uninitialized ONU, it learns its MAC address and round-trip time. The method for round-trip time measurement is explained in Sec. 5.3.3.

Step 3.   Upon parsing and verifying the REGISTER_REQ message, the OLT issues the REGISTER message sent directly to an initializing ONU using the MAC address received during the previous step. The REGISTER message contains a unique identification value called the *logical link* ID (LLID) that the OLT assigns to each ONU. The use of LLID is explained in Chap. 6.

Following the REGISTER message, the OLT sends a normal GATE (nondiscovery or unicast GATE) to the same ONU.

Step 4.   Finally, after receiving both the REGISTER and the normal GATE messages, the ONU sends the REGISTER_ACK message to acknowledge to the OLT that it has successfully parsed the REGISTER message. The REGISTER_ACK should be sent in the timeslot granted by the previously received GATE message.

Since multiple uninitialized ONUs may respond to the same discovery GATE message, the REGISTER_REQ messages may collide. In that case, the ONUs whose REGISTER_REQ messages have collided will not get the REGISTER message. If an ONU does not receive the REGISTER message before it receives another discovery GATE, it will infer that a collision has occurred and will attempt to initialize again.

5.3.2.1 Discovery slot and discovery window. Discovery slot is a length of the grant advertised to all uninitialized ONUs in the discovery GATE. The discovery window is an interval reserved by the discovery

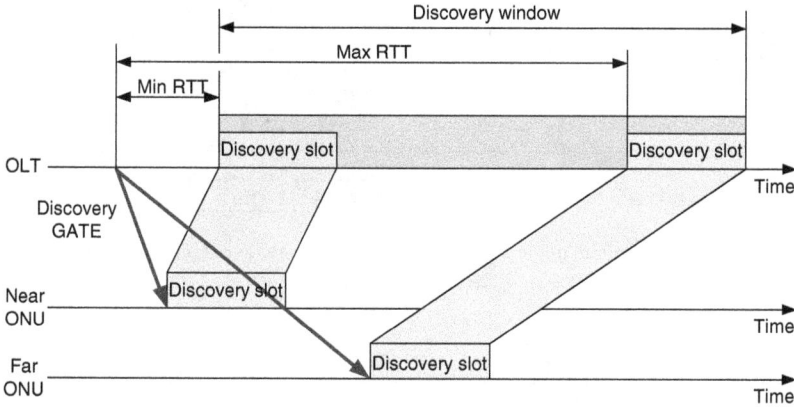

Figure 5.8 Relationship of discovery slot and discovery window.

agent. No data traffic should be scheduled during the discovery window. As shown in Fig. 5.8, the discovery window size and discovery slot size are related. The discovery window should be at least as large as the discovery slot. In addition, since the distance to an uninitialized ONU is not known yet, the discovery window should accommodate the entire range of possible round-trip times (RTTs). Thus, the relationship between discovery slot and discovery window can be expressed as

$$discovery Window \geq discovery Slot + maxRTT - minRTT$$

Often, either for simplicity or because it is not known, $minRTT$ is taken as 0. Considering the maximum PON distance of 20 km (per IEEE 802.3ah specification), the following relationship should hold:

$$discovery Window \geq discovery Slot + 200 \text{ µs}$$

5.3.2.2 Avoiding persistent collisions. Since more than one ONU may attempt initialization at the same time, autodiscovery is a contention-based procedure. If two or more uninitialized ONUs happen to be at the same distance from the OLT, their REGISTER_REQ messages will persistently collide and such ONUs will never be discovered by the OLT. To avoid the persistent collisions, the OLT allocates the discovery slot larger than the necessary time to transmit a single REGISTER_REQ message.

Each uninitialized ONU will apply some random delay to offset the transmission of the REGISTER_REQ message within the discovery slot (Fig 5.9).

Two or more REGISTER_REQ messages still may collide in this configuration; however, given a sufficiently large discovery slot, such collisions will not be persistent. During the next discovery opportunity,

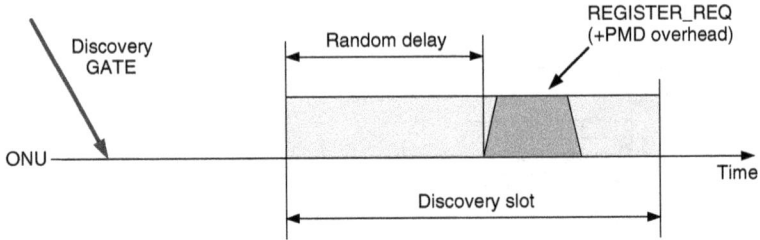

Figure 5.9   Applying random delay during discovery process to avoid persistent collisions.

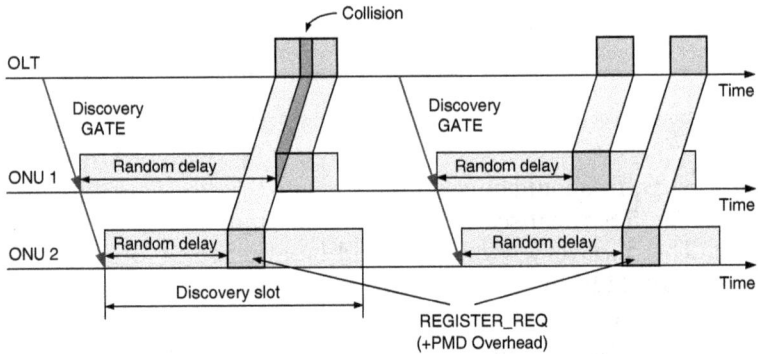

Figure 5.10   Discovery attempt with and without REGISTER_REQ collision.

the ONUs will choose different random delays, possibly avoiding the collision (Fig 5.10).

Clearly, the discovery is an invasive process, since no normal traffic can be carried upstream by EPON during the discovery window allocation. In Chap. 13 we investigate how the size of the discovery slot may be chosen to minimize overall loss of bandwidth caused by the discovery procedure.

It is interesting to note that the IEEE 802.3ah task force has also considered an alternative method for avoiding the persistent collisions, namely, a back-off mechanism. Using this method, an ONU, whose REGISTER_REQ message has collided, will skip a random number of discovery attempts. This random number could be chosen from a range that doubles in size after each collision, resulting in a *binary exponential back-off* (BEB) algorithm, akin to the method used in CSMA/CD.

Even though in the BEB method the discovery slot may be smaller (only long enough to transmit one REGISTER_REQ message), the number of required discovery attempts will be significantly higher than in the *random delay* (RD) method described above. This is so because, in the BEB method, there is a significant probability that all collided

ONUs will back off by more than one and that none of them will respond during the next discovery opportunity. This will consume EPON bandwidth without any reduction in the number of undiscovered ONUs. In the RD method, all undiscovered ONUs transmit the REGISTER_REQ message in every discovery slot until they succeed. The group has also considered a combination of BEB and RD, called BEB+RD, in which the ONUs would apply random delay to the REGISTER_REQ messages and would back off in case these messages collided. Intuitively, the efficiency of the BEB+RD method was between those of the BEB and RD methods. In the end, the efficiency argument prevailed, and the task force voted against the BEB and BEB+RD methods.

5.3.3   Round-trip time measurement

Probably, a simplest way to measure RTT is to send a message from the OLT to an ONU and request an ONU to echo it back right away. Then the RTT is simply the time difference between sending the message and receiving the response at the OLT. However, this simple scheme suffers from three issues:

1. Possibly the varying time to generate a response at the ONU is counted as part of RTT.

2. Downstream and upstream transmission timing is coupled.

3. If two ONUs happen to be at the same distance from the OLT, they would receive discovery GATE messages simultaneously and generate REGISTER_REQ messages at the same time. Without the possibility of applying random delay, these messages would persistently collide.

To resolve the above issues, the IEEE 802.3ah standard defines a more sophisticated RTT measurement scheme. The timing diagram of RTT measurement mechanisms is shown in Fig. 5.11.

When the discovery GATE is passed through the control multiplexer at the OLT, it is timestamped with OLT's MPCP clock (t_0). The time-stamp reference point is the first byte of the discovery GATE message. In other words, the timestamp value should be equal to the MPCP clock value at the moment when the first byte of *destination address* (DA) is transmitted, i.e., passed from MAC control to MAC.

When this discovery GATE arrives at the ONU, the ONU sets its local MPCP counter to the value of the received timestamp. Here as well, the reference point should be the first byte of DA, as received by the ONU. After the initial value of the local MPCP clock is set, this clock continues

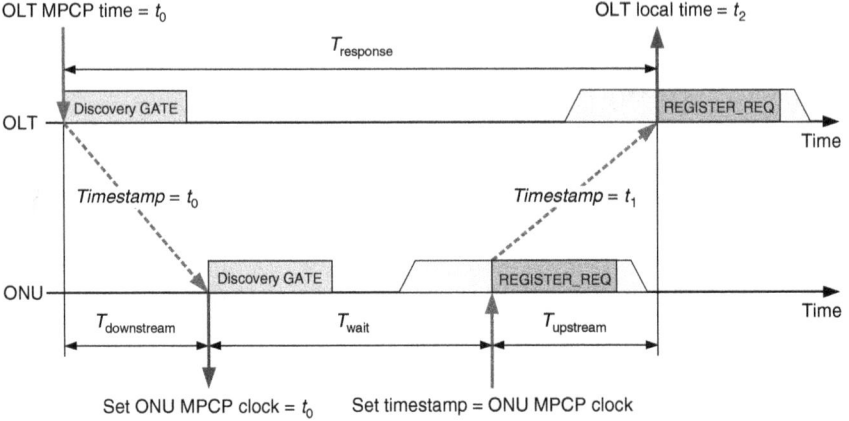

Figure 5.11   Round-trip time measurement.

running synchronously with the clock recovered from the received data stream.

When the value of the MPCP clock reaches the timeslot start time, the ONU applies an additional random delay, after which it starts transmitting the REGISTER_REQ message. When the REGISTER_REQ is passed through control multiplexer at the ONU, it is time-stamped with the ONU's MPCP clock (t_1). The timestamp reference point is the first byte of the REGISTER_REQ message. In Fig 5.11, the interval of time between receiving the discovery GATE message and transmitting the REGISTER_REQ is denoted T_{wait} and is equal $t_1 - t_0$. This interval provides sufficient time for the ONU to generate the REGISTER_REQ message.

Finally, when the REGISTER_REQ message arrives at the OLT, the OLT notes the value of its MPCP clock corresponding to the first byte of the DA field. In Fig 5.11, this value is denoted t_2. The time elapsed at the OLT between sending the discovery GATE and receiving the REGISTER_REQ is denoted T_{response} and is equal to $t_2 - t_0$. From the timing diagram it is clear that T_{response} is equal to $T_{\text{downstream}} + T_{\text{wait}} + T_{\text{upstream}}$. Thus, we have

$$RTT = T_{\text{downstream}} + T_{\text{upstream}} = T_{\text{response}} - T_{\text{wait}} = (t_2 - t_0) - (t_1 - t_0) = t_2 - t_1 \quad (5.1)$$

Equation (5.1) shows that the RTT equals exactly the difference between the REGISTER_REQ arrival time and the timestamp contained in the REGISTER_REQ message. Of course, this equation is only valid if T_{response} and T_{wait} are measured in the same time domain, i.e., if the ONU's MPCP clock is synchronized to the OLT's clock.

5.3.3.1 Timestamp reference. Interestingly enough, the timestamp reference point being the first byte of the DA is not listed in the IEEE 802.3ah standard as a *mandatory* requirement. That means that standard-compliant devices are allowed to use different timestamp reference points.

Below we consider several examples of RTT measurement in the EPON system where the OLT and an ONU assume different timestamp reference points. Recall that the RTT is measured only to enable pipelined granting (see Sec. 5.3.1.1). In pipelined granting, the OLT must be able to precalculate the arrival time of the data burst from a given ONU. In the following examples, we will analyze how the choice of a reference point may affect the OLT's ability to precalculate the burst arrival time.

We first consider a case when the OLT and an ONU choose distinct reference points for the GATE message (Fig. 5.12). For example, the OLT will read MPCP clock Δ_{OLT} TQ ahead of transmitting the first byte of DA, and the ONU will set its MPCP clock to the received timestamp value Δ_{ONU} TQ after receiving the first byte of DA. As shown in Fig. 5.12*a*, in this situation, the calculated RTT value will be

$$RTT = T_{\text{downstream}} + T_{\text{upstream}} + \Delta_{OLT} + \Delta_{ONU} \qquad (5.2)$$

Figure 5.12*b* illustrates a cycle of bandwidth assignment performed after the autodiscovery completes. In pipelined granting mode, if the OLT expects to receive ONU's data at time S, it will send to this ONU a GATE message with timeslot start time equal to $S - RTT$. As can be seen from the diagram, the actual arrival time A is equal to

$$A = t_0 + \Delta_{OLT} + T_{\text{downstream}} + \Delta_{ONU} + S - RTT - t_0 + T_{\text{upstream}} \qquad (5.3)$$

Expanding RTT per Eq. (5.2), we get $A = S$; that is, the actual data arrival time A exactly corresponds to the expected arrival time S. This example demonstrated that any time delta between actual GATE timestamp reference points used by the OLT and an ONU is indistinguishable from a downstream propagation delay. *The actual timestamp reference points for downstream MPCP messages do not need to coincide for the OLT and ONUs. The location of reference points is irrelevant as long as these points remain the same during autodiscovery and during the normal granting process.*

In our next example we consider a case when the OLT and an ONU use different reference points for the REGISTER_REQ message (Fig. 5.13). As before, we assume that the timestamp value is prepared before a frame transmission begins, and that the receiving device gets the timestamp value after the entire message is received and parsed. Therefore, in this example, the ONU will read MPCP clock Δ_{ONU} TQ

OLT GATE reference point is advanced by Δ_{OLT} from 1st byte of DA

OLT REGISTER_REQ reference point corresponds to 1st byte of DA

$T_{response}$

OLT

Discovery GATE

REGISTER_REQ

Time

Timestamp = t_0

Timestamp = t_1

Discovery GATE

REGISTER_REQ

Time

ONU Δ_{OLT} $T_{downstream}$ Δ_{ONU} T_{wait} $T_{upstream}$

ONU GATE reference point is delayed by Δ_{ONU} from 1st byte of DA

ONU REGISTER_REQ reference point corresponds to 1st byte of DA

(a) Measured RTT: $RTT = T_{response} - T_{wait} = T_{downstream} + T_{upstream} + \Delta_{OLT} + \Delta_{ONU}$

OLT MPCP time = t_0

Actual data arrival time A

OLT

GATE

Time

Timestamp = t_0
StartTime = $S - RTT$

Timestamp = t_1

GATE

Time

ONU Δ_{OLT} $T_{downstream}$ Δ_{ONU} $T_{wait} = S - RTT - t_0$ $T_{upstream}$

Set ONU MPCP clock = t_0

StartTime

(b) Data Arrival: $A = t_0 + \Delta_{OLT} + T_{downstream} + S - RTT - t_0 + \Delta_{ONU} + T_{upstream} = S$

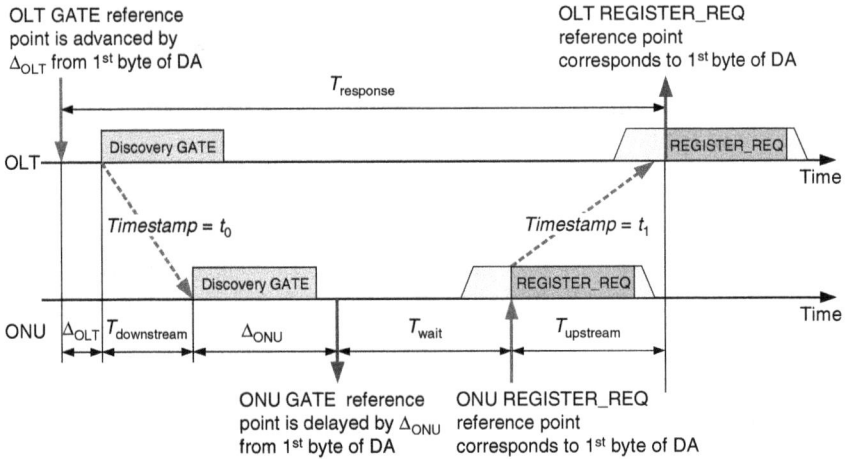

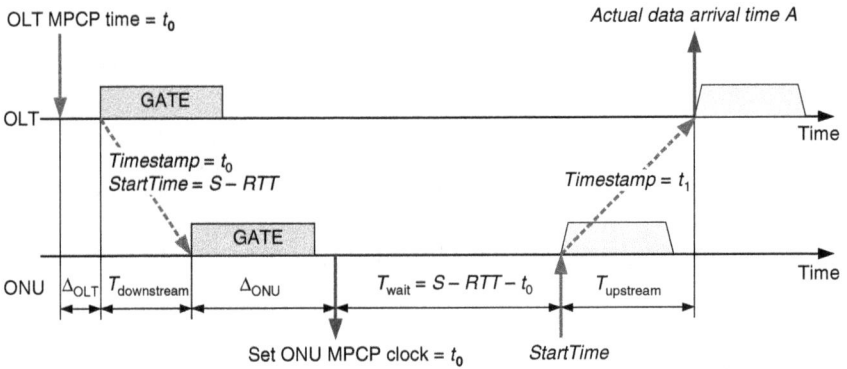

Figure 5.12 Precalculation of arrival time when the OLT and ONUs use different GATE timestamp reference points.

ahead of transmitting the first byte of DA, and the OLT will latch message arrival time Δ_{OLT} TQ after receiving the first byte of DA. As shown in Fig. 5.13a, the calculated RTT value will be

$$RTT = T_{downstream} + T_{upstream} + \Delta_{OLT} + \Delta_{ONU} \tag{5.4}$$

Fig. 5.13b illustrates a cycle of bandwidth assignment performed after the autodiscovery completes. As can be seen from the diagram, the actual arrival time A is

OLT GATE reference point
corresponds to 1st byte of DA

OLT REGISTER_REQ
reference point delayed by
Δ_{OLT} from 1st byte of DA

$T_{response}$

OLT — Discovery GATE   REGISTER_REQ   Time

Timestamp = t_0 Timestamp = t_1

ONU   Discovery GATE   REGISTER_REQ   Time

$T_{downstream}$ T_{wait} Δ_{ONU} $T_{upstream}$ Δ_{OLT}

ONU GATE reference point
corresponds to 1st byte of DA

ONU REGISTER_REQ
reference point is advanced
by Δ_{ONU} from 1st byte of DA

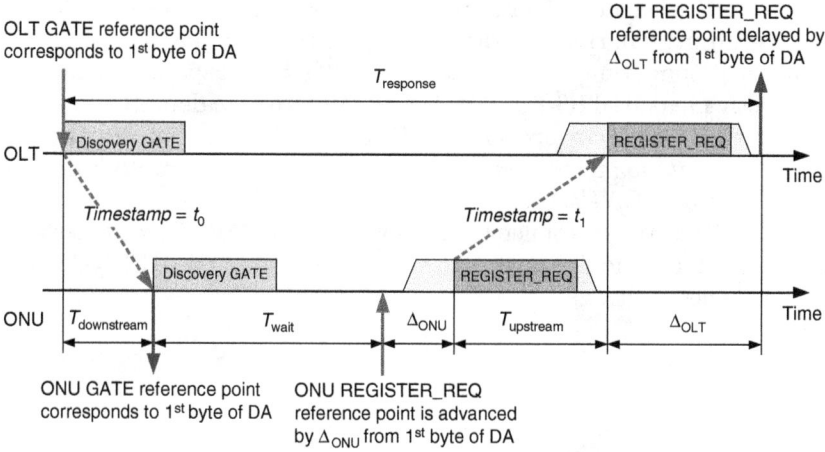

(a) Measured RTT: $RTT = T_{response} - T_{wait} = T_{downstream} + T_{upstream} + \Delta_{OLT} + \Delta_{ONU}$

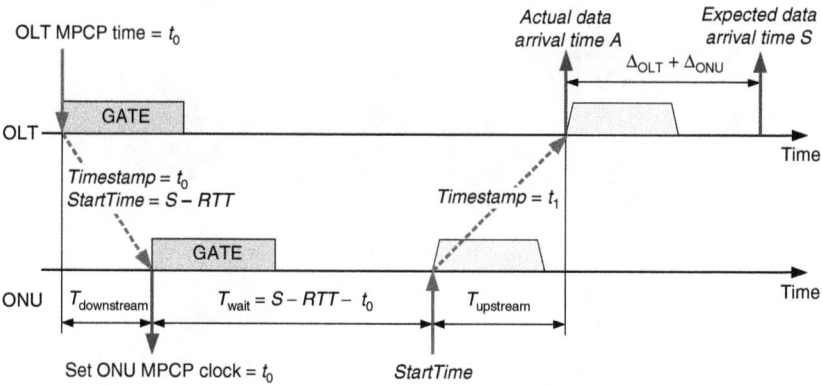

OLT MPCP time = t_0

Actual data
arrival time A

Expected data
arrival time S

$\Delta_{OLT} + \Delta_{ONU}$

OLT — GATE   Time

Timestamp = t_0
StartTime = $S - RTT$

Timestamp = t_1

ONU   GATE   Time

$T_{downstream}$ $T_{wait} = S - RTT - t_0$ $T_{upstream}$

Set ONU MPCP clock = t_0 StartTime

(b) Data Arrival: $A = t_0 + T_{downstream} + S - RTT - t_0 + T_{upstream} = S - \Delta_{OLT} - \Delta_{ONU}$

Figure 5.13   Precalculation of arrival time when the OLT and ONUs use different REGISTER_REQ timestamp reference points.

$$
\begin{aligned}
A &= t_0 + T_{\text{downstream}} + S - RTT - t_0 + T_{\text{upstream}} \\
&= t_0 + T_{\text{downstream}} + S - T_{\text{downstream}} - T_{\text{upstream}} \\
&\quad - \Delta_{\text{OLT}} - \Delta_{\text{ONU}} - t_0 + T_{\text{upstream}} \\
&= S - \Delta_{\text{OLT}} - \Delta_{\text{ONU}}
\end{aligned}
\tag{5.5}
$$

The actual data arrival time A is earlier than the expected arrival time S by $\Delta_{\text{OLT}} + \Delta_{\text{ONU}}$. During the autodiscovery, the time delta between REGISTER_REQ timestamp reference points used by the OLT and an ONU is accounted for as part of the upstream propagation delay. However, during normal operation, the upstream propagation delay is simply a propagation time of the signal. This discrepancy results in incorrectly measured RTT and OLT's inability to precalculate the exact data arrival times. *For proper MPCP operation, the actual timestamp reference points for upstream MPCP messages should exactly coincide for the OLT and ONUs.*

The fact that the reference points for upstream MPCP messages are not required to coincide appears to be an oversight in the IEEE 802.3ah standard specification.

6

Logical Topology Emulation

The IEEE 802 architecture [802] assumes all communicating stations in a LAN segment to be connected to a shared medium. In a shared medium, all stations are considered as belonging to a single *access domain*, where at most one station can transmit at a time and all stations can receive all the time.

Multiple access domains can be interconnected by a device called a *bridge*. Bridges selectively forward packets to create an appearance of a LAN consisting of all access domains. The selective forwarding prevents transmitting a packet into the domains that do not include any destination stations for this packet. Bridging of multiple LANs is widely used to provide administrative isolation of access domains, to increase the number of stations or the physical reach of the network beyond the limitations of individual LAN segments, and to improve the throughput.

In an extreme case, an access domain may consist of just one station. Typically, many such single-station domains are connected by *point-to-point* (P2P) links to a bridge, forming a *switched LAN*.

Relying on the notion of access domains, bridges never forward a frame back to its ingress port. In case the access domain consists of multiple stations, it is assumed that all the stations connected to the same port on the bridge can communicate with one another without the bridge's help. In the case of a switched LAN, there can be no recipients in the access domain of the sender, so no frames are ever forwarded back.

This bridge behavior has led to an interesting problem: Users connected to different ONUs in the same PON cannot belong to the same LAN and are unable to communicate with one another at layer 2 (data link layer). The reason is that the PON medium does not allow ONUs to communicate with one another directly, due to the directivity of passive splitters/combiners. Yet, the OLT has only a single port connecting to all ONUs, and a bridge located in the OLT would never forward a data frame back to its ingress port. In IEEE 802.3ah task force, this issue raised a question of EPON compliance with IEEE 802 architecture, particularly with P802.1D bridging.

The above example is illustrative of the conflict endured by EPON throughout its entire development cycle in the IEEE 802.3ah work group. On one hand, to be part of the IEEE 802.3 Ethernet standard, EPON specification must comply with all the requirements put forward by the 802 architectural model. Specifically, all stations interconnected by a shared medium should form an access domain and be able to communicate with one another. On the other hand, EPON was being developed for subscriber access networks with requirements drastically different from those of private LANs. Subscriber access networks serve noncooperating, independent users who, for various security, regulatory, and economical reasons, may not be allowed to communicate to one another, except when provisioned by a network operator to do so.

To resolve this issue and to ensure seamless integration with other Ethernet networks, devices attached to the PON medium implement a *logical topology emulation* (LTE) function that, based on its configuration, may emulate either a shared medium or a point-to-point medium.

To preserve the existing Ethernet MAC operation defined in the IEEE 802.3 standard, the LTE function should reside below the MAC sublayer. Operation of this function relies on tagging of Ethernet frames with tags unique for each ONU. These tags are called *logical link identifiers* (LLIDs) and are placed in the preamble at the beginning of each frame. To guarantee uniqueness of LLIDs, each ONU is assigned one or more tags by the OLT during the initial registration (autodiscovery) phase.

6.1   Point-to-Point Emulation (P2PE)

The objective of P2P emulation mode is to achieve the same physical connectivity as in switched LAN, where all the stations are connected to a central switch using point-to-point links.

In P2P emulation mode, the OLT must have N MAC ports (interfaces), one for each ONU (Fig. 6.1). During ONUs registration, a unique LLID value will be assigned to each ONU. Each MAC port at the OLT will be assigned the same LLID as its corresponding ONU.

When sending a frame downstream (from the OLT to an ONU), the emulation function in the OLT inserts the LLID associated with a particular MAC port that the frame arrived from (Fig. 6.1*a*). Even though the frame will pass through a splitter and reach each ONU, only one P2PE function will match that frame's LLID with the value assigned to the ONU and will accept the frame and pass it to its MAC layer for further verification. LTE functions in all other ONUs will discard this frame, so the MAC sublayers will never see that frame. In this

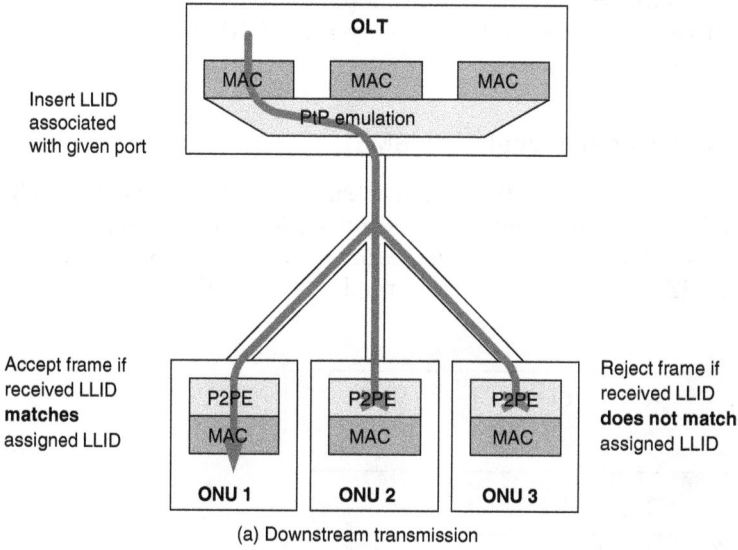

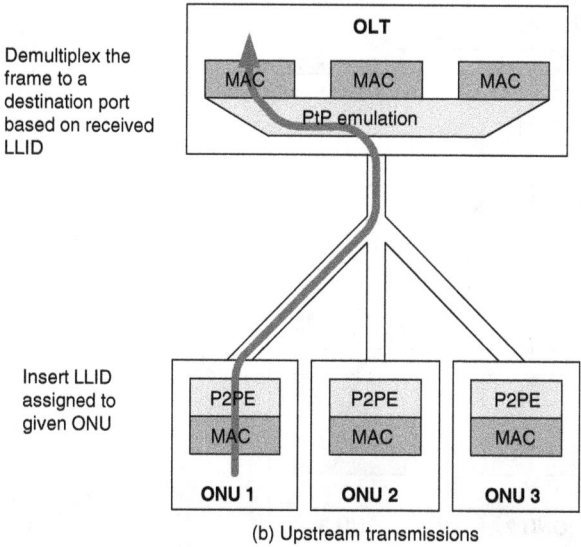

(a) Downstream transmission

(b) Upstream transmissions

Figure 6.1   Point-to-point virtual topology emulation.

sense, from the MAC sublayer perspective, it appears as if the frame was sent on a point-to-point link to only one ONU.

In the upstream direction, the ONU will insert its assigned LLID in the preamble of each transmitted frame. The P2PE function in the OLT will demultiplex the frame to the proper MAC port based on this unique LLID (Fig. 6.1*b*).

The P2PE configuration is clearly compatible with bridging, as each ONU is virtually connected to an independent bridge port. The bridge placed in the OLT (Fig. 6.2) will relay inter-ONU traffic between its ports.

6.2 Shared-Medium Emulation (SME)

In shared-medium emulation, frames transmitted by *any* node (OLT or any ONU) should be received by *every* node (OLT and every ONU), except the sender. In the downstream direction, the OLT inserts a *broadcast* LLID, which will be accepted by every ONU (Fig. 6.3*a*).

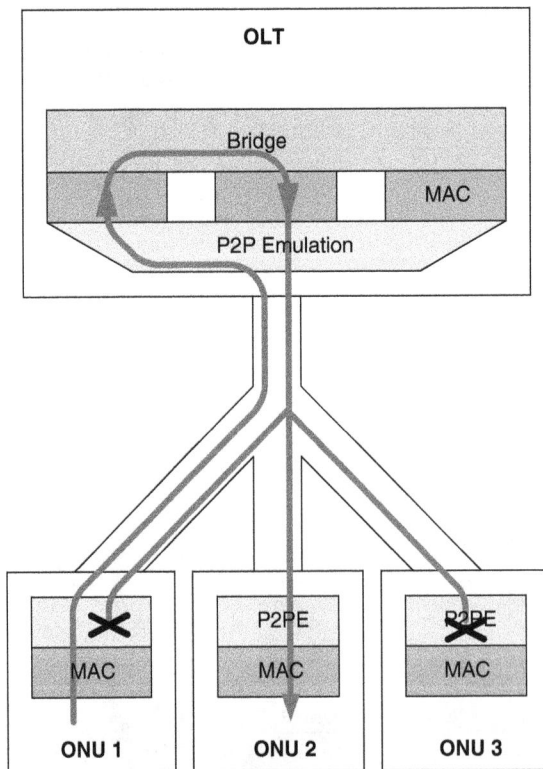

Figure 6.2   Bridging between ONU 1 and ONU 2 using the point-to-point emulation.

To ensure shared-medium operation for upstream data (frames sent by ONUs), the LTE function in the OLT must mirror all frames back downstream to be received by all other ONUs (Fig. 6.3*b*). To avoid frame duplication, when an ONU receives its own frame, the LTE function in an ONU accepts a frame only if the frame's LLID is different from the LLID assigned to that ONU. Thus, in SME mode, the ONU's filtering rules are opposite to those of P2PE mode. While in P2PE mode an ONU only accepts frames whose LLIDs match ONU's own LLID, in the SME

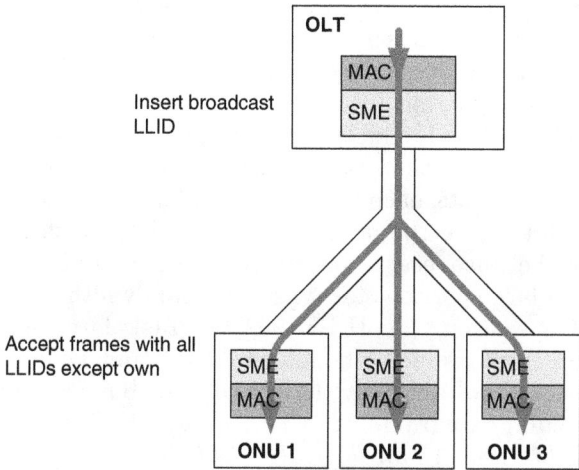

(a) Downstream transmission

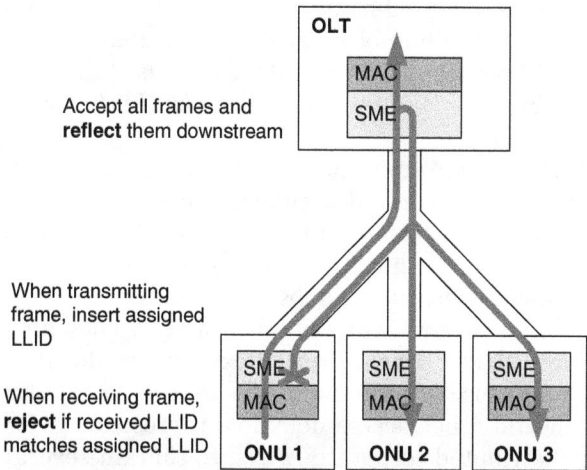

(b) Upstream transmissions

Figure 6.3   Shared-medium emulation.

mode an ONU accepts frames whose LLIDs are different from the ONU's assigned LLID.

The shared-medium emulation requires only one MAC port in the OLT and presents PON to a bridge as a single access domain. Physical-layer functionality (LTE function) provides the ONU-to-ONU communicability, eliminating the need for a bridge.

6.3   Combined P2PE and SME Mode

While both P2PE and SME options provide solutions for P802.1 standards compliance issues, both also have drawbacks, specifically when considered for an application in a subscriber access network. The P2PE mode precludes a single-copy multicast/broadcast when a single frame sent by the OLT is received by several ONUs. This feature is very important for services such as video broadcast or any real-time broadcast services. To support such services, the OLT operating in the P2PE mode must duplicate broadcast packets, each time with a different LLID.

Shared-medium emulation, on the other hand, provides broadcast capabilities. However, because *every* upstream frame is reflected downstream, it wastes a large portion of downstream bandwidth.

To achieve optimal operation, the IEEE 802.3ah task force has considered the possibility of using both point-to-point and shared-medium emulation modes simultaneously. To identify which mode is to be used with each particular data frame, the 16-bit-wide LLID field was divided into a *mode bit* and 15-bit *logicalLinkId*. The mode bit represents the emulation mode with a 0 indicating the point-to-point emulation and a value of 1 indicating the shared-medium emulation. The basic idea was that if the received mode bit is 0, the LTE function at the ONU will accept the frame only if *logicalLinkId* matches its assigned *logicalLinkId*. If, however, the received mode bit is 1, the LTE function will accept the frame only if the received *logicalLinkId* does not match the assigned value.

The idea of combining different emulation modes did not work out quite well. The SME mode only allowed a single access domain per EPON, which means that a data frame sent by any ONU will reach every ONU. Yet, it was recognized that broadcasting user's frames to all other ONUs is not a desirable feature in subscriber access networks. What is needed is the ability to specify any number of access domains between 1 (SME mode) and N (P2PE mode). Such flexibility would allow some access domains, representing individual subscribers, to contain only a single ONU each and other access domains, representing, for example, campuses or distributed corporate LANs, to contain several ONUs.

At various times, the task force considered different ideas to achieve this. One proposal called for the LLID to be a bitmap with every bit mapped to a particular ONU. This would allow 2^N access domains with any combination of ONUs being able to form an access domain. Clearly, because only a limited number of bits are available in the preamble, this solution is not scalable with the number of ONUs.

A more flexible solution proposed splitting LLID into three fields: mode bit, *logicalGroupId*, and *logicalLinkId* [Cho04]. The *logical-GroupId* essentially identified an access domain. An ONU would only accept frames belonging to the same access domain, i.e., having a matching *logicalGroupId*. Should this proposal be accepted, the ONU's filtering rules would look like the following:

Accept frame only if

1. *logicalLinkId* is equal to broadcast LLID, or

2. Mode bit is 0 and the received *logicalGroupId* is equal to the assigned *logicalGroupId* and the received *logicalLinkId* is equal to the assigned *logicalLinkId*, or

3. Mode bit is 1 and the received *logicalGroupId* is equal to the assigned *logicalGroupId* and the received *logicalLinkId* is not equal to the assigned *logicalLinkId*.

This solution, as proposed, was limited to only 8 access domains and only 2047 logical links, which was a point of concern. And while technically this solution could be improved, it was proposed too late to be included in the standard.

6.4   Final Solution

Whether it was due to lack of interest or its uselessness in the access environment, the idea of shared emulation died and was buried without a ceremony. The compromise was to retain only the point-to-point emulation and add an auxiliary *single-copy broadcast* (SCB) port at the OLT. In such a configuration, in an EPON with N ONUs, the OLT will contain $N + 1$ MACs: one for each ONU (P2PE) and one for broadcasting to all ONUs (Fig. 6.4). To optimally separate the traffic, higher layers (above MAC) will decide which port to send the data to.

If the SCB and unicast ports are connected to a 802.1D bridge, it is possible that the *spanning tree protocol* (STP) will detect a loop, since the same ONU will be reachable through virtual P2P link and through virtual broadcast channel. To avoid STP disabling one of the ports that formed the loop, the standard recommends that the SCB port not be connected to a 802.1D bridge. The SCB channel is to be used for

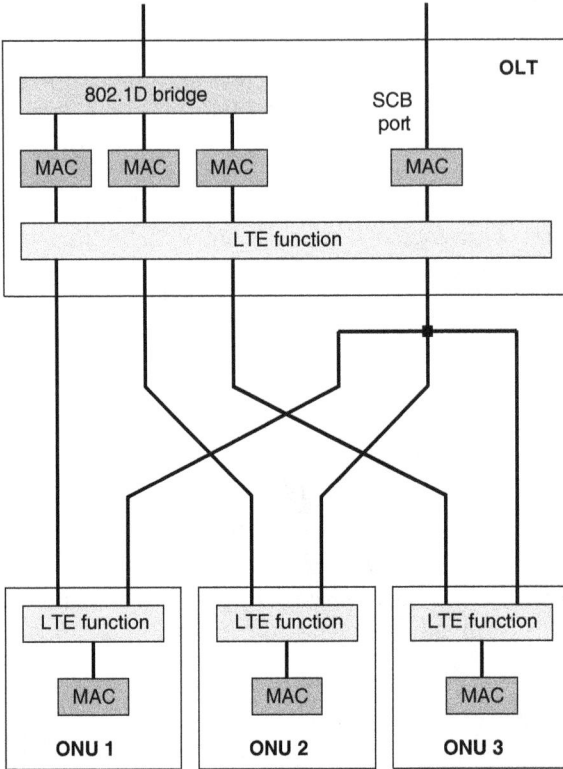

Figure 6.4   Combined point-to-point and shared-medium emulation mode.

downstream broadcast only. ONUs are not allowed to send upstream frames with broadcast LLID. The exceptions are several special control frames used for ONU's autodiscovery and registration.

6.4.1   LLID filtering rules

The following LLID filtering rules for the ONU are specified in the standard:

1. If the received mode bit is 0 and the received *logicalLinkId* value matches the assigned *logicalLinkId*, then the frame is accepted.

2. If the received mode bit is 1 and the received *logicalLinkId* value doesn't match the assigned *logicalLinkId*, then the frame is accepted.

3. If the received *logicalLinkId* is a broadcast *logicalLinkId* (has value 0x7FFF), then the frame is accepted.

4. All other frames are discarded by the LTE function.

At the OLT, the filtering rules are the following:

1. If the received *logicalLinkId* is a broadcast *logicalLinkId* (has value 0x7FFF) and a virtual port exists with an assigned broadcast *logicalLinkId*, then the frame is accepted and is transferred to this port.

2. If the received *logicalLinkId* is any value other than the broadcast *logicalLinkId* and a virtual port exists with an assigned mode bit 0 and an assigned *logicalLinkId* matching the received *logicalLinkId*, then the frame is accepted and is transferred to this port.

3. All other frames are discarded by the LTE function.

Even though the shared emulation mode died, its legacy in the form of the mode bit lived on. One may reasonably argue that the final specification is awkward, as neither is the mode bit necessary nor is its usage well defined. For example, if the OLT transmits a frame with a mode bit set to 1 and the *logicalLinkId* different from any assigned *logicalLinkId*, such frame will be accepted by every ONU. The same effect will be achieved if the frame is transmitted with the broadcast *logicalLinkId*. In effect, the mode bit simply reduces the available LLID address space almost by one half.

6.5 Preamble Format

The frame preamble is a relic from the early days of CSMA/CD networks. Because, in the CSMA/CD protocol, the channel goes silent between transmissions, the receiving station would need to resynchronize on each individual frame. The preamble simply consisted of alternating 0s and 1s (pattern 0101 . . . or octet values 0x55) and provided a periodic waveform of highest frequency for the given line rate. These days, Ethernet matured into higher speeds and CSMA/CD almost universally gave way to full-duplex Ethernet MAC. In full-duplex mode, even if the sender has no data to transmit, idle characters are being transmitted and the receiver remains synchronized at all times. Even though the CSMA/CD mode is gone, preambles in front of Ethernet frames remained as unsightly pimple scars reminding of Ethernet's puberty years. Not surprisingly, the IEEE 802.3ah task force decided to put the preamble to good use.

To allow additional information to be carried in the frame preamble, its format is modified as shown in Fig. 6.5. At the sending device, the LTE function, located in the reconciliation sublayer, replaces some of the octets of the preamble with several fields: *start of LLID delimiter* (SLD), LLID consisting of mode bit and *logicalLinkId*, and 8-bit *cyclic redundancy check* (CRC-8). The LTE function on the receiving side will

Size (octets) :| 1 | 8 | 6 | 6 | 2 | 46 – 1500 | 4 |

| 802.3 frame | S P D | Preamble | DA | SA | Len / type | Payload | ⟩⟨ | FCS |

Frame preamble | Reserved | SLD | Reserved | LLID | CRC-8 |

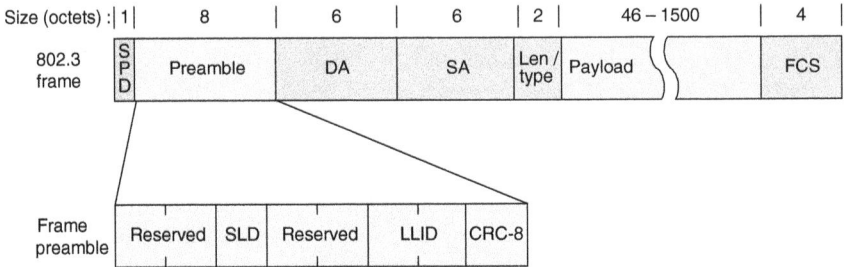

Figure 6.5   Format of frame preamble in EPON.

extract these fields and replace them with conventional preamble pattern before passing this preamble with the frame following it to the MAC sublayer.

6.5.1 Start-of -LLID delimiter

The SLD field has value 0xD5 and is located either immediately following the *start of packet delimiter* (SPD) field or one octet from the SPD. The reason for this is the gigabit Ethernet specification, which requires that the SPD code-group always be located at an even byte position. If the frame transmission starts (i.e., the TX_EN signal is asserted by GMII) at the even byte position, the SPD field (/S/ code-group) replaces the first octet of the preamble. If the frame transmission starts at an odd byte position, the first preamble octet is replaced by the idle and the second preamble octet is replaced by the SPD field. Figure 6.6 illustrates these two possibilities.

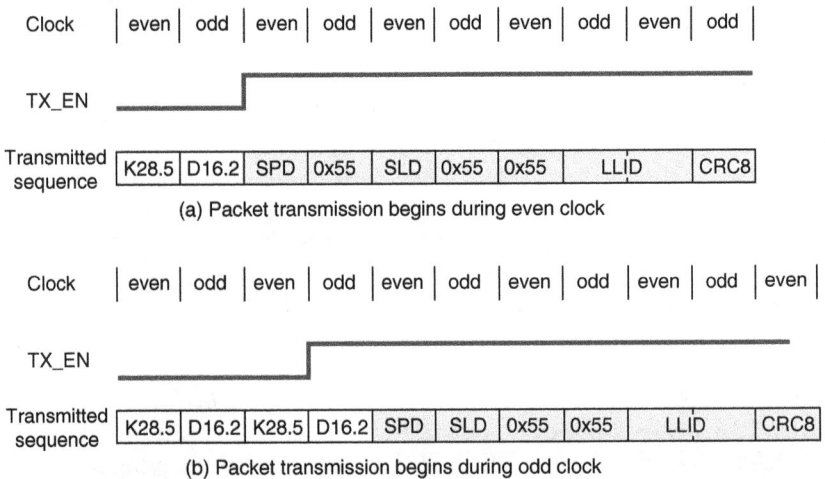

Clock | even | odd | even | odd | even | odd | even | odd | even | odd |

TX_EN

Transmitted sequence | K28.5 | D16.2 | SPD | 0x55 | SLD | 0x55 | 0x55 | LLID | CRC8 |

(a) Packet transmission begins during even clock

Clock | even | odd | even | odd | even | odd | even | odd | even | odd | even |

TX_EN

Transmitted sequence | K28.5 | D16.2 | K28.5 | D16.2 | SPD | SLD | 0x55 | 0x55 | LLID | CRC8 |

(b) Packet transmission begins during odd clock

Figure 6.6   Position of SLD field depending on odd/even byte alignment.

6.5.2 Cyclic redundancy check

Because in EPON the preamble carries useful information, the receiving device should be able to verify the preamble's integrity. This is achieved by adding the cyclic redundancy check field to the preamble. The transmitting device calculates the CRC over the fields of the preamble, starting with the SLD field and ending with the LLID field, a total of five octets. The receiving device also calculates the CRC value over the same fields and compares it with the received value. The nonmatching CRC indicates one or more transmission errors. Some of these errors may possibly be in the LLID field; this means that the LTE function cannot reliably determine whether the given frame is really destined to this device or not. Therefore, if the received and calculated CRC values do not match, the frame that follows this preamble should be discarded.

The standard specifies the following generating polynomial to calculate the CRC value:

$$G(x) = x^8 + x^2 + x + 1$$

The generating polynomial has degree 8 and generates 8-bit-wide CRC codes, hence the name of the method: CRC-8. The CRC-8 checksums will detect all single-bit errors and all errors with an odd number of erroneous bits, and burst errors less than 8 bits long.

The CRC calculation can be performed using the shift register, as shown in Fig. 6.7. This register should be initialized to zero. The residual register value, after all the data is shifted through this register, represents the CRC-8 checksum. The data octets are shifted through the register in order of transmission, the least-significant bit first.

Figure 6.8 represents a function that calculates the CRC-8 according to the above method. This function may be used to generate CRC-8 test values. Please note that this function requires a large number of shift operations and may not be efficient for some implementations. In practice, more advanced methods based on table lookup are often used.

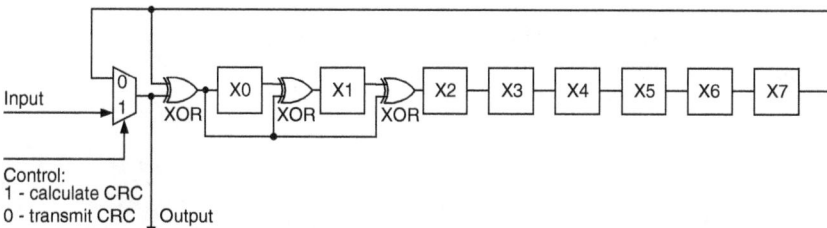

Figure 6.7 Shift register generating CRC-8.

```
typedef unsigned char   octet_t;

#define CRC8_POLYNOMIAL = 0x07;
#define LSB( X )         ( X & 0x01 )

octet_t CRC8( octet_t* data, int num_octets )
{
    octet_t shift_reg = 0;  // register holding the crc value
    octet_t octet;

    for( int i = 0; i < num_octets; i++ )
    {
        octet = data[i];
        for( int offset = 0; offset < 8; offset++, octet >>= 1 )
        {
            if( LSB(octet) ^ LSB(shift_reg) )
            {
                shift_reg >>= 1;
                shift_reg ^= CRC8_POLYNOMIAL;
            }
            else
                shift_reg >>= 1;
        }
    }
    return shift_reg;
}
```

Figure 6.8 Function calculating CRC-8 value.

Reference

[Cho04]   Su-il Choi, "Multicasting in EPON," presented at IEEE 802.3 plenary meeting in Orlando, FL, March 2004. Presentation is available at http://www.ieee802.org/3/efm/public/comments/d3_1/pdfs/choi_p2mp_1_0304.pdf.

7

Laser Control Function

Even in the absence of data transmissions, lasers generate *spontaneous emission noise*. This noise, accumulated over all nontransmitting ONUs, can easily obscure the data signal from a distant ONU. Thus, the ONU's lasers must be turned off between transmissions.

The MPCP framework developed by IEEE 802.3ah originally considered an ONU's laser controlled by a signal generated by the MAC control sublayer [Gum+01]. This seemed a logical decision since only the MPCP located in MAC control had knowledge of the assigned transmission windows and could turn the laser on and off at the precise moments of time.

However, in subsequent discussions it was recognized that such an approach, while technically feasible, nevertheless presents a violation of the protocol layering model, since it would require the laser control signal to bypass multiple sublayers: MAC, RS, GMII, PCS, and PMA (see Fig. 4.1). A new solution was found only in September 2003 [KM03]. This solution would allow the PCS sublayer to monitor passing data units and decide when the laser should be turned on and off. This new functionality is defined in a new PCS function called *data detector*.

7.1 Data Detector Function

On the data path, the data detector is located after 8b/10b encoding; therefore, it operates on 10-bit words, further referred to as *code-groups*. In essence, data detector is simply a delay line, which imposes a constant delay on data passing through the PCS (see Fig. 7.1).

The delay line can be implemented as a FIFO buffer with matching input and output rates. The purpose of introducing a delay line is to

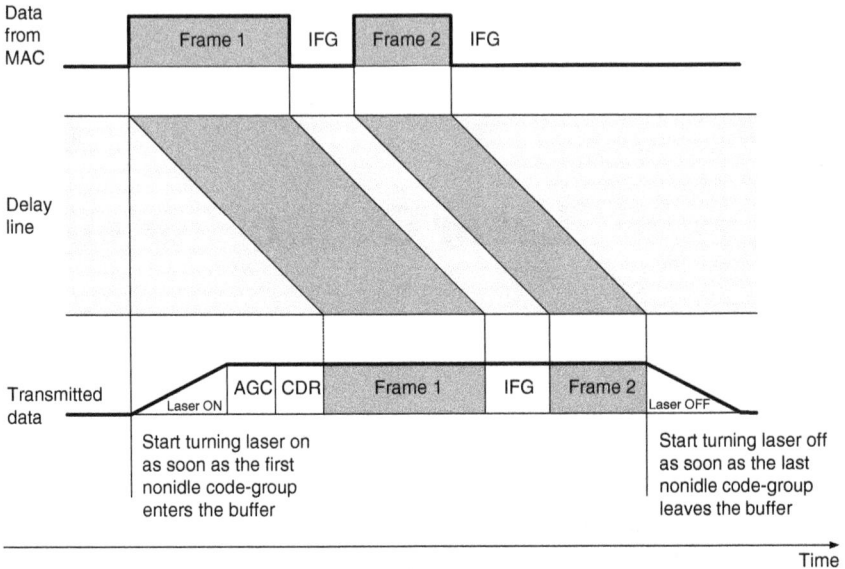

Figure 7.1   Timing diagram of data detector function.

provide the physical layer enough time to turn the laser on and generate the necessary synchronization sequence before transmitting the data. This synchronization sequence consists of idle code-groups required by the receiver to perform gain adjustment (AGC interval) and synchronize clock (CDR interval).

Upon initialization, the buffer is filled with idle code-groups. When the first nonidle code-group arrives to the buffer, the buffer immediately generates a signal to turn on the laser. By the time the first data code-group reaches the head of buffer, the laser will have been completely turned on and the necessary synchronization sequence will have been transmitted.

When the last nonidle code-group leaves the buffer, the data detector generates the signal to turn off the laser. Since it is known that the buffer is empty of data, the data detector is guaranteed to have enough time to turn on the laser when the next nonidle character arrives.

Figure 7.2 illustrates that if an ONU leaves a large idle gap in its upstream transmission, the data detector may partially or completely shut down the laser in the middle of a granted timeslot. This behavior does not introduce any undesirable effects, since data detector will always guarantee sufficient time for the laser to turn back on and the synchronization sequence to be generated.

Another externally observable feature of the data detector function is that if an ONU is granted a timeslot but it has no data frames to

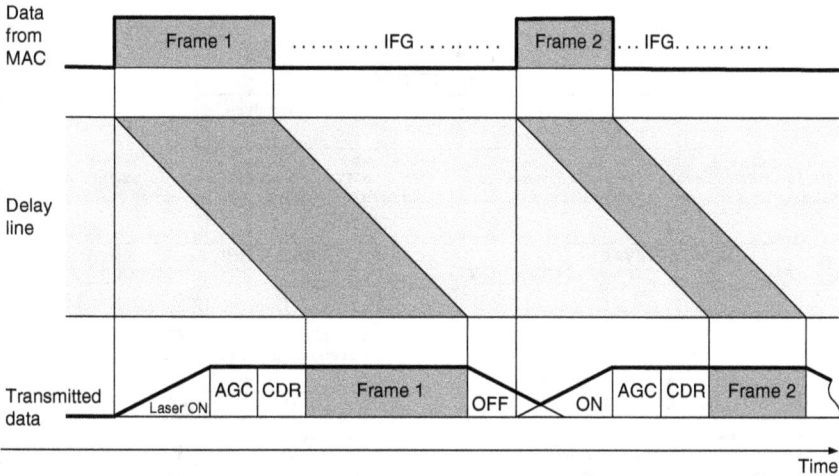

Figure 7.2 Illustration of partial laser shutdown during ONU's transmission.

transmit, its laser will not turn on. It is important that the OLT be designed in such a way that it understands such ONU's behavior. Some OLT implementations rely on detecting received optical power within a timeslot granted to a particular ONU to ensure that the ONU is alive. Such implementations may not operate properly if the ONU is allowed not to turn on the laser.

7.2 Data Detector State Diagram

The data detector state diagram is shown in Fig. 7.3. To monitor whether the FIFO buffer is empty (i.e., contains only idle code-groups), the data detector maintains a variable called IdleLength, which represents the continuous run of idles ending with the most recently received code-group. If the most recently received code group is not idle, the IdleLength is reset to 0.

7.2.1 WAIT_FOR_CODE_GROUP state

Upon initialization, the data detector enters the WAIT_FOR_CO-DE_GROUP state and remains in this state until the 8b/10b encoder issues the service primitive PMA_UNITDATA.request(tx_code_group), requesting the data decoder to transmit the next code-group. If this next code-group is idle, as determined by the IsIdle(...) function, the data detector transitions to IDLE_ARRIVAL state; otherwise, it enters DATA_ARRIVAL state.

BEGIN

```
┌─────────────────────────────────────────────────────────┐
│                   WAIT_FOR_CODE_GROUP                    │
├─────────────────────────────────────────────────────────┤
│   // Receive next tx_code_group from 8b/10b encoder     │
└─────────────────────────────────────────────────────────┘
```

PMA_UNITDATA.request(tx_code_group) AND   PMA_UNITDATA.request(tx_code_group) AND
IsIdle(tx_code_group) == false   IsIdle(tx_code_group) == true

```
┌──────────────────────────┐              ┌──────────────────────────┐
│       DATA_ARRIVAL       │              │       IDLE_ARRIVAL       │
├──────────────────────────┤              ├──────────────────────────┤
│   IdleLength = 0         │              │   IdleLength++           │
└──────────────────────────┘              └──────────────────────────┘
```

                                          LaserIsOn == true AND   else
     LaserIsOn == false   else   IdleLength ≥ DelayBound

```
┌──────────────────────────┐              ┌──────────────────────────┐
│      TURN_LASER_ON       │              │      TURN_LASER_OFF       │
├──────────────────────────┤              ├──────────────────────────┤
│   LaserIsOn = true       │              │   LaserIsOn = false       │
│   PMD_SIGNAL.Request(true)│              │   PMD_SIGNAL.Request(false)│
└──────────────────────────┘              └──────────────────────────┘
```

          UCT   UCT

```
┌─────────────────────────────────────────────────────────┐
│                   TRANSMIT_CODE_GROUP                    │
├─────────────────────────────────────────────────────────┤
│   dtx_code_group = FIFO.RemoveHead()                    │
│   FIFO.Append(tx_code_group)                            │
└─────────────────────────────────────────────────────────┘
```

                          UCT

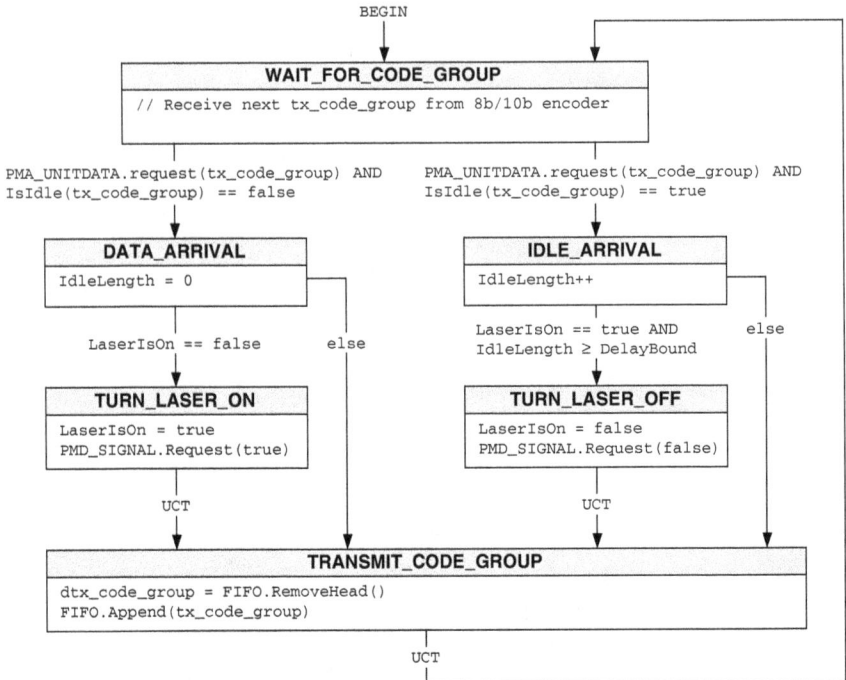

Figure 7.3   Data detector state diagram. (*Reprinted from IEEE Standard 802.3ah with permission from IEEE.*)

7.2.1.1 IsIdle(..) function.

The IsIdle(tx_code_group) function checks whether the next code-group is idle. Recall that in the IEEE 802.3 standard, idle is represented by two code-groups, also called an *ordered set*. The first code-group in the ordered set is /K28.5/, and the second code-group is either /D5.6/ or /D16.2/ (refer to [802.3] for information of how the idle ordered sets are generated). The /D5.6/ and /D16.2/ code-groups may occur in data frames; however, a combination of /K28.5/ followed by /D5.6/ or /D16.2/ represents an idle. In addition, these code-groups can take one of two possible 10-bit values depending on the current state of running disparity.

The standard provides the following definition for the IsIdle(...) function:

> IsIdle(tx_code_group). This function is used to determine whether tx_code_group is /T/, /R/, /K28.5/, or any code-group other than /D21.5/ or /D2.2/ that follows a /K28.5/.

Turning on the laser by an ONU at a wrong time can have undesirable consequences in EPON, leading to data corruption or even ONU deregistration. The `IsIdle(...)` function should be robust enough, and even conservative in preventing erroneous activation of the laser. Therefore, this function treats a combination of /K28.5/ followed by any other code-group as an idle, and encountering such a combination will not trigger the laser activation process. The exception to this rule is two ordered sets: /K28.5/D21.5/ and /K28.5/D2.2/. These ordered sets represent configuration messages used for line rate autonegotiation, and generally, they should not be encountered in EPON. It would be a safer approach for the `IsIdle(...)` function not to differentiate these two sets from all other combinations and to treat them as idles as well.

⚠️

Please note that in the IEEE 802.3ah standard, the definition for the `IsIdle` (tx_code_group) function is incorrect.

The /T/ and /R/ code-groups together form a sequence called the *end-of-packet delimiter* (EPD), which is used to delineate the ending boundary of a packet. The EPD can be represented by either combination /T/R/ or /T/R/R/. The particular sequence is chosen such that the idle following it started in an even code-group position.

Following this definition, the `IsIdle` (...) function would return true for /T/ and /R/ code-groups. Thus, even though these code-groups may still remain in the buffer, the data detector may start shutting down the laser. This would result in an ONU cutting short its transmission, truncating the EPD of the last packet in a burst.

The definition of the `IsIdle` (...) function is expected to be corrected in a future maintenance revision of the standard.

Considering both the above error and the improved robustness if configuration ordered sets are treated as idles, we recommend the following definition of the `IsIdle` (...) function to be used for practical implementations:

`IsIdle(tx_code_group)`. This function is used to determine whether tx_code_group is /K28.5/ or any code-group that follows a /K28.5/.

The pseudocode for the corrected `IsIdle` (tx_code_group) function is shown below:

```
prev_code_group = /K28.5/; // start with empty buffer
bool IsIdle( tx_code_group )
{
    if(tx_code_group == /K28.5/ OR prev_code_group == /K28.5/)
    {
        prev_code_group = tx_code_group;
```

```
    return TRUE;
}
prev_code_group = tx_code_group;
return FALSE;
}
```

7.2.2 DATA_ARRIVAL state

If a nonidle code-group is received, the variable IdleLength is set to 0, even though this variable may already have value 0 if the previous code-group was not idle either. In this state, the data detector checks the current state of the laser, represented by the boolean variable IsLaserOn. If the laser is currently turned off, as indicated by condition IsLaserOn == false, the data detector enters state TURN_LASER_ON, where it will turn on the laser; otherwise, it transitions directly to state TRANSMIT_CODE_GROUP.

7.2.3 TURN_LASER_ON state

In this state, the data detector function turns on the laser. The service primitive to control the laser is called PMD_SIGNAL.request(..), and it should be given the argument true in order to turn on the laser. From this state, the data detector unconditionally transitions to state TRANSMIT_CODE_GROUP.

7.2.4 IDLE_ARRIVAL state

An arrival of idle code-group increments the variable IdleLength by 1. Further, the value of this variable is compared to the DelayBound constant. The DelayBound represents the FIFO buffer depth. The value of IdleLength being equal to or greater than DelayBound means that the entire FIFO buffer is filled with idle ordered sets, and therefore, it is safe to turn off the laser.

If the new value of IdleLength is greater than or equal to DelayBound and the laser is currently turned on, the data detector will enter the state TURN_LASER_OFF. Otherwise, if the laser is already turned off, the state machine will transition directly to the TRANSMIT_CODE_GROUP state.

7.2.5 TURN_LASER_OFF state

In this state, the data detector instructs the PMD to turn off the laser by issuing the service primitive PMD_SIGNAL.request(false). From this state, the data detector unconditionally transitions to state TRANSMIT_CODE_GROUP.

7.2.6 TRANSMIT_CODE_GROUP state

In this state, the data detector shifts all stored code-groups one position. The head-of-line group is removed from the buffer and is transferred to transmit the function of the PMA sublayer (`dtx_code_group` = `FIFO.Re-moveHead()`). The code-group that just arrived to the data detector from the 8b/10b encoder is added at the end of the FIFO buffer (indicated by command `FIFO.Append(tx_code_group)`).

7.3 FIFO Buffer Size

The delay introduced by the FIFO buffer shall be sufficient to turn on the laser, and to generate the synchronization sequence needed by the OLT's receiver to adjust the gain and synchronize the clocks. The time to turn on laser T_{on} is specified by the IEEE 802.3ah standard to be 512 ns. The time to adjust the gain T_{AGC} and to synchronize clocks T_{CDR} can vary, but should not exceed 400 ns each. In addition, 32 ns should be allocated to the receiver to align code-groups ($T_{code\_group\_align}$). Therefore, the maximum delay introduced by the FIFO buffer will not exceed 512 + 400 + 400 + 32 = 1344 ns. The `DelayBound` value is expressed in terms of buffer capacity to store code-groups. Since one code-group arrives every 8 ns, the FIFO buffer should be able to store a maximum of 168 code-groups.

It was explained in Sec. 4.1.2 that the T_{AGC} or T_{CDR} times are negotiable; if the OLT has faster optical components or the signal quality is high, it may not require a 400-ns T_{AGC} or T_{CDR}. The OLT has the ability to convey the actual required combined T_{AGC}, T_{CDR}, and $T_{code\_group\_align}$ time to ONUs in the discovery GATE message in a field called `sync-Time`. Upon receiving the new synchronization time, the ONUs should change the FIFO buffer depth as well as the value of `DelayBound` to `syncTime` plus the T_{on} time to turn on the laser. The implementer should be mindful of the different units used for `DelayBound` and `syncTime`. `DelayBound` is expressed in 8-ns units equal to the transmission time of one code-group, whereas `syncTime` is expressed in units of time quanta equal to 16 ns each. Therefore, the following formula for setting the `Delay-Bound` should be used:

```
DelayBound = 2 × syncTime + 64
```

where 64 represents laser-on time T_{on} expressed in code-group transmission times (512 ns / 8 ns).

References

[802.3]   *IEEE Standard for Information Technology — Telecommunications and information exchange between systems—Local
and metropolitan area networks—Specific Requirements. Part 3:
Carrier sense multiple access with collision detection (CSMA/
CD) access method and physical layer specification*, ANSI/IEEE
Std. 802.3-2002, 2002 edition. Available at http://stand-
ards.ieee.org/getieee802/download/802.3-2002.pdf.

[Gum+01]   A. Gummalla et al., "Multi-Point Control Protocol: Common
Framework," IEEE 802.3ah task force meeting, Austin, TX,
November 2001. Available at http://www.ieee802.org/3/efm/
public/nov01/kramer_1_1101.pdf.

[KM03]   G. Kramer and A. Maislos, "LaserControl problem statement,"
IEEE 802.3ah task force meeting, Portonovo, Italy, September
2003. Available at http://www.ieee802.org/3/efm/public/sep03/
p2mp/kramer_1_0903.pdf.

8

Multi-Point Control Protocol: A Formal Specification

This chapter provides detailed explanations of *multi-point control protocol* (MPCP) control messages and state machines specified in IEEE 802.3ah standard, as it was approved by the Standards Board of IEEE Standards Association on June 24, 2004.

Each MPCP function may be implemented in a number of ways; quite often implementations may be more optimal and efficient than what is presented in the standard. However, all devices that claim conformance with the standard must implement the MPCP functions such that the externally observable behavior of a specific implementation is indistinguishable from behavior exhibited by the state machines specified in the standard.

8.1 MPCP Frame Structure

MPCP frames are commonly referred to as *MPCP data units* (MPCP-DUs). MPCP defines five messages used to exchange information between the OLT and ONUs: GATE, REPORT, REGISTER_REQ, REGISTER, and REGISTER_ACK. All MPCPDUs are 64-byte MAC control frames consisting of the following fields (Fig. 8.1):

1. *Destination address (DA)*. The destination address field of a MAC control frame contains the 48-bit address of the station(s) for which the frame is intended. With the exception of the REGISTER message, all MPCPDUs use a globally assigned 48-bit multicast address 01-80-C2-00-00-01$_{16}$. The REGISTER frames use the individual MAC address of the destination ONU.

Fields	Octets
Destination address (DA)	6
Source address (SA)	6
Length/type = $88\text{-}08_{16}$	2
Opcode	2
Timestamp	4
Opcode-specific fields/pad	40
Frame check sequence (FCS)	4

Figure 8.1   Format of generic MPCP frame.

2. *Source address (SA).* The source address field of a MAC control frame contains the 48-bit individual address of the station sending the frame. In the OLT, the LTE function interfaces a single *gigabit media-independent interface* (GMII) with multiple MAC instances, and these instances may be assigned unique addresses. The frames originated at the OLT should use the source address associated with the MAC instance which transmitted the frame.

3. *Length/type.* The length/type field of a MAC control frame is a 2-octet field that contains the hexadecimal value 88-08. This value has been universally assigned to identify MAC control frames.

4. *Opcode.* The opcode field identifies the specific MAC control frame as follows:

   $00\text{-}01_{16}$:   PAUSE
   $00\text{-}02_{16}$:   GATE
   $00\text{-}03_{16}$:   REPORT
   $00\text{-}04_{16}$:   REGISTER_REQ
   $00\text{-}05_{16}$:   REGISTER
   $00\text{-}06_{16}$:   REGISTER_ACK

5. *Timestamp.* The timestamp field carries the value of MPCP clock corresponding to the transmission of the first byte of the DA. The timestamp values are used to synchronize MPCP clocks in the OLT and ONUs.

6. *Opcode-specific fields.* These fields carry information pertinent to specific MPCP functions. The portion of the payload not used by the opcode-specific fields should be padded with zeros.

7. *Frame check sequence* (FCS). The FCS field carries a CRC-32 value used by the MAC to verify integrity of received frames.

Fields	Octets
Destination address (DA)	6
Source address (SA)	6
Length/Type = 88–08$_{16}$	2
Opcode = 00–03$_{16}$	2
Timestamp	4
Number of queue sets	1
Report bitmap	[1]
Queue #1 report	[2]
Queue #2 report	[2]
Queue #3 report	[2]
Queue #4 report	[2]
Queue #5 report	[2]
Queue #6 report	[2]
Queue #7 report	[2]
Queue #8 report	[2]
Pad = 0	0–39
Frame check sequence (FCS)	4

Repeated *n* times as indicated by *Number of queue sets*

Figure 8.2   Format of REPORT frame.

8.1.1   REPORT control frame

REPORT messages are used by ONUs to report local queue status to the OLT. The format of REPORT frame is shown in Fig. 8.2.

8.1.1.1   Queue #*n* report.   A REPORT message conveys queue lengths of up to 8 queues, represented by *queue #n report* fields. The reported queue lengths should be adjusted to account for the necessary frame preamble, interframe spacing, and FEC parity overhead, even though this additional data may not be physically present in the queue. In other words, this field represents not the actual queue length, but the transmission window size required to transmit the data stored in the queue.

8.1.1.2   Report bitmap.   The reported queues are identified by a *report bitmap*—an 8-bit field where a bit in position *n* represents queue #*n*. The bit *n* being set to 1 indicates that queue #*n* report field is present; otherwise the queue #*n* report is not present.

8.1.1.3   Number of queue sets.   If the ONU reports the total queue length, it is very likely that the OLT will grant a timeslot smaller than what was requested by the ONU, or else risk allocating an unproportionally large timeslot to an ONU. Lacking any additional knowledge about the queue's composition, the OLT is not able to grant a smaller timeslot that exactly fits some number of frames. Since Ethernet frames cannot be fragmented, a frame that does not fit in the slot remainder will have to be deferred to the next timeslot, leaving an unused remainder in the current timeslot.

To cope with this situation, the REPORT message may contain multiple *queue sets*. Each queue set reports a cumulative length (including the overhead) of a subset of queued packets, always starting from the head of the queue. If the OLT is unable to grant the entire queue length, it may choose the slot length equal to any of the reported values, and such timeslot will not have wasted bandwidth, due to packet delineation.

The standard is somewhat vague about how the subsets of packets should be determined. The original idea, discussed and adopted by the EFM task force, assumed that several *thresholds* would be specified for each queue. Then the reported queue length would be equal to the length of all packets (including the overhead) not exceeding the threshold (Fig. 8.3 shows an example of reported values for three queues and two thresholds). However, since the REPORT messages are generated by DBA client, which is outside the scope of the standard, the formal definition for multiple thresholds was never given.

The maximum number of queue sets in a REPORT message depends on how many queues are to be reported. Given the 64-byte limit on the length of the REPORT message, an ONU that reports 8 queues may have up to 2 thresholds per queue, as shown in Fig. 8.4a. If only one queue is available at the ONU, it may report up to 13 thresholds, as shown in Fig. 8.4b.

8.1.2   GATE control frame

The GATE control frame serves a dual role: A discovery GATE message is used to advertise a discovery slot for which all uninitialized ONUs may contend, and a normal GATE message is used to grant transmission opportunity to a single, already discovered ONU. The GATE message could easily be defined as two distinct messages (with different opcodes). For various reasons, the task force decided to keep it as a single message and differentiate the intended use of the message (discovery versus normal) using an additional field within the payload. Figure 8.5 presents the formats of discovery GATE and normal GATE messages.

8.1.2.1   Number of grants / flags.   As was explained in Sec. 5.3.1.2, each transmission window or grant is represented by a pair {*startTime*, *length*}. One GATE message can contain up to 4 grants. The number

Threshold T1 = 2000   Threshold T2 = 4000

Queue 0 | 1000 | 500 | 1200 | 1000

Queue 1 | 700 | 500

Queue 2 | 1500 | 1000 | 1000 | 1000

2000

4000

(a) Queue composition and thresholds

- Preamble = 4 TQ

- Inter-frame gap (IFG) = 6 TQ

Number of queue sets = 2

Report bitmap = 00000111$_2$

Queue #0 report = **770** (1540 bytes)   1000   500

Queue #1 report = **620** (1240 bytes)   700   500

Queue #2 report = **760** (1520 bytes)   1500

Report bitmap = 00000111$_2$

Queue #0 report = **1890** (3780 bytes)   1000   500   1200   1000

Queue #1 report = **620** (1240 bytes)   700   500

Queue #2 report = **1780** (3560 bytes)   1500   1000   1000

(b) Reported values

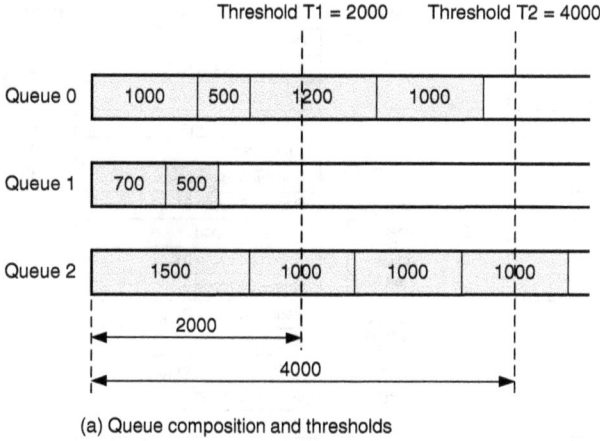

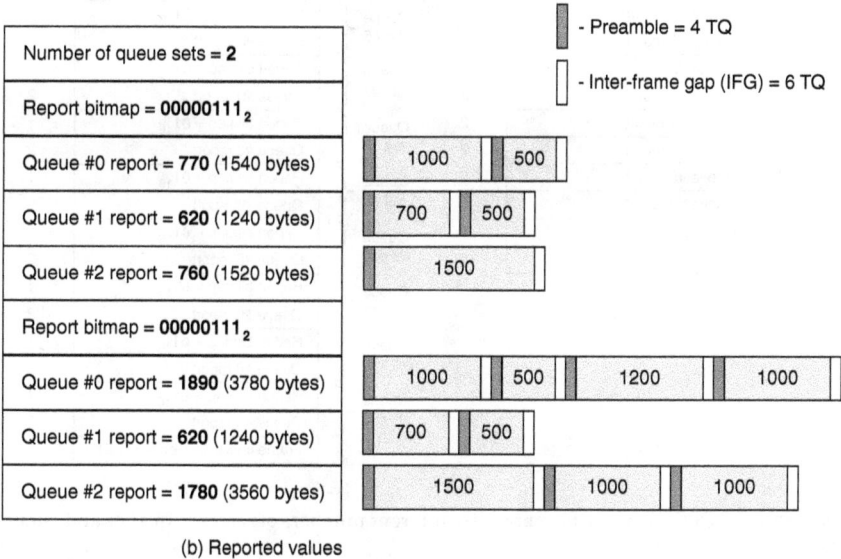

Figure 8.3   Example of queue's composition and reported values.

of grants/flag field indicates the exact number of grants in the given GATE message, as well as some additional information, summarized in Table 8.1.

The *number of grants* subfield indicates, as its name suggests, the number of grants in the GATE message. The valid values are 0 through 4. A GATE message with 0 grants does not assign a transmission window to an ONU and is only used as a keep-alive mechanism, explained later in this chapter.

Fields	Octets
Destination address (DA)	6
Source address (SA)	6
Length/type = 88–08$_{16}$	2
Opcode = 00–03$_{16}$	2
Timestamp	4
Number of queue sets = 2	1
Report bitmap = FF$_{16}$	1
Queue #0 report	2
Queue #1 report	2
Queue #2 report	2
Queue #3 report	2
Queue #4 report	2
Queue #5 report	2
Queue #6 report	2
Queue #7 report	2
Report bitmap = FF$_{16}$	1
Queue #0 report	2
Queue #1 report	2
Queue #2 report	2
Queue #3 report	2
Queue #4 report	2
Queue #5 report	2
Queue #6 report	2
Queue #7 report	2
Pad = 0	5
Frame check sequence (FCS)	4

Queue set #1 spans Report bitmap = FF$_{16}$ through Queue #7 report (first group). Queue set #2 spans the second Report bitmap = FF$_{16}$ through Queue #7 report.

(a)

Fields	Octets
Destination address (DA)	6
Source address (SA)	6
Length/type = 88–08$_{16}$	2
Opcode = 00–03$_{16}$	2
Timestamp	4
Number of queue sets = 13	1
Report bitmap = 01$_{16}$	1
Queue #0 report	2
Report bitmap = 01$_{16}$	1
Queue #0 report	2
Report bitmap = 01$_{16}$	1
Queue #0 report	2
Report bitmap = 01$_{16}$	1
Queue #0 report	2
Report bitmap = 01$_{16}$	1
Queue #0 report	2
Report bitmap = 01$_{16}$	1
Queue #0 report	2
Report bitmap = 01$_{16}$	1
Queue #0 report	2
Report bitmap = 01$_{16}$	1
Queue #0 report	2
Report bitmap = 01$_{16}$	1
Queue #0 report	2
Report bitmap = 01$_{16}$	1
Queue #0 report	2
Report bitmap = 01$_{16}$	1
Queue #0 report	2
Report bitmap = 01$_{16}$	1
Queue #0 report	2
Report bitmap = 01$_{16}$	1
Queue #0 report	2
Frame check sequence (FCS)	4

Each Report bitmap = 01$_{16}$ / Queue #0 report pair forms Queue set #1 through Queue set #13 respectively.

(b)

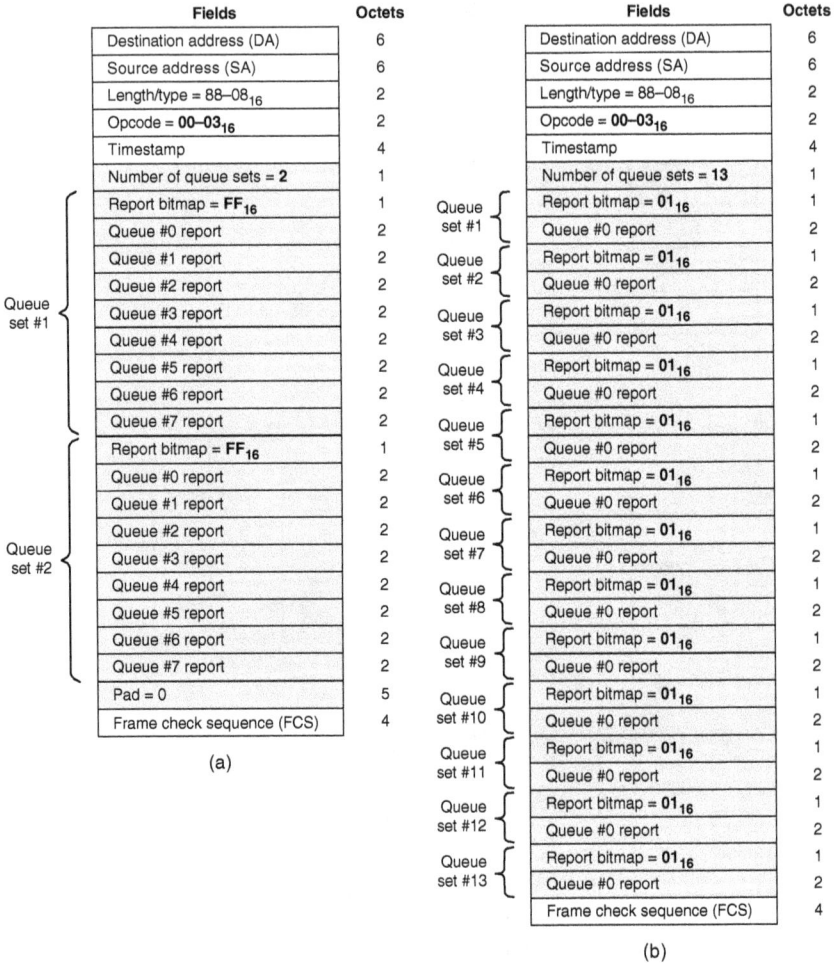

Figure 8.4 REPORT frame formats with different numbers of reported queues and queue sets.

The *discovery/normal GATE* bit indicates the purpose and payload format of the message. When set to 1, this field indicates that the frame is the discovery GATE; otherwise, it is a normal GATE.

Force report is a bitmap indicating whether the OLT requests the ONU to transmit the REPORT message in any of the assigned grants. Each bit in the force report bitmap corresponds to an individual grant as follows: bit 4 corresponds to grant 1, bit 5 corresponds to grant 2, bit 6 corresponds to grant 3, and bit 7 corresponds to grant 4. Typically, if the OLT requests a REPORT from an ONU in grant #n, the OLT should increase the length of this grant to accommodate the additional

Fields	Octets
Destination address (DA)	6
Source address (SA)	6
Length/type = 88–08$_{16}$	2
Opcode = **00–02$_{16}$**	2
Timestamp	4
Number of grants/flags = **09$_{16}$**	1
Grant start time	4
Grant length	2
Sync time	2
Pad = 0	31
Frame check sequence (FCS)	4

(a)

Fields	Octets
Destination address (DA)	6
Source address (SA)	6
Length/type = 88–08$_{16}$	2
Opcode = **00–02$_{16}$**	2
Timestamp	4
Number of grants/flags	1
Grant #1 start time	[4]
Grant #1 length	[2]
Grant #2 start time	[4]
Grant #2 length	[2]
Grant #3 start time	[4]
Grant #3 length	[2]
Grant #4 start time	[4]
Grant #4 length	[2]
Pad = 0	15/39
Frame check sequence (FCS)	4

(b)

Figure 8.5   Format of GATE message: (*a*) discovery GATE and (*b*) normal GATE.

TABLE 8.1   Contents of Number of Grants / Flags Field

Bits	Subfield name
0–2	Number of grants
3	Discovery/normal GATE
4–7	Force report bitmap

REPORT message. If the OLT grants exactly the length that was previously requested by the ONU, the REPORT message from the ONU will have to preempt one of data frames. If the preempted frame has lengths larger than the length of the REPORT frame, the granted slot will have an unused remainder and some bandwidth will be wasted. ONUs are also allowed to generate REPORT messages on their own, without the OLT requesting it.

The discovery GATE message always carries a single grant. This message should have the discovery/normal GATE bit set to 1, and should have none of the force report bits set. Therefore, in discovery GATEs, the number of grants/flags field always has the value of 09$_{16}$.

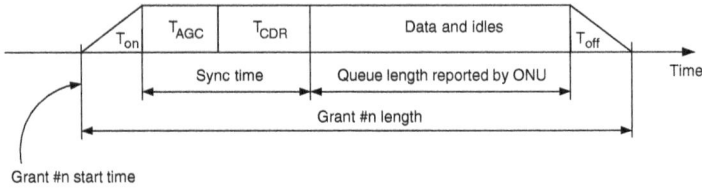

Figure 8.6   Grant structure.

8.1.2.2   Grant #n start time.   The *grant #n start time* field represents a value of ONU's local MPCP clock at which the ONU is supposed to start turning on the laser. The value of this field is represented in units of TQ. If a GATE message contains more than one grant, all the grants within the message should be ordered by their start times. However, no such requirement exists for the grants that are delivered in separate GATE messages, so the OLT may assign grants to ONUs in an order different from the order in which these grants will become active at ONUs.

In discovery GATE messages, the interpretation of the grant start time value is slightly different: rather than being the time when the laser should turn on, this field is interpreted as the time when the laser *may* turn on. As was explained in Sec. 5.3.2.2, uninitialized ONUs will wait some random delay before responding in the discovery slot. Thus, unless this random delay happened to be zero, the laser will be turned on sometime after the advertised grant start time.

8.1.2.3   Grant #n length.   The *grant #n length* field represents the length of ONU's transmission. This length is measured in TQ and includes the necessary time intervals to turn the laser on (T_{on}), generate the necessary synchronization sequence for the OLT's receiver to adjust its gain (T_{AGC}) and synchronize its receive clock (T_{CDR}), and finally, to turn off the laser (T_{off}). Figure 8.6 illustrates the grant structure and its relation to the grant length value.

Once again, in discovery GATE messages, the interpretation of the grant length field is different. In responding to the discovery GATE message, the uninitialized ONUs will keep the laser turned on only long enough to transmit the REGISTER_REQ message. The *grantLength* value is used by ONU to calculate the maximum allowed range for random delay applied to its transmission. For example, if the REGISTER_REQ transmission time, including T_{on}, T_{AGC}, T_{CDR}, and T_{off}, is T TQ, then the random delay may be chosen from the interval [0:*grantLength*–*T*]. Clearly, the goal here is to guarantee that no ONU ever transmits past the timeslot end time *grantStartTime* + *grantLength*.

Fields	Octets
Destination address (DA)	6
Source address (SA)	6
Length/type = $88\text{--}08_{16}$	2
Opcode = **$00\text{--}04_{16}$**	2
Timestamp	4
Flags	1
Pending grants	1
Pad = 0	38
Frame check sequence (FCS)	4

Figure 8.7   Format of REGISTER_REQ frame.

8.1.2.4 Sync time. The *sync time* field is present only in the discovery GATEs. Different burst-mode receivers at the OLT may require different AGC and/or CDR intervals. To improve EPON efficiency when faster receivers are used, the OLT uses the sync time field to advertise to ONUs the total of AGC and CDR times its receiver requires. According to the IEEE 802.3ah standard, the T_{AGC} time should not exceed 400 ns. The T_{CDR} interval consists of two components—bit synchronization and code-group alignment. The bit synchronization interval shall be no longer than 400 ns. The code-group alignment interval is specified not to exceed 32 ns. Thus, the maximum allowed sync time is 832 ns, or 52 TQ. The ONU should transmit only idle code-groups during the sync time period.

8.1.3 REGISTER_REQ control frame

REGISTER_REQ message is used by uninitialized ONUs to respond to discovery GATEs. When the OLT receives the REGISTER_REQ message from an ONU, it learns two key pieces of information: the round-trip time to the ONU and the ONU's MAC address.

Already registered ONUs may also issue the REGISTER_REQ message to request deregistration by the OLT. The format of this message is shown in Fig. 8.7.

8.1.3.1 Flags. The value of the *flags* field indicates whether the REGISTER_REQ message is requesting registration (flags = 1) or deregistration (flags = 3). All other values of the flags field are reserved. The OLT should ignore all received REGISTER_REQ messages in which the flags field takes one of the reserved values.

Fields	Octets
Destination address (DA)	6
Source address (SA)	6
Length/type = 88–08$_{16}$	2
Opcode = **00–05**$_{16}$	2
Timestamp	4
Assigned port	2
Flags	1
Sync time	2
Echoed pending grants	1
Pad = 0	34
Frame check sequence (FCS)	4

Figure 8.8   Format of REGISTER frame.

8.1.3.2 Pending grants. When an ONU receives a GATE message, it should store the grant parameters, such as start time, length, force report indication, and discovery flag, until the local MPCP clock reaches the grant start time value. The *pending grants* field indicates to the OLT the ONU's buffering capacity to store future grants. At the OLT, the received value of the pending grants field serves as an advisory to the DBA client to not issue more outstanding grants than the ONU can buffer. All outstanding grants in excess of the pending grants value will be discarded by the ONU.

8.1.4 REGISTER control frame

The REGISTER message is used by the OLT to assign a unique LLID to a newly discovered ONU. Among all MPCPDUs, the REGISTER message is the only one that uses the ONU's MAC address as the DA. This message is intended to only a single ONU, yet it is transmitted before logical link to this ONU is established. Therefore, the REGISTER message is transmitted with broadcast LLID, but with individual MAC address.

Additionally, the OLT may send the REGISTER message to an already registered ONU, to deregister this ONU or to request the ONU to repeat its registration procedure. Such messages will be sent on already established unicast logical links and may use a globally assigned 48-bit multicast address 01-80-C2-00-00-01$_{16}$. The format of the REGISTER message is shown in Fig. 8.8.

8.1.4.1 Assigned port. The assigned port field carries the LLID value being assigned to the given ONU. This value is unique per EPON. Once the LLID is assigned to the LTE function in the ONU and corresponding port at the OLT, the unicast logical link between the OLT and ONU is created.

8.1.4.2 Flags. The *flags* field identifies the specific registration instructions to the ONU. This field can take the following values:

1. *Reregister.* An already registered ONU is asked to reregister.

2. *Deregister.* An already registered ONU is asked to deallocate the LLID and transition to an uninitialized state. After deregistration, the ONU may participate in autodiscovery procedure again.

3. *Ack.* The discovery agent at the OLT confirms successful registration of an ONU.

4. *Nack.* The discovery agent at the OLT denies registration to an ONU. The ONU will remain in an uninitialized state.

All other values of the flags field are reserved. The ONU should ignore all received REGISTER messages in which the flags field takes one of the reserved values.

8.1.4.3 Sync time. The definition of this field is similar to the sync time field defined for the discovery GATE message (see Sec. 8.1.2.4); however, the standard does not mandate the sync time value in the REGISTER message to match the sync time value in the discovery GATE. Quite conceivably, the OLT will require a larger sync time in the discovery GATE. Once the first response from an ONU (i.e., the REGISTER_REQ message) is received, the OLT may reassess the quality of the received signal, and possibly reduce the sync time, if, for example, a good quality signal allows for a shorter T_{CDR} time.

8.1.4.4 Echoed pending grants. The *echoed pending grants* field acknowledges to an ONU that the DBA agent in the OLT is aware of the maximum number of outstanding grants that the ONU is able to store. Strictly speaking, this field is not necessary, as it carries no useful information.

8.1.5   REGISTER_ACK control frame

The REGISTER_ACK message serves as an ONU's final registration acknowledgment. It is the first MPCPDU transmitted by an ONU on a newly established logical link. The format of REGISTER_ACK message is shown in Fig. 8.9.

Fields	Octets
Destination address (DA)	6
Source address (SA)	6
Length/type = 88–08$_{16}$	2
Opcode = 00–06$_{16}$	2
Timestamp	4
Flags	1
Echoed assigned port	2
Echoed sync time	2
Pad = 0	35
Frame check sequence (FCS)	4

Figure 8.9 Format of REGISTER_ACK frame.

8.1.5.1 Flags. The *flags* field identifies the specific registration response from an ONU. This field can take the following values:

1. *Nack.* The discovery agent at the ONU refused registration. The ONU will remain in an uninitialized state.

2. *Ack.* The discovery agent at the ONU confirms successful registration.

All other values of the flags field are reserved. The OLT should ignore all received REGISTER_ACK messages in which the flags field takes one of the reserved values.

8.1.5.2 Echoed assigned port. The *echoed assigned port* field carries a copy of the assigned port value received in the REGISTER message.

8.1.5.3 Echoed sync time. The *echoed sync time* field carries a copy of the sync time value received in the REGISTER message.

8.2 Opcode-Independent Processes

Traditionally, processes specified in the MAC control sublayer were separated into opcode-independent processes and opcode-dependent processes. In pre-EFM MAC control, the opcode-independent processes included control parser and control multiplexer, and the only opcode-dependent process was flow control.

In the *multi-point* MAC control, specified by the EFM task force, the opcode-independent processes include control parser, control multiplexer, and multi-point transmission control (in OLT only). The opcode-dependent processes, in addition to flow control, include discovery process, gating process, and reporting process.

As the name suggests, opcode-independent processes perform identical operations on all MAC control frames, independently of the opcode value. These operations include parsing the received frames and determining whether a frame is a MAC control frame or any other type of frame; MAC control frames are further verified to have a correct opcode value. In the transmission direction, the opcode-independent operations involve serializing frames received from multiple interfaces and prioritizing MAC control frames over data frames.

In the multi-point MAC control, some more operations have been added to the control parser and the control multiplexer. The control multiplexer is responsible for timestamping all outgoing MPCPDUs, and the control parser is responsible for latching the local MPCP clock when an MPCPDU arrives. These time-related operations are performed only if the frame's opcode is recognized to belong to one of the MPCPDUs; therefore, strictly speaking, the control parser and the control multiplexer do not remain opcode-independent processes.

8.2.1 Control parser

The control parser is responsible for parsing the received frames and demultiplexing them to opcode-specific control functions, such as gating, reporting, or discovery processes. The operations of the control parsers in the OLT and an ONU are very similar, the only difference being the PARSE TIMESTAMP state. The OLT control parser state diagram is shown in Fig. 8.10*a*. Figure 8.10*b* shows only the PARSE TIMESTAMP state from the control parser state diagram in the ONU.

8.2.1.1 WAIT FOR RECEIVE state. Upon initialization, the control parser enters the WAIT FOR RECEIVE state. It remains in this state until a frame is received, as indicated by a receiveStatus return code from the ReceiveFrame(...) function call. Further, the length/type field of the received frame is checked to see whether the received frame is a MAC control frame (i.e., Length/Type = 88-08$_{16}$) or data frame. If the received frame is a data frame, the control transitions to the PASS TO MAC CLIENT state. If the received frame is a MAC control frame, the PARSE OPCODE state is entered.

8.2.1.2 PASS TO MAC CLIENT state. In this state, the received frame is simply passed to an upper-layer entity and the control parser unconditionally returns to the WAIT FOR RECEIVE state.[1]

8.2.1.3 PARSE OPCODE state. All MAC control frames are further parsed in the PARSE OPCODE state. If the control parser recognizes

---

[1]Transition label UCT stands for *unconditional transition.*

BEGIN

```
                              WAIT FOR RECEIVE
              ReceiveFrame(DA, SA, Length/Type, data_rx):receiveStatus
```

Length/Type == MAC_Control_type   Length/Type != MAC_Control_type

```
        PARSE OPCODE                          PASS TO MAC CLIENT
    opcode_rx = data_rx[0:15]     MA_DATA.Indication(DA, SA, Length/Type|data_rx, receiveStatus)
```

UCT

opcode_rx ∉
{supported opcode}

opcode_rx ∈ {timestamp opcode}

```
                                              PARSE TIMESTAMP
opcode_rx ∈ {supported opcode} AND    timestamp = data_rx[16:47]
opcode_rx ∉ {timestamp opcode}        newRTT = localTime - timestamp
                                      timestampDrift = abs(newRTT - RTT) > guardThresholdOLT
                                      RTT = newRTT
```

UCT

```
        INITIATE MAC CONTROL FUNCTION
    Perform opcode-specific operation
```

UCT

(a)

opcode_rx ∈ {timestamp opcode}

```
                              PARSE TIMESTAMP
    timestamp = data_rx[16:47]
    timestampDrift = abs(timestamp - localTime) > guardThresholdONU
    localTime = timestamp
```

UCT

(b)

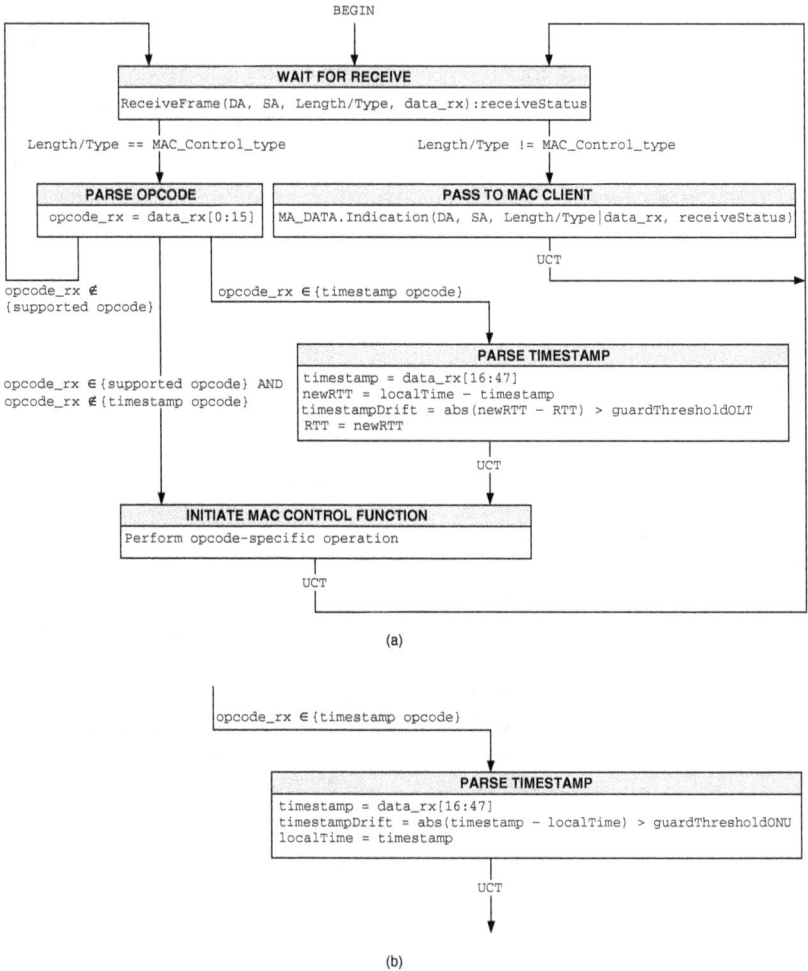

Figure 8.10 (a) OLT control parser state diagram. (b) PARSE TIMESTAMP state in ONU control parser. (*Reprinted from IEEE Standard 802.3ah with permission from IEEE.*)

the opcode as belonging to one of the MPCPDUs (i.e., opcode ∈ {02_{16}, 03_{16}, 04_{16}, 05_{16}, 06_{16}}), then the PARSE TIMESTAMP state is entered. If the opcode is supported, but does not belong to any of the MPCPDU opcodes, the control parser transitions to the INITIATE MAC CONTROL FUNCTION state. At the time of this writing, there exists only one non-MPCPDU opcode—opcode 01_{16}, which identifies a PAUSE frame. Finally, if the control parser does not recognize the opcode of a received MAC control frame (i.e., if opcode_rx ∉ {supported opcode}), it discards this frame and returns to the WAIT FOR RECEIVE state.

8.2.1.4 PARSE TIMESTAMP state. The PARSE TIMESTAMP state is the only control parser state that is different between the OLT and an ONU. We first consider the OLT version of this state.

In the OLT, in this state, the received MPCPDU is further parsed and its timestamp field is extracted. Once the value of the timestamp is known, the new RTT value to the ONU is calculated as the difference between the local MPCP clock (represented by the localTime variable) and the timestamp. The mechanism behind the RTT measurement is explained in Sec. 5.3.3. Note that, although it is not explicitly shown in this state, the localTime value should correspond to the time when the first byte of DA was received.

After the new RTT value is calculated, this value is compared to the previously calculated RTT. If the absolute difference between the two measurements exceeds the guardThresholdOLT constant, the timestamp-Drift error is asserted and the OLT proceeds to deregister such ONU, as shown in Fig. 8.26. Finally, the new RTT value is saved in the RTT variable for future comparison with the newRTT value.

In the ONU, in the PARSE TIMESTAMP state (Fig. 8.10b), the timestamp field is extracted. Once the value of timestamp is known, it is compared to the local MPCP clock represented by the localTime variable. If the absolute difference between the two values exceeds the guardThresholdONU constant, the timestampDrift error is asserted and the ONU deregisters itself, as shown in Fig. 8.27. Finally, the MPCP clock is adjusted by setting localTime = timestamp.

8.2.1.5 Timestamp drift tolerance. It has been explained in Sec. 5.3.1.3 that the MPCP timestamping mechanism relies on constant propagation delay between the control multiplexer of the sending device and the control parser of the receiving device. Yet, the standard allows 1-TQ delay variability for data (including MPCPDUs) passing through the MAC sublayer and an additional 1 TQ while passing through the PHY sublayer. As shown in Fig. 8.11, in the worst case, these delay variabilities could accumulate for transmitting and receiving devices. In this figure, the constant delay component through MAC and PHY is ignored; only the delay variability is shown. It can be seen that in the downstream direction (from the OLT to an ONU), the accumulated delay variability at the ONU becomes 4 TQ. In the round-trip path (OLT to ONU to OLT), the accumulated variability becomes 8 TQ.

The IEEE 802.3ah task force has added an additional margin of 4 TQ to account for any propagation delay variability introduced outside MAC and PHY, such as variability of propagation delay in fiber. As the result, the guardThresholdONU value is specified as 8 TQ, and guardThresholdOLT is specified as 12 TQ.

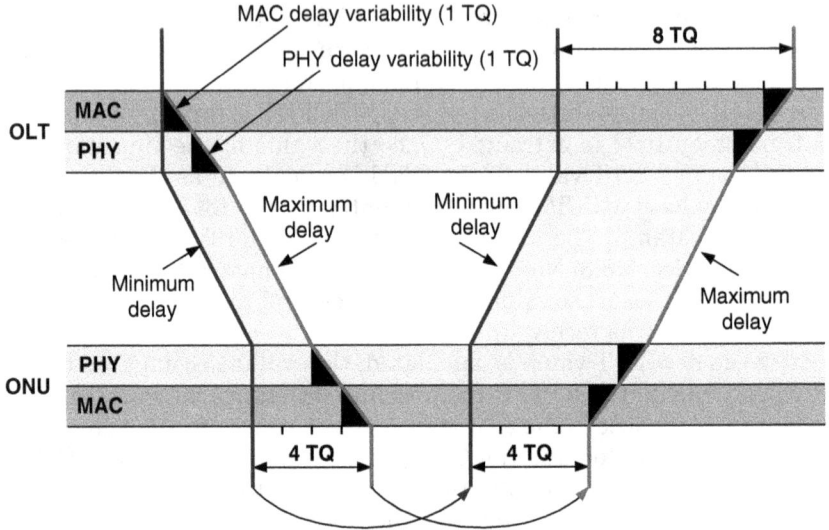

Figure 8.11   Accumulation of delay variabilities in MAC and PHY sublayers.

Please, note that the final version of IEEE 802.3ah has the values of guardThresholdOLT and guardThresholdONU mixed. The correct values should be 8 TQ for guardThresholdONU and 12 TQ for guardThresholdOLT. This mixup is expected to be corrected in one of future maintenance revisions of the standard.

8.2.1.6   INITIATE MAC CONTROL FUNCTION state.   In this state, an opcode-specific function is invoked. The opcode-specific processes run in parallel to control parser. In other words, the INITIATE MAC CONTROL FUNCTION state is a nonblocking state, and upon initiating an opcode-specific function, the control parser immediately returns to the WAIT FOR RECEIVE state and becomes ready to process the next frame.

8.2.2   ONU Control Multiplexer

The control multiplexer is responsible for gating the transmission data path, i.e., for allowing the data to pass only inside the transmission window specified by a previously received grant.

It can be seen in Figs. 5.3 and 5.7 that, in ONUs, frames can be generated by MAC client as well as several independent processes, such as the discovery process or the reporting process. Therefore, the second

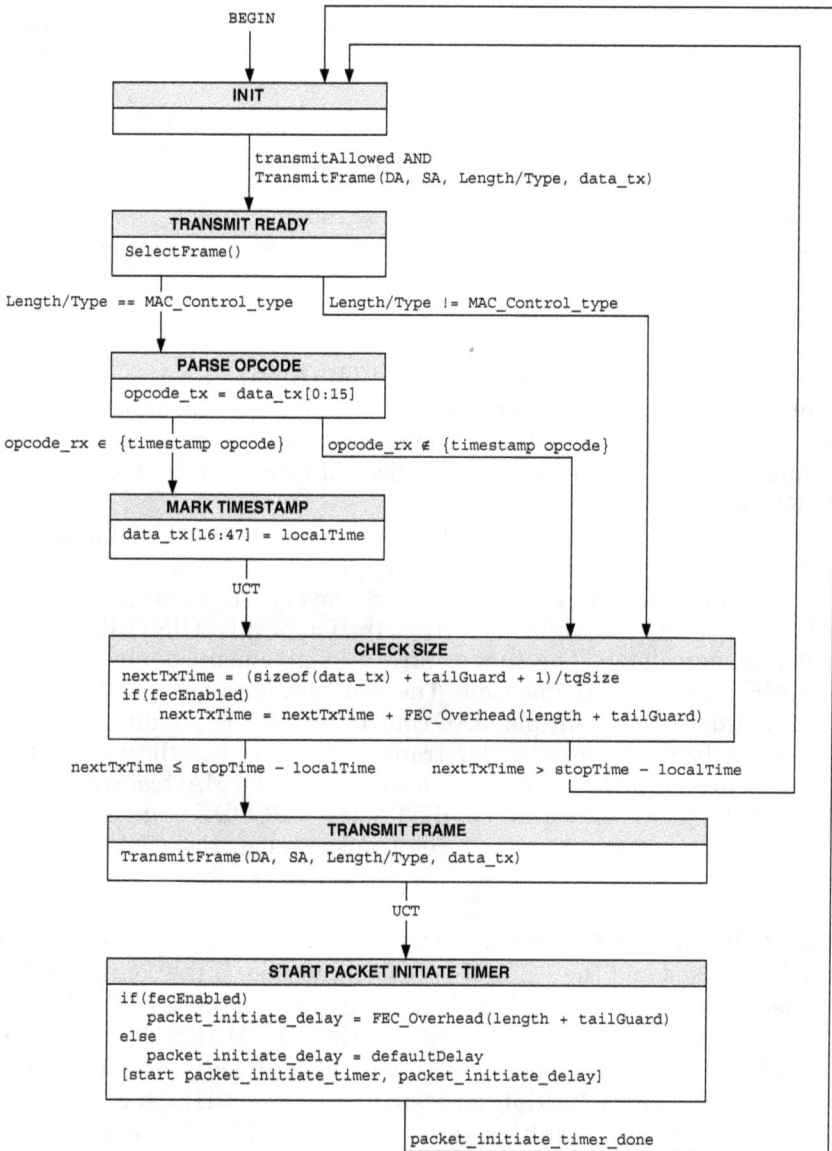

Figure 8.12   ONU control multiplexer state diagram. (*Reprinted from IEEE Standard 802.3ah with permission from IEEE.*)

function of the control multiplexer is to prioritize and serialize concurrent frames issued by multiple processes, and forward these frames to the MAC sublayer for further transmission. If a frame being forwarded is an MPCPDU, the control multiplexer is also responsible for timestamping this frame. The ONU control multiplexer state diagram is shown in Fig. 8.12.

8.2.2.1   INIT state.   Upon initialization, the control multiplexer enters the INIT state. It remains in this state until a frame becomes available for transmission and transmission is allowed by the gating process. The latter condition is indicated by boolean variable `transmitAllowed`, which is set by the gating process (see Sec. 8.3.2). When transmission is allowed and the frame is available, the control multiplexer transitions to TRANSMIT READY state.

8.2.2.2   TRANSMIT READY state.   In the TRANSMIT READY state, the control multiplexer invokes function `SelectFrame()`. This function is responsible for selecting only one frame if multiple frames are available simultaneously. This function prioritizes MAC control frames over the MAC client frames.

   In ONU, multiple MAC control frames may become available simultaneously. For example, the reporting process may issue a REPORT frame simultaneously with the discovery process issuing a REGISTER_ACK to confirm the registration or a REGISTER_REQ in order to deregister. The flow control, if implemented, may issue a PAUSE frame at the same time. The standard does not specify the selection order when multiple MAC control frames are pending.

   Once a frame is selected for transmission, its length/type field is checked to determine whether it is a data frame or a MAC control frame. In case of a MAC control frame, the PARSE OPCODE state is entered; otherwise, the control multiplexer transitions to the CHECK SIZE state.

8.2.2.3   PARSE OPCODE state.   In this state, the MAC control frame is further parsed and the value of opcode is checked. If the control multiplexer recognizes the opcode as belonging to one of the MPCPDUs (i.e., $opcode \in \{02_{16}, 03_{16}, 04_{16}, 05_{16}, 06_{16}\}$), then the MARK TIMESTAMP state is entered. If the opcode does not belong to any of the MPCPDU opcodes, the control multiplexer transitions to the CHECK SIZE state. Note that the control multiplexer will forward all MAC control frames, even if opcode cannot be recognized or is not valid.

8.2.2.4   MARK TIMESTAMP state.   In the MARK TIMESTAMP state, the value of the local MPCP clock, represented by variable `localTime`, is copied into the timestamp field of the outgoing MPCPDU. Note, that for the proper RTT measurement, the MPCP clock should correspond to the time when the first byte of the DA is transmitted from the MAC control sublayer to the MAC sublayer. The control multiplexer then unconditionally transitions to the CHECK SIZE state.

> ⚠️
>
> Please note that the end-of-packet delimiter /T/R/R/, which is accounted for in the `tailGuard` constant, is also a part of the extended end-of-packet delimiter accounted for in `FEC_Overhead (...)` function; i.e., it is counted twice. As such, the fitness test in state CHECK SIZE overestimates the required timeslot size by 3 bytes. This overestimation may result in control multiplexer unnecessarily deferring the frame to the next timeslot, thus wasting a significant portion of the upstream bandwidth.

8.2.2.5   CHECK SIZE state.   In the CHECK SIZE state, a fitness test is performed to verify that the remaining timeslot size is large enough to transmit the pending frame.

First, the frame transmission time `nextTxTime` in units of TQ is calculated. The transmission length in bytes includes the frame's payload `data_tx` and additional overhead, represented by a constant `tailGuard`. The `tailGuard` adds to 29 bytes and consists of the following components: preamble + SFD (8 bytes), DA field (6 bytes), SA field (6 bytes), length/type field (2 bytes), FCS (4 bytes), and end-of-packet delimiter /T/R/R/ (3 bytes). The result of this calculation is the transmission length in bytes; it is then converted to units of TQ by dividing its value by `tqSize`, which has value of 2 bytes/TQ. To round up the result of integer division, an additional 1 (which is `tgSize - 1`) is added to the transmission length.

In the next step, if the optional *forward error correction* (FEC) is enabled, the transmission time `nextTxTime` is increased to account for the FEC parity overhead. The FEC mechanism adds 16 bytes of overhead for each 239-byte block of data. In addition, as explained in Sec. 12.3, FEC uses extended start-of-frame and end-of-frame delimiters, which add up to 26 bytes of overhead. Therefore, given the frame length x in bytes, the `FEC_Overhead (...)` function calculates an additional FEC overhead in TQ as

$$FEC\_Overhead\ (x) = 13 + \left\lceil \frac{x}{239} \right\rceil \times 8 \qquad (8.1)$$

The timeslot end time is represented by variable `stopTime`, which is set by the gating process (see Sec. 8.3.2). If the total frame transmission time `nextTxTime` is less than the remaining transmission time (`stopTime - localTime`), the TRANSMIT STATE is entered; otherwise, the frame is deferred and the control multiplexer returns to the INIT state.

The IEEE 802.1D standard disallows frame reordering; therefore the deferred frame should retain its place at the head of its queue. However, if multiple queues are present, it seems reasonable if the SelectFrame() function chooses a possibly shorter frame from a different queue, to utilize otherwise wasted timeslot remainder.

8.2.2.6   TRANSMIT FRAME state.   In the TRANSMIT FRAME state, the control multiplexer passes the frame to the MAC, by invoking the TransmitFrame(...) function. Readers should not confuse this function with the TransmitFrame(...) function listed in transition from the state INIT to the state TRANSMIT READY. The former represents passing the frame from the MAC control to the MAC, while the latter represents passing the frame from one of the MPCP processes or the MAC client to the control multiplexer, or in other words, receiving a frame by the control multiplexer.

The state diagram assumes the TransmitFrame(...) function is blocking; i.e., it will not return until the frame is transmitted. Upon the function return, the START PACKET INITIATE TIMER state is entered.

8.2.2.7   START PACKET INITIATE TIMER state.   Recall that the MPCP requires a constant propagation delay between the control multiplexer and the control parser, or more precisely, between writing the timestamp value in MPCPDU at the transmitting side and reading it at the receiving side. However, the full-duplex MAC, as currently defined in IEEE 802.3 standard, does not guarantee a constant delay. Immediately after the function TransmitFrame(...) returns, the MAC is ready to accept another frame. Yet, at the same time, an internal process inside the MAC (called the deferral process) enforces minimum *interframe gap* (IFG) by delaying the transmission of the second frame. The delay experienced by the second frame will depend on the time of the second TransmitFrame(...) function call and can range from 0 to 96 ns (minimum IFG).

To avoid this additional delay variability, the control multiplexer also defers the timestamping of the second frame by applying a packet_initiate_delay. In the absence of FEC overhead, this delay is equal to 6 TQ or 96 ns and is represented by a constant defaultOverhead. If the FEC is enabled, the gap between adjacent frames should increase even more to allow insertion of FEC parity data and extended delimiters. This overhead is calculated using the FEC_Overhead(...) function described above.

The calculated value of packet_initiate_delay is used to start packet_initiate_timer. When this timer expires (condition `packet_initiate_timer_done`), the control multiplexer returns to the INIT state.

8.2.3 Multi-point transmission control

Unlike the ONUs, the control multiplexer in the OLT does not perform data path gating functions. Indeed, because the OLT is the only device transmitting in the downstream direction, it has full access to the channel and does not need to gate its transmission.

However, the control multiplexer in the OLT faces a different challenge. Recall that to implement the point-to-point emulation, the OLT creates a logical port for each ONU. Each of these logical ports is represented by an independent instance of MAC and MAC control, including separate instances of control multiplexers. Because each instance of MPCP can generate MPCPDUs or data frames independently of other instances, there is a need for a transmission arbitration mechanism to avoid data loss or contention for the downstream channel.

The data streams generated by multiple logical ports are combined at the reconciliation sublayer (RS); so it may seem logical that RS would be responsible for multiplexing multiple data streams. However, the handicap of this approach is that, if any arbitration is performed below the MAC control, the propagation delay between the control multiplexer and the control parser may not be constant. The IEEE 802.3ah task force had to come up with a solution that would perform the arbitration between different MPCP instances at the same MAC control sublayer. As a result, the *multi-point transmission control* process was defined, as shown in Fig. 8.13. This process has a global view of all the MPCP instances and is responsible for allowing only one MPCP instance to transmit at any given time.

The multi-point transmission control process communicates with each instance of the control multiplexer by using three signals (boolean variables): `transmitPending`, `transmitInProgress`, and `transmitEnable`.

The state diagram of the multi-point transmission control is shown in Fig. 8.14.

8.2.3.1 INIT state. Upon initialization, the multi-point transmission control enters the INIT state, in which it sets all `transmitEnable[i]` variables to false. Then an unconditional transition to the WAIT PENDING state occurs.

8.2.3.2 WAIT PENDING state. The multi-point transmission control remains in the WAIT PENDING state until at least one instance of the

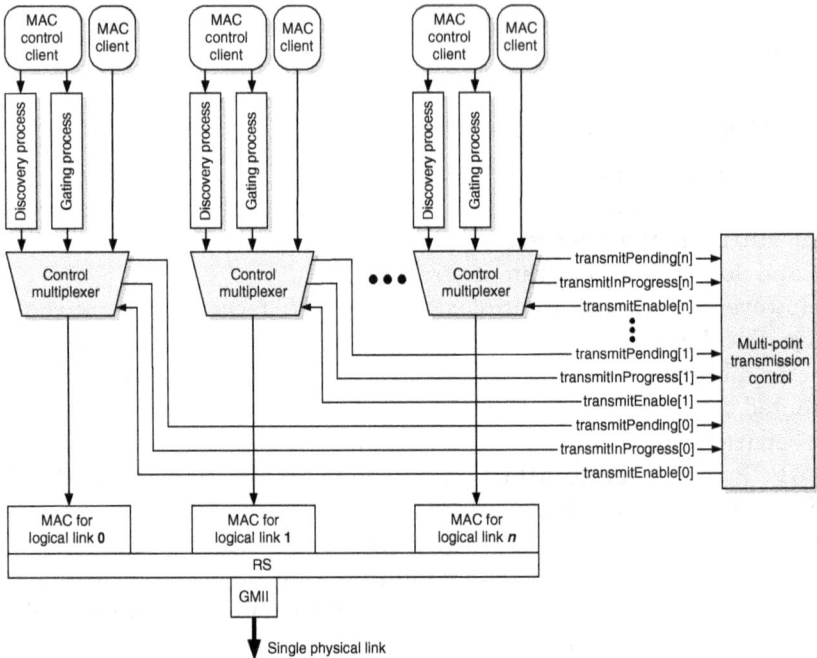

Figure 8.13 Relationship of multiple MPCP instances and multi-point transmission control.

control multiplexer reports a pending frame. The function `transmissionPending()` is simply an OR of all `transmitPending[i]` variables associated with n instances of the control multiplexer:

```
transmissionPending = transmitPending[0] OR
                      transmitPending[1] OR
                      ...
                      transmitPending[n-1]
```

When at least one frame is waiting, the multi-point transmission control enters the ENABLE state.

8.2.3.3 ENABLE state. In the ENABLE state, the multi-point transmission control selects one instance of the control multiplexer to transmit a frame. The `select()` function returns an index of the control multiplexer instance which has a pending frame. The standard does not specify the selection criteria, if multiple instances have waiting frames. It is reasonable for an implementation of this function to guarantee fairness for each MPCP instance in accessing the channel. Also, it may be justified if, in the selection process, instances with pending control

```
                        BEGIN
                          |
                          v
 +------------------------------------------------+
 |                    INIT                         |
 +------------------------------------------------+
 |  transmitEnable[0..n-1] = false                 |
 +------------------------------------------------+

                          |
                         UCT
                          |          +----------+
                          v          v          |
 +------------------------------------------+    |
 |               WAIT PENDING                |    |
 +------------------------------------------+    |
 |                                           |    |
 +------------------------------------------+    |

                          |                       |
               transmissionPending()              |
                          |                       |
                          v                       |
 +------------------------------------------+    |
 |                  ENABLE                   |    |
 +------------------------------------------+    |
 |  j = select()                             |    |
 |  transmitEnable[j] = true                 |    |
 +------------------------------------------+    |

                          |                       |
         transmitInProgress[j] = false            |
                          |                       |
                          v                       |
 +------------------------------------------+    |
 |                 DISABLE                   |    |
 +------------------------------------------+    |
 |  transmitEnable[j] = false                |    |
 +------------------------------------------+    |

                          |                       |
                         UCT----------------------+
```

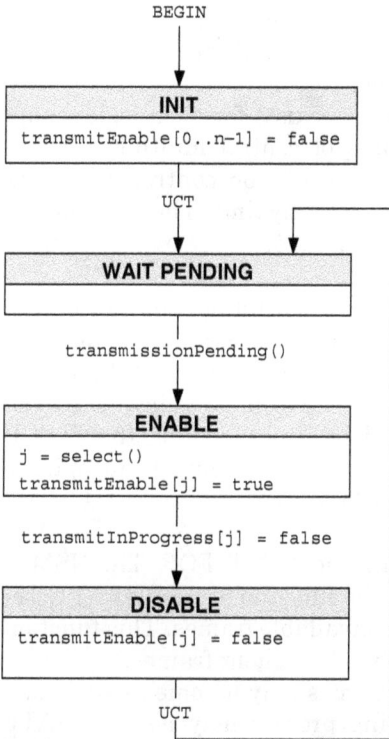

Figure 8.14   Multi-point transmission control state diagram. (*Reprinted from IEEE Standard 802.3ah with permission from IEEE.*)

messages get higher weight than the instances with waiting user data frames.

Once an instance j is selected, its transmitEnable[j] variable is set to true, thus allowing this instance to transmit the waiting frame. The multi-point transmission control remains in the ENABLE state until frame transmission completes, as indicated by the transmitInProgress[j] variable, after which the DISABLE state is entered.

8.2.3.4 DISABLE state. In the DISABLE state, the multi-point transmission control disallows instance j any further transmissions and returns to the WAIT PENDING state.

8.2.4 OLT control multiplexer

Similarly to an ONU, in the OLT, the concurrent frames can be generated by the MAC client as well as several independent processes, such as the discovery process or the gating process. The control multiplexer is responsible for prioritizing and serializing such frames and forwarding them to the MAC sublayer for further transmission. If the frame being forwarded is an MPCPDU, the control multiplexer is also responsible for timestamping this frame.

A separate control multiplexer instance exists for each logical port created at the OLT. Multiple instances of control multiplexers are arbitrated by a common multi-point transmission control. Only one control multiplexer is allowed to transmit at any time. The OLT control multiplexer state diagram is shown in Fig. 8.15.

8.2.4.1 INIT state. Upon initialization, the control multiplexer enters the INIT state, where the variables `transmitInProgress` and `transmitPending` are set to false. The control multiplexer remains in this state until a frame becomes available for transmission, as indicated by the `TransmitFrame(...)` function call received from one of the MPCP processes. When a frame becomes available, the control multiplexer transitions to the WAIT FOR TRANSMIT state.

8.2.4.2 WAIT FOR TRANSMIT state. In the WAIT FOR TRANSMIT state, the control multiplexer invokes the function `SelectFrame()` to select only one frame of possibly multiple available frames. This function prioritizes MAC control frames over the MAC client frames.

In the OLT, multiple MAC control frames may become available simultaneously. For example, the gating process may issue a GATE frame simultaneously with the discovery process issuing a REGISTER message requesting ONU's deregistration. The standard does not specify the selection order when multiple MAC control frames are pending.

After a frame is selected, the `transmitPending[j]` variable is asserted to notify the multi-point transmission control that the jth instance of the control multiplexer has a pending frame. The control multiplexer remains in the WAIT FOR TRANSMIT state until the multi-point transmission control authorizes the transmission, i.e., until `transmitEnable` becomes true. When transmission is allowed, the state TRANSMIT READY is entered.

8.2.4.3 TRANSMIT READY state. In the TRANSMIT READY state, the length/type field of the selected frame is checked to determine whether it is a data frame or a MAC control frame. In the case of a MAC control frame, the PARSE OPCODE state is entered; otherwise, the control multiplexer transitions to the SEND FRAME state.

8.2.4.4 PARSE OPCODE state. In this state, the MAC control frame is further parsed and the value of opcode is checked. If the control multiplexer recognizes the opcode as belonging to one of the MPCPDUs (i.e.,

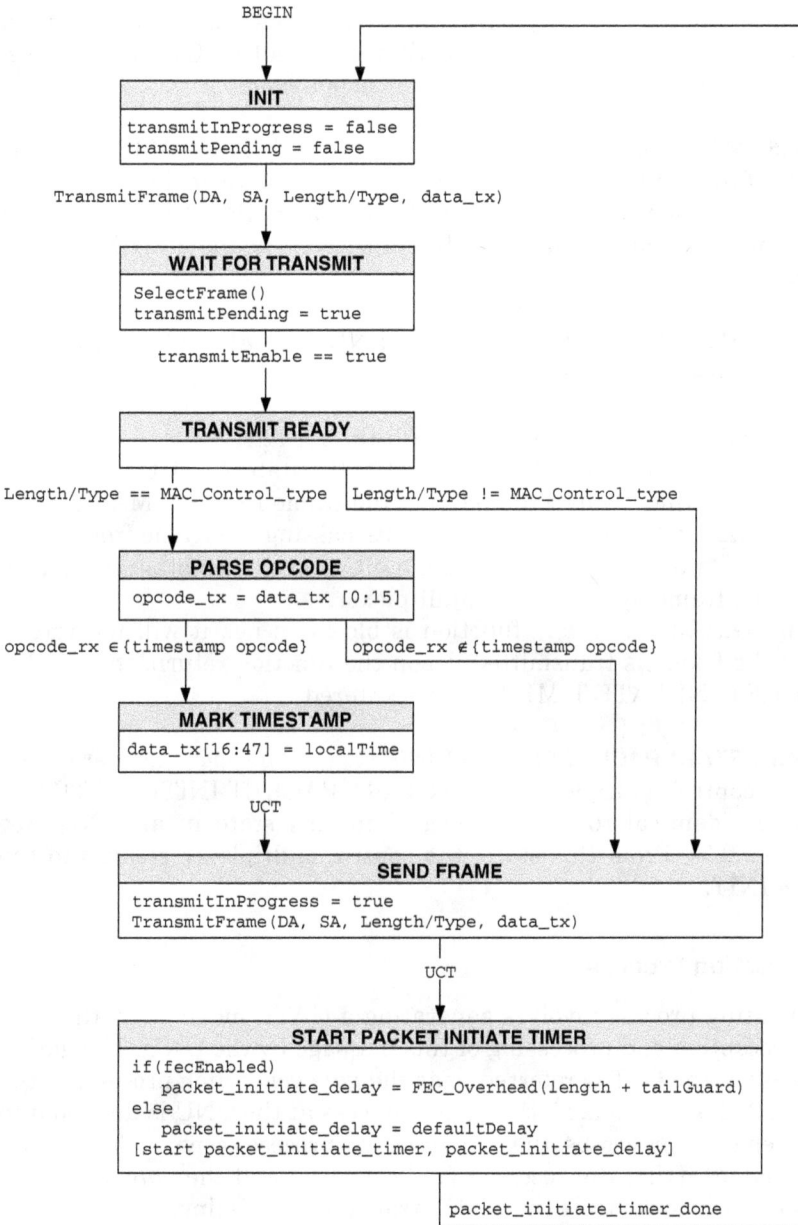

Figure 8.15 OLT control multiplexer state diagram. (*Reprinted from IEEE Standard 802.3ah with permission from IEEE.*)

opcode $\in \{02_{16},\ 03_{16},\ 04_{16},\ 05_{16},\ 06_{16}\}$), then the MARK TIMESTAMP state is entered. If the opcode does not belong to any of the MPCPDU opcodes, the control multiplexer transitions to the SEND FRAME state. Note that the control multiplexer will forward all MAC control frames, even if opcode cannot be recognized or is not valid.

8.2.4.5 MARK TIMESTAMP state. In the MARK TIMESTAMP state, the value of the local MPCP clock, represented by the variable `localTime`, is copied into the timestamp field of the outgoing MPCPDU. The control multiplexer then unconditionally transitions to the SEND FRAME state.

8.2.4.6 SEND FRAME state. In the SEND FRAME state, the control multiplexer asserts the `transmitInProgress` signal and passes the frame to the MAC, by invoking the `TransmitFrame(...)` function. Readers should not confuse this function with the `TransmitFrame(...)` function listed in transition from the state INIT to the state WAIT FOR TRANSMIT. The former represents passing the frame from the MAC control to the MAC, while the latter represents passing the frame from one of the MPCP processes to the control multiplexer, or in other words, receiving a frame by the control multiplexer.

The `TransmitFrame(...)` function is blocking; i.e., it will not return until the frame is transmitted. Upon the function return, the START PACKET INITIATE TIMER state is entered.

8.2.4.7 START PACKET INITIATE TIMER state. The operation performed by the control multiplexer in the START PACKET INITIATE TIMER state is identical to that performed in this state in an ONU (see Sec. 8.2.2.7). From this state, the control multiplexer returns to the state INIT.

8.3 Gating Process

The gating process involves generating a GATE message by the OLT and reception and processing of this message by the ONU. The gating process at the OLT is referred to as the *gate generation* process. In the IEEE 802.3ah standard, the gating process at the ONU is divided into two separate processes: the *gate reception* process responsible for parsing and verifying the received GATE frames and the *gate activation* process, which controls the ONU's transmission timing.

8.3.1 Gate generation at the OLT

A separate instance of the gate generation process exists for each logical port at the OLT (or for each registered ONU). The gate generation process is driven by the DBA agent, which determines the start time and

BEGIN

```
┌─────────────────────────────────────────────────────────────┐
│                            WAIT                               │
└─────────────────────────────────────────────────────────────┘
```

registered == true

```
┌─────────────────────────────────────────────────────────────┐
│                        WAIT FOR GATE                         │
│ [start gate_periodic_timer, gate_timeout]                    │
└─────────────────────────────────────────────────────────────┘
```

registered == true AND
MACR (DA,
     GATE,
     grant_number,
     start[4],
     length[4],
     force_report[4])

registered == false

registered == true AND
gate_periodic_timer_done

```
┌──────────────────────────────────────┐   ┌──────────────────────────────────────────────┐
│              SEND GATE               │   │           PERIODIC TRANSMISSION              │
│ data_tx[0:15]      = GATE            │   │ data_tx[0:15] = GATE                         │
│ data_tx[48:50]     = grant_number    │   │ data_tx[48:55] = 0                           │
│ data_tx[52:55]     = force_report[0:3]│  │ TransmitFrame(DA, SA, MAC_Ctrl_type, data_tx)│
│ data_tx[56:87]     = start[0]        │   └──────────────────────────────────────────────┘
│ data_tx[88:103]    = length[0]       │
│ data_tx[104:135]   = start[1]        │   UCT
│ data_tx[136:151]   = length[1]       │
│ data_tx[152:183]   = start[2]        │
│ data_tx[184:199]   = length[2]       │
│ data_tx[200:231]   = start[3]        │
│ data_tx[232:247]   = length[3]       │
│                                      │
│ TransmitFrame(DA, SA, MAC_Ctrl_type, data_tx) │
└──────────────────────────────────────┘
```
UCT

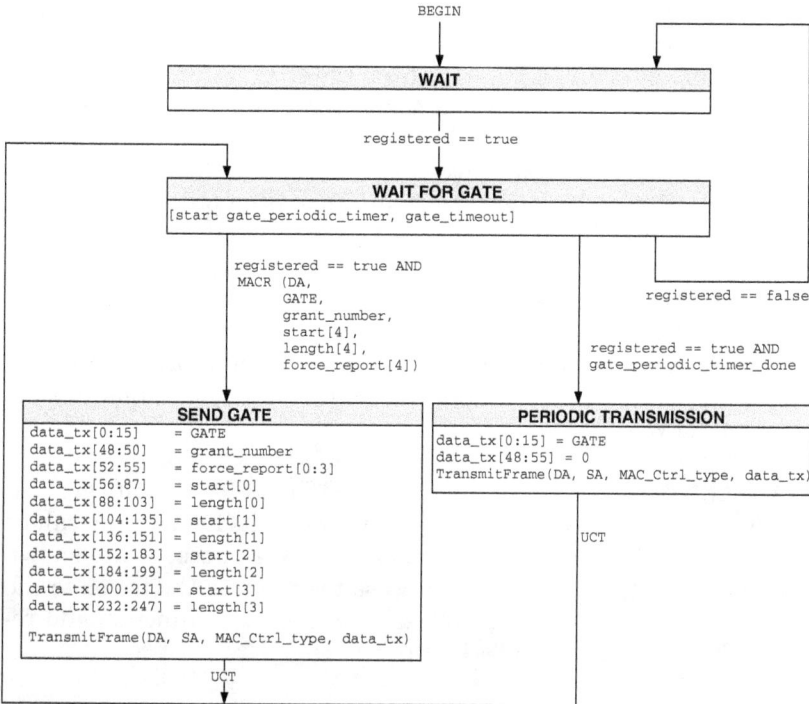

Figure 8.16   Gate generation state diagram. (*Reprinted from IEEE Standard 802.3ah with permission from IEEE.*)

length for each grant issued to an ONU. Upon receiving a request from the DBA agent, the gate generation process forms a GATE message and transmits it to the ONU.

The GATE messages are also used as a keep-alive mechanism, informing the ONUs that the corresponding logical port at the OLT is functioning properly. If the DBA agent does not issue a request to send a grant to a particular ONU for a predefined period of time, the gate generation process will autonomously issue an empty GATE message (with grant_number = 0) to this ONU.

The gate generation state diagram is shown in Fig. 8.16.

8.3.1.1 WAIT state. Upon initialization, the gate generation process enters the WAIT state. It remains in this state until the corresponding ONU becomes registered. When the ONU successfully completes the autodiscovery, i.e., the variable registered becomes true, the gate generation process transitions to the WAIT FOR GATE state.

8.3.1.2 WAIT FOR GATE state. In the WAIT FOR GATE state, the gate generation process starts the `gate_periodic_timer` and waits for a request to transmit a GATE message from the DBA agent.

To request a GATE transmission, the DBA agent issues a service primitive `MACR(DA, GATE, grant_number, start[4], length[4], force_report[4])`, where MACR stands for MA_CONTROL.request. Upon reception of such request, the gate generation process enters the SEND GATE state.

The expiration interval of the `gate_periodic_timer` is set to 50 ms. If the timer expires before a request from DBA agent arrives, the PERIODIC TRANSMISSION state is entered.

Finally, if, while waiting for the request from the DBA agent or the timeout, the ONU becomes unregistered, the gate generation process returns to the WAIT state.

8.3.1.3 SEND GATE state. In the SEND GATE state, the gate generation process forms a GATE message by setting all its fields to the values issued by the DBA agent. The fields of the GATE message are shown in Fig. 8.5b. The gate generation process then passes the frame to the control multiplexer by calling the `TransmitFrame(...)` function and returns to the WAIT FOR GATE state.

8.3.1.4 PERIODIC TRANSMISSION state. In this state, the gate generation process forms an empty GATE message (with `grant_number = 0`). This frame is then passed to control multiplexer by calling the `TransmitFrame(...)`, after which the process returns to the WAIT FOR GATE state.

8.3.2 Gate reception at the ONU

The gate reception process at the ONU verifies the integrity of each received grant and stores all valid grants in a structure called `grantList`. The state diagram of the gate reception process is shown in Fig. 8.17.

8.3.2.1 WAIT state. Upon initialization, the gate reception process enters the WAIT state. It remains in this state until the ONU becomes registered or a GATE frame is received, as indicated by the `opcode_rx == GATE` condition. When the ONU successfully completes the autodiscovery, the gate reception process transitions to the WAIT FOR GATE state. If a GATE frame arrives before the discovery is completed, the gate reception process enters the PARSE GATE state.

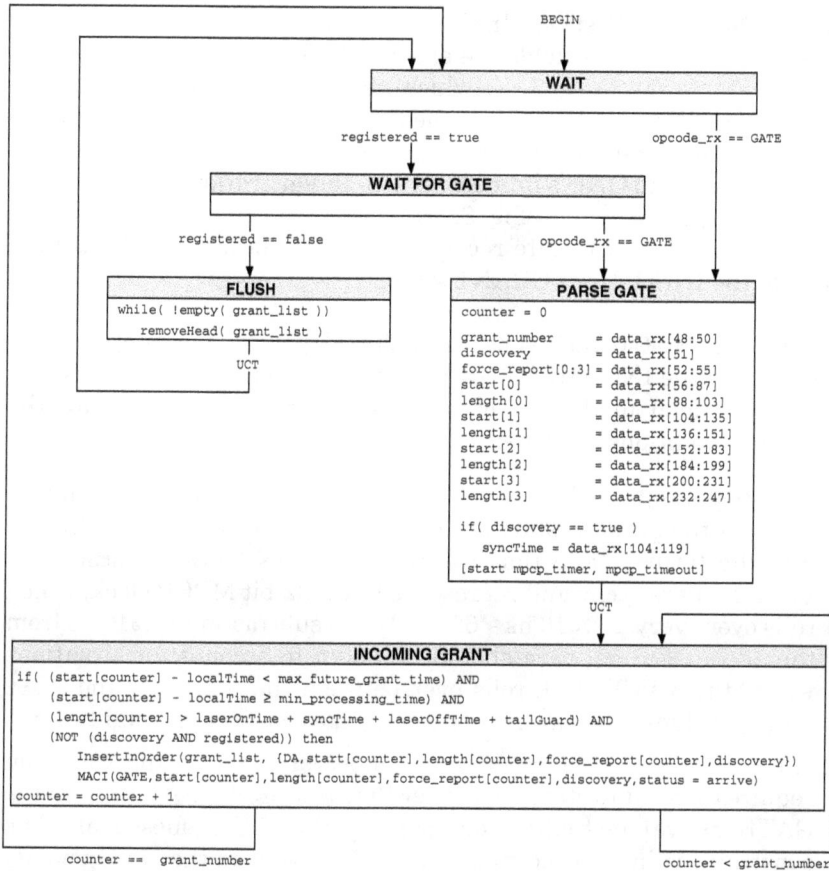

```
                                          BEGIN
                                            │
                                            ▼
                          ┌─────────────────────────────────────┐
                          │                WAIT                  │
                          └─────────────────────────────────────┘
                            registered == true          opcode_rx == GATE
                          ┌─────────────────────────────────────┐
                          │            WAIT FOR GATE             │
                          └─────────────────────────────────────┘
            registered == false                    opcode_rx == GATE
     ┌───────────────────────────────┐  ┌──────────────────────────────────────┐
     │            FLUSH              │  │            PARSE GATE                 │
     │ while( !empty( grant_list ))  │  │ counter = 0                           │
     │    removeHead( grant_list )   │  │                                       │
     │                               │  │ grant_number     = data_rx[48:50]     │
     │             UCT               │  │ discovery        = data_rx[51]        │
     └───────────────────────────────┘  │ force_report[0:3] = data_rx[52:55]    │
                                        │ start[0]         = data_rx[56:87]     │
                                        │ length[0]        = data_rx[88:103]    │
                                        │ start[1]         = data_rx[104:135]   │
                                        │ length[1]        = data_rx[136:151]   │
                                        │ start[2]         = data_rx[152:183]   │
                                        │ length[2]        = data_rx[184:199]   │
                                        │ start[3]         = data_rx[200:231]   │
                                        │ length[3]        = data_rx[232:247]   │
                                        │                                       │
                                        │ if( discovery == true )               │
                                        │    syncTime = data_rx[104:119]        │
                                        │ [start mpcp_timer, mpcp_timeout]      │
                                        └──────────────────────────────────────┘
                                                        UCT
┌──────────────────────────────────────────────────────────────────────────────────┐
│                              INCOMING GRANT                                         │
│ if( (start[counter] - localTime < max_future_grant_time) AND                       │
│    (start[counter] - localTime ≥ min_processing_time) AND                          │
│    (length[counter] > laserOnTime + syncTime + laserOffTime + tailGuard) AND       │
│    (NOT (discovery AND registered)) then                                            │
│        InsertInOrder(grant_list, {DA,start[counter],length[counter],force_report[counter],discovery}) │
│        MACI(GATE,start[counter],length[counter],force_report[counter],discovery,status = arrive) │
│ counter = counter + 1                                                              │
└──────────────────────────────────────────────────────────────────────────────────┘
    counter == grant_number                              counter < grant_number
```

Figure 8.17   ONU gate reception state diagram. (*Reprinted from IEEE Standard 802.3ah with permission from IEEE.*)

8.3.2.2 WAIT FOR GATE state.   In the WAIT FOR GATE state, the gate reception process continues to wait for a GATE frame. When a GATE frame arrives, as indicated by a condition opcode_rx == GATE being true, the gate reception process enters the PARSE GATE state. If the ONU becomes unregistered, the state FLUSH is entered.

8.3.2.3 FLUSH state.   When an ONU becomes unregistered, it is not allowed to transmit any data. When the ONU becomes unregistered after being registered, the gate reception process enters the FLUSH state. In this state, all the pending grants are deleted by successively calling the function RemoveHead(grantList) until the list becomes empty. The process then returns to the WAIT state.

8.3.2.4   PARSE GATE state. In this state, the incoming GATE frame is parsed and its various fields are extracted. Following the parsing, the process starts the `mpcp_timer`, which measures the interval between arriving GATE MPCPDUs. The timeout interval for `mpcp_timer` is set to 1 s. Failure to receive a GATE frame before `mpcp_timer` expires is a fatal fault that leads to ONU's immediate self-deregistration (see transition labeled `mpcp_timer_done` in Fig. 8.27).

From this state, the gate reception process unconditionally transitions to the INCOMING GRANT state.

8.3.2.5   INCOMING GRANT state. A GATE MPCPDU may contain up to 4 grants. In the INCOMING GRANT state, each received grant is verified and added to the list of pending grants (`grantList`) if all the following conditions are true:

1. `start[counter] − localTime < max_future_grant_time`. This condition requires the grant to start not more than `max_future_grant_time` TQ into the future. Note that the timing values (`start[counter]` and `localTime`) are cyclic and represented by a 32-bit MPCP clock, which rolls over every $2^{32} \times 16$ ns ≈ 68.7 s. When subtracting `localTime` from the `start[counter]`, care should be taken to account for situations where the MPCP clock rolls over between the `localTime` and `start[count]` values. The `max_future_grant_time` constant is equal to 1 s.

2. `start[counter] − localTime ≥ min_processing_time`. This condition requires a grant to start not sooner than `min_procesing_time` from the GATE arrival (actually from parsing the GATE message). This constraint is introduced to give the DBA agent in the ONU enough time to prepare data for transmission, including generating the REPORT message, if requested by an asserted `force_report` flag. The value of the `min_processing_time` constant is 1024 TQ or 16.384 μs.

3. `length[counter] > laserOnTime + syncTime + laserOffTime + tailGuard`. A check is made of whether the grant length is sufficient to turn the laser on and off and send the necessary synchronization sequence. This check is somewhat inconsistent as the ability to fit `tailGuard` in the timeslot does not mean that a minimum-size frame can be transmitted. It would be more appropriate to specify a minimum frame size including the preamble and IFG instead of `tailGuard`. However, this issue is not significant and would not affect the overall system behavior, because an additional check will be made by the ONU's control multiplexer to ensure that a specific frame fits in the effective slot length left after the physical-level overhead has been subtracted.

4. NOT (discovery AND registered). This condition simply discards all discovery gates after ONU has registered.

If all the above conditions are satisfied, the grant is added to the pending grant list (grantList). All grants are stored in the grantList in order of their start times. The gate reception process also informs the DBA agent of the new pending grant by invoking the MACI(...) service primitive (MACI stands for MA_CONTROL.indication).

The code in this state is executed separately for each grant, identified by an index variable counter. When all grants received in the last GATE message are processed (i.e., when counter == grants_num), the gate reception process returns to the WAIT state.

8.3.3   Gate activation

The main function of the gate activation process is to control the timing of ONU's transmission by setting and clearing the variable transmit-Allowed, which is used by ONU's control multiplexer. The state diagram of the gate activation process is shown in Fig. 8.18.

8.3.3.1   WAIT FOR GRANT state.   Upon initialization, the gate activation process enters the WAIT state and sets the transmitAllowed variable to false. The gate activation process remains in this state until a pending grant is added to the grantList, upon which it enters the WAIT FOR START TIME state.

8.3.3.2   WAIT FOR START TIME state.   In the WAIT FOR START TIME state, the gate activation process removes the head-of-line grant from the grantList and stores it in the currentGrant structure. This structure consists of the following members:

DA	GATE destination address (48 bits)
start	Grant start time (32 bits)
length	Grant length (16 bits)
force_report	Flag indicating that a REPORT message is requested (1 bit)
discovery	Flag indicating that this is a discovery grant (1 bit)

When the local MPCP clock reaches the value specified as the grant start time, the gate activation process transitions to the CHECK GATE TYPE state.

8.3.3.3   CHECK GATE TYPE state.   In the CHECK GATE TYPE state, the gate activation process makes a decision whether random delay should be applied to this grant or not. Recall from Sec. 5.3.2.2 that

BEGIN

WAIT FOR GRANT

transmitAllowed = false

NOT empty(grantList)

WAIT FOR START TIME

currentGrant = RemoveHead(grantList)

localTime == currentGrant.start

CHECK GATE TYPE

(registered == false) AND
(currentGrant.discovery == true) AND
(IsBroadcast(currentGrant) == true)

else

RANDOM WAIT

maxDelay = currentGrant.length
    - laserOnTime - laserOffTime
    - syncTime - discoveryGrantLength

if(fecEnabled)
    maxDelay = maxDelay
        - FEC_Overhead(discoveryGrantLength*tqSize)

[start rndDlyTmr, Random(maxDelay)]

((registered == true) AND
(currentGrant.discovery == false))
OR
((registered == false) AND
(currentGrant.discovery == true)) AND
(IsBroadcast(currentGrant) == false)

rndDlyTmr_done

START TX

stopTime = currentGrant.start + currentGrant.length - laserOnTime - laserOffTime - syncTime
transmitAllowed = true

if(currentGrant.discovery == true)
    insideDiscoveryWindow = true
    effectiveLength = discoveryGrantLength
else
    effectiveLength = stopTime - localTime

[start gntWinTmr, effectiveLength]

MACI(GATE, localTime, effectiveLength, currentGrant.force_report, currentGrant.discovery,
    status=active)

gntWinTmr_done

STOP TX

insideDiscoveryWindow = false
MACI(GATE, status = deactivate)

empty(grantList)

NOT empty(grantList)

CHECK NEXT GRANT

nextGrant = PeekHead(grantList)
nextStopTime = nextGrant.start + nextGrant.length - laserOffTime

else

(nextStopTime ≤ stopTime)
OR
(nextGrant.discovery == true) AND
(nextGrant.start ≤ stopTime + laserOffTime)

(nextStopTime > stopTime) AND
(nextGrant.discovery == false) AND
(nextGrant.start ≤ stopTime + laserOffTime)

HIDDEN GRANT

RemoveHead(grantList)

UCT

BACK TO BACK GRANT

currentGrant = RemoveHead(grantList)

UCT

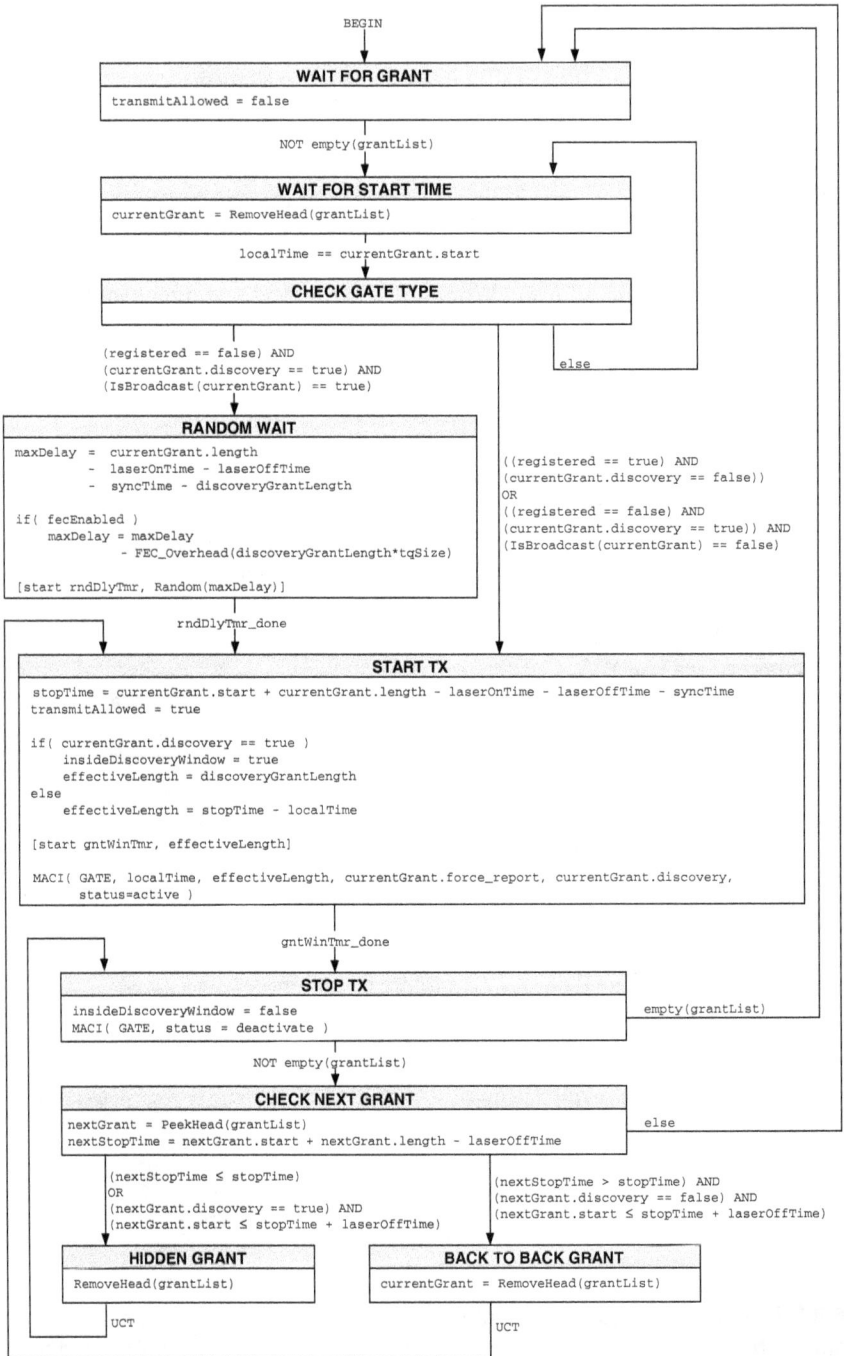

Figure 8.18   ONU gate activation state diagram. (*Reprinted from IEEE Standard 802.3ah with permission from IEEE.*)

random delay is used to avoid persistent collisions when multiple un-registered ONUs transmit their REGISTER_REQ messages. There-fore, the random delay should be applied only if (1) the grant is a discovery grant, (2) the ONU is not registered yet, and (3) this grant was sent to multiple ONUs, i.e., the GATE's DA is a globally assigned MAC control address. If all these conditions are true, the gate activation process enters the RANDOM DELAY state.

However, if the discovery agent in the OLT somehow knows the MAC address of an uninitialized ONU, it may send a discovery GATE with a unicast DA address. Then the ONU does not need to apply the random delay. Thus, if (1) the grant is a discovery grant, (2) the ONU is not registered yet, and (3) this grant was sent to a single ONU, i.e., the GATE message had a unicast DA, then the state START TX is entered. The START TX is also entered if (1) the ONU is already registered and (2) the current grant is a normal grant (i.e., currentGrant.discovery == false).

In all other cases, the current grant is discarded and the gate activation process returns to the WAIT FOR START TIME state, where it waits for and extracts the next grant from the grantList.

8.3.3.4 RANDOM DELAY state.   In this state, the random delay is cal-culated and applied. The maximum allowed delay maxDelay is calculated first. This delay should be chosen such that the ONU's transmission does not extend beyond the end of the timeslot. The discovery-GrantLength represents the length of transmitted REGISTER_REQ message, including start-of-packet delimiter, preamble, MPCPDU, and end-of-packet delimiter—a total of 38 TQ. The maxDelay is obtained by subtracting the discoveryGrantLength and optical overhead from the currentGrant.length.

If FEC is enabled, the ONU's transmission length will further increase by the size of FEC parity data and extended delimiters. To account for this, maxDelay should be decreased by FEC_Overhead (discoveryGrantLength * tgSize).

Finally, a rndDlyTmr timer is started with an expiration interval chosen randomly from an interval [0, maxDelay]. When this timer expires, the gate activation process enters the START TX state.

8.3.3.5 START TX state.   In the START TX state, the gate activation process calculates the stopTime of the current grant—a time when data transmission should cease in order not to extend beyond the granted timeslot.

The process also asserts the value of the transmitAllowed variable, thus allowing the control multiplexer to start data transmission. It may seem confusing that the transmitAllowed variable is set to true at a time corresponding to the grant start time, whereas in Fig. 8.6 it was shown that the grant start time corresponds to a time when the laser should just begin to turn on and the actual data should be sent laserOnTime + syncTime later. The precise timing relationship shall become clear if we consider the operation of the data detector function discussed in Chap. 7. The data detector introduces an additional delay equal to laserOnTime + syncTime. Thus, if transmitAllowed is asserted at time T (or, correspondingly, the first byte of preamble is transmitted by MAC at time T), the data detector will start turning on laser at time T, and the first byte of preamble will be transmitted out at time $T +$ laserOn-Time + syncTime, as expected.

If the current grant is a discovery grant, the variable insideDiscov-eryWindow is set to true (this variable is used by discovery process explained in Sec. 8.5) and the length of transmission effective-Length is set to discoveryGrantLength.

For normal grants, the effectiveLength is calculated to cover the entire available timeslot, excluding the physical layer overhead.

Finally, the gate activation process informs the DBA agent about the activated grant and starts the gntWinTmr timer, which is set to expire after an interval equal to the effectiveLength value. When the timer gntWinTmr expires, the STOP TX state is entered.

8.3.3.6 STOP TX state. In the STOP TX state, the insideDiscoveryWin-dow is reset to false, and the DBA agent is informed about the grant's deactivation. If there are more pending grants in the grantList, the gate activation process enters the CHECK NEXT GRANT state; otherwise, it returns to the WAIT FOR GRANT state.

8.3.3.7 CHECK NEXT GRANT state. The reason to the check the next grant before resetting the transmitAllowed variable to false is to make sure that the next grant does not overlap with the current grant. If it does overlap, then the transmission may continue uninterrupted until the end of the next grant.

In this state, the gate activation process peeks at the next head-of-line grant without actually removing it from the grantList. It calculates nextStopTime and determines if the next grant overlaps with the current grant. As shown in Fig. 8.19, there could be two possibilities. In the first case, referred to as *back-to-back* grants, the next grant extends beyond the current grant. In this case, the gate activation process enters the BACK TO BACK GRANT state. The second case, referred to as *hidden* grant, occurs when the next grant does not extend beyond the current grant. In this situation, the process transitions to the HIDDEN GRANT state.

A transition to the HIDDEN GRANT state occurs also in the case when the next grant partially overlaps the current grant and the next

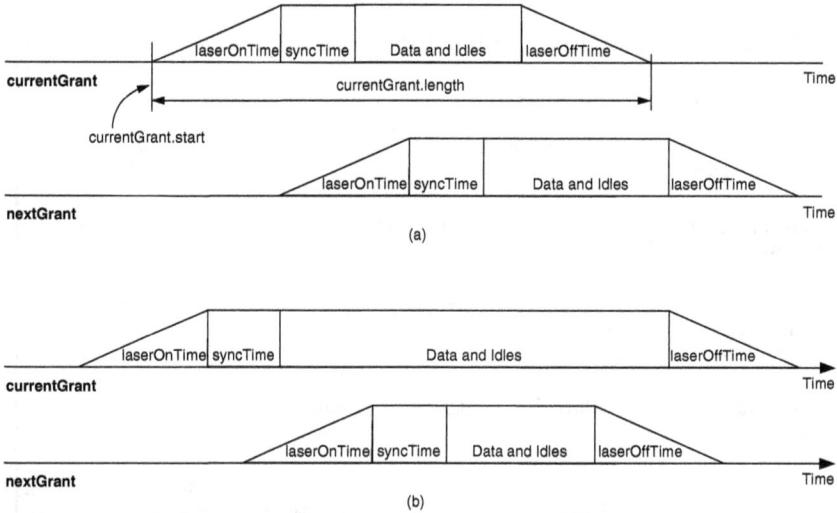

Figure 8.19   Examples of (a) back-to-back grant and (b) hidden grant.

grant is a discovery grant. From the gate receiving process (Sec. 8.3.2) we know that a registered ONU discards all discovery GATE MPCPDUs. If an ONU becomes unregistered, it discards all pending grants. Thus, it seems the only situation in which a discovery grant may overlap with the normal grant occurs if the ONU manages to get deregistered, the ONU receives a discovery GATE, and the random delay elapses, all during the current grant being active, i.e., while the gate activation process remains in the START TX state.

> ⚠️ According to the IEEE 802.3ah, an ONU that becomes deregistered while it is in a transmission state will not cease the transmission until the current grant completes. This represents a significant flaw in the specification. It is recommended that implementations of gate activation state diagrams be modified such that the transmitAllowed variable is reset to false immediately upon ONU's deregistration.

A transition to the BACK TO BACK GRANT state occurs only if the next grant is not a discovery grant.

> ⚠️ Please note that, in the gate activation state diagram, the conditions for transitions from the CHECK NEXT GRANT state to states HIDDEN GRANT and BACK TO BACK GRANT are incorrect.

The transition to the HIDDEN GRANT state should be as follows:

```
(nextStopTime ≤ stopTime)
OR
(nextGrant.discovery == true) AND
(nextGrant.start ≤ currentGrant.start + currentGrant.length)
```

The condition for transition to the BACK TO BACK GRANT state should be as follows:

```
(nextStopTime > stopTime) AND
(nextGrant.discovery == false) AND
(nextGrant.start ≤ currentGrant.start + currentGrant.length)
```

In addition, the calculation of nextStopTime value is incorrect and should be changed to

```
nextStopTime = nextGrant.start + nextGrant.length
             - laserOnTime - laserOffTime - syncTime
```

8.3.3.8 HIDDEN GRANT state. If a grant is determined to be a hidden grant, it is simply removed from the grantList and the process enters the STOP TX state.

8.3.3.9 BACK TO BACK GRANT state. In this state, the gate activation process extracts the next grant from the grantList; this grant becomes the new currentGrant. The process then transitions to the START TX state where a new stopTime is calculated.

8.4   Reporting Process

The reporting process is responsible for passing queue status information from an ONU to the OLT. Reports are generated by the DBA agent at the ONU and are sunk by the DBA agent in the OLT.

8.4.1   Report generation at an ONU

The report generation process is driven by the DBA agent, which determines the number of queue sets to report as well as queue length values for each of the reported queues. Upon receiving a request from the DBA agent, the report generation process forms a REPORT message and transmits it to the OLT.

REPORT messages are also used as a keep-alive mechanism, informing the OLT that the reciprocal ONU is functioning properly. If the DBA agent does not issue a request to send a report for a predefined period of time, the report generation process will autonomously issue

BEGIN

WAIT

registered == true

WAIT FOR REPORT
[start report_periodic_timer, report_timeout]

registered == false

registered == true AND
MACR (DA,
    REPORT,
    report_number,
    report_list)

registered == true AND
report_periodic_timer_done

SEND REPORT	PERIODIC TRANSMISSION
data_tx[0:15] = REPORT data_tx[48:55] = report_number data_tx[56:311] = report_list TransmitFrame(DA, SA, MAC_Ctrl_type, data_tx)	data_tx[0:15] = REPORT data_tx[48:55] = 0 TransmitFrame(DA, SA, MAC_Ctrl_type, data_tx)

UCT

UCT

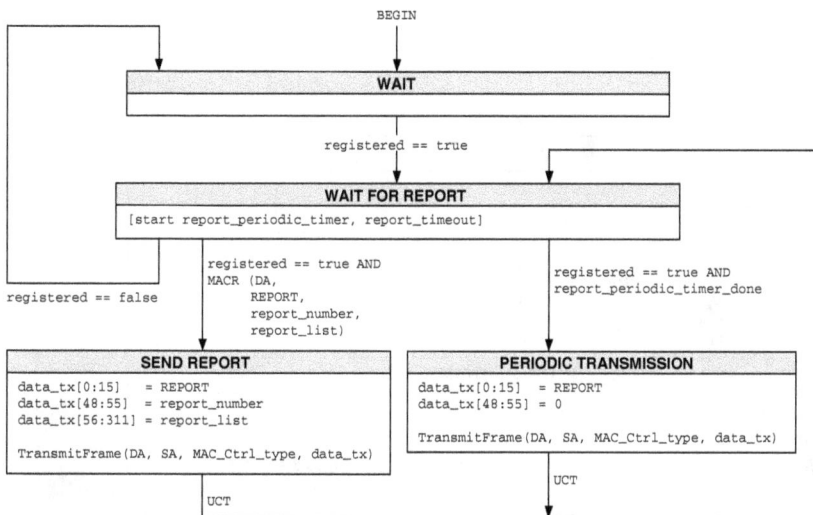

Figure 8.20   REPORT generation state diagram. (*Reprinted from IEEE Standard 802.3ah with permission from IEEE.*)

an "empty" REPORT message, i.e., a REPORT with the number of queue sets, represented by variable report_number, being 0. The report generation state diagram is shown in Fig. 8.20.

8.4.1.1 WAIT state. Upon initialization, the report generation process enters the WAIT state. It remains in this state until the ONU becomes registered. When the ONU successfully completes the autodiscovery, i.e., the variable registered becomes true, the report generation process transitions to the WAIT FOR REPORT state.

8.4.1.2 WAIT FOR REPORT state. In the WAIT FOR REPORT state, the report generation process starts the report_periodic_timer and waits for a request to transmit a REPORT message from the DBA agent. The report_periodic_timer's expiration interval is set to 50 ms. If the timer expires before a request from the DBA agent arrives, the PERIODIC TRANSMISSION state is entered.

To request a REPORT transmission, the DBA agent issues a service primitive MACR (DA, REPORT, report_number, report_list), where MACR stands for MA_CONTROL.request. The report_number variable represents the number of queue sets. The report_list structure contains report bitmap and queue length fields, as shown in several examples in Fig. 8.4. The report_list is generated and sunk by DBA agents at the ONU and the OLT, and is not processed by the MPCP processes. Upon reception of such request from the DBA agent, the report generation process enters the SEND REPORT state.

BEGIN

```
                                      ┌──────────────┐
                                      │              │
              │                       │              │
              ▼                       ▼              │
┌─────────────────────────────────────────────┐     │
│                     WAIT                     │     │
├─────────────────────────────────────────────┤     │
│                                             │     │
└─────────────────────────────────────────────┘     │
              │                                       │
              │                                       │
       opcode_rx == REPORT                            │
              │                                       │
              ▼                                       │
┌─────────────────────────────────────────────┐     │
│               RECEIVE REPORT                 │     │
├─────────────────────────────────────────────┤     │
│ report_number = data_rx[48:55]              │     │
│ report_list   = data_rx[56:311]             │     │
│ MACI( REPORT, RTT, report_number, report_list ) │ │
│                                             │     │
│ [start mpcp_timer, mpcp_timeout]            │     │
└─────────────────────────────────────────────┘     │
              │                                       │
             UCT                                      │
              │                                       │
              └───────────────────────────────────────┘
```

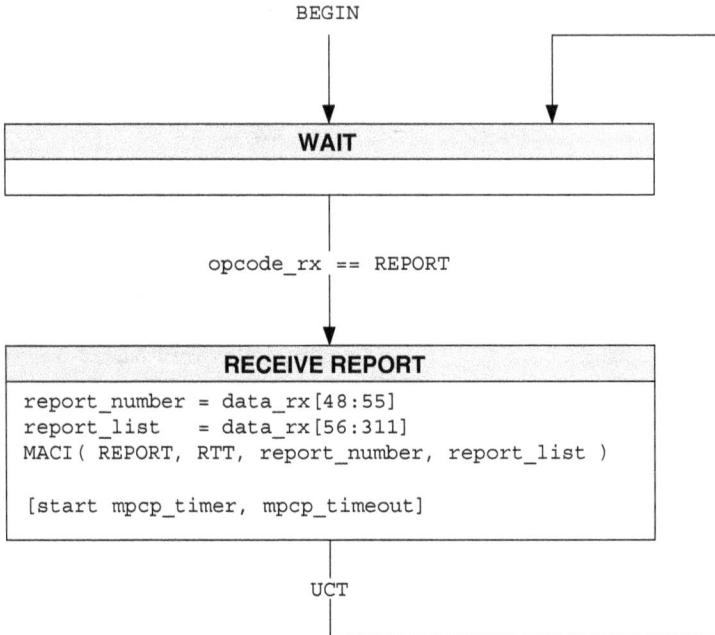

Figure 8.21   OLT report reception state diagram. (*Reprinted from IEEE Standard 802.3ah with permission from IEEE.*)

Finally, if, while waiting for the request from the DBA agent or the timeout, the ONU becomes unregistered, the report generation process returns to the WAIT state.

8.4.1.3 SEND REPORT state.   In the SEND REPORT state, the report generation process creates a REPORT frame and sets all its fields to the values issued by the DBA agent. The fields of the REPORT message are shown in Fig. 8.2. The process then passes the frame to the control multiplexer by calling the `TransmitFrame(...)` function and returns to the WAIT FOR REPORT state.

8.4.1.4 PERIODIC TRANSMISSION state.   In this state, the report generation process forms an empty REPORT message (with `report_number = 0`). This frame is then passed to the control multiplexer by calling the `TransmitFrame(...)` function, after which the process returns to the WAIT FOR REPORT state.

8.4.2 Report reception at the OLT

The report reception process at the OLT is responsible for receiving the REPORT messages and passing the received data to the DBA agent in

`MACI(...)` service primitive, where MACI stands for MA_CONTROL.indication. A separate instance of the report reception process exists for each logical port at the OLT (or for each registered ONU). The state diagram of the report reception process is shown in Fig. 8.21.

8.4.2.1 WAIT state. Upon initialization, the report reception process enters the WAIT state. It remains in this state until a REPORT frame is received, as indicated by the `opcode_rx == REPORT` condition, upon which the report reception process enters the RECEIVE REPORT state.

8.4.2.2 RECEIVE REPORT state. In this state, the incoming REPORT frame is parsed and its various fields are extracted. Following the parsing, the process starts the `mpcp_timer`. The `mpcp_timer` measures the interval between arriving REPORT MPCPDUs. The timeout interval for `mpcp_timer` is set to 1 s. Failure to receive a REPORT frame before the `mpcp_timer` expires is a fatal fault that leads to ONU's immediate deregistration (see transition labeled `mpcp_timer_done` in Fig. 8.26).

From this state, the report reception process unconditionally transitions to the WAIT state.

8.5 Discovery Process

The autodiscovery mechanism is used to detect newly connected ONUs and learn the round-trip delays and MAC addresses of these ONUs. For simplicity, Fig. 5.7 presented a single discovery process at the OLT. In reality, the standard breaks this into four separate processes: *discovery gate generation* process, *request reception* process, *register generation* process, and *final registration* process. All four processes are driven by the discovery agent. The main reason for such break-out is the fact that discovery gate generation, request reception, and register generation state machines are implemented only for the instance of MPCP associated with the broadcast logical port, while the final registration is implemented for each of multiple MPCP instances associated with unicast logical ports.

ONUs typically have one instance of MPCP which responds to both broadcast and unicast LLIDs. For this reason, only one state machine is required at the ONU.

Figure 8.22 illustrates the interaction of various processes involved in autodiscovery. The autodiscovery handshake procedure consists of the following steps:

1. The discovery agent in the OLT instructs the gate generation process to send a discovery GATE message. This message is addressed to a

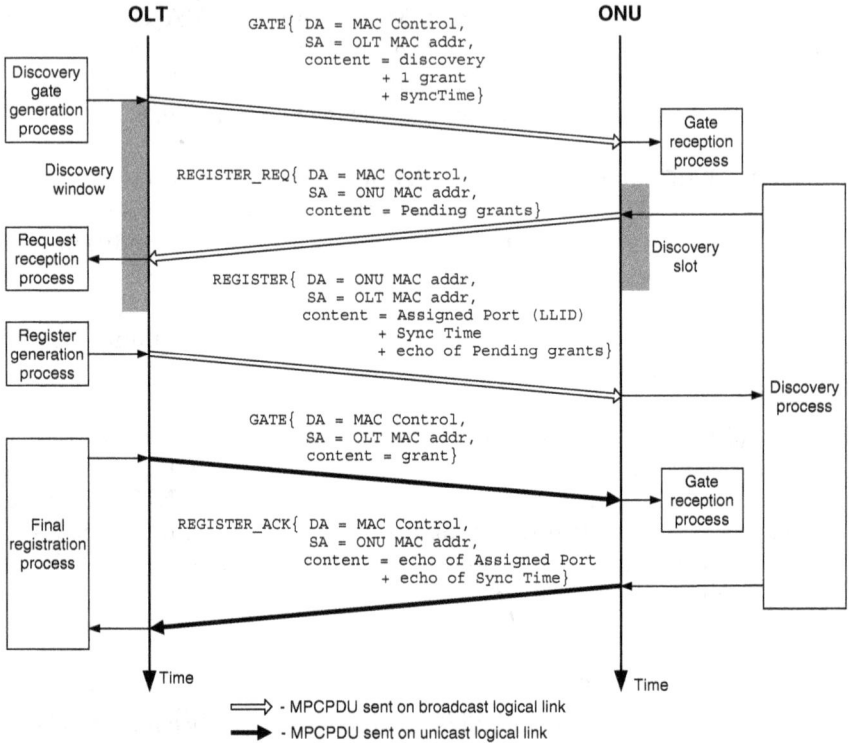

Figure 8.22   Autodiscovery message exchange.

group MAC control address and is transmitted on broadcast channel (with LLID = 7F FF$_{16}$).

   The discovery GATE MPCPDU is received and verified by the gate reception process at the ONU. The received discovery grant is stored for future activation.

2. Upon initialization, the discovery process in the ONU generates a REGISTER_REQ message. This message remains buffered until the discovery grant activates, i.e., until the transmission window opens. Then the REGISTER_REQ is transmitted upstream to the OLT on the broadcast channel. At the OLT, the REGISTER_REQ is passed to the request reception process, which further forwards it to the discovery agent.

3. Upon processing the REGISTER_REQ message from the ONU, the discovery agent issues a unique LLID value and requests the register generation process to transmit a REGISTER MPCPDU to the ONU. This message is addressed to an individual ONU but is transmitted

on the broadcast channel, because the unique LLID has not been assigned to the ONU yet.

At the ONU, the REGISTER message is forwarded to the discovery process. It is expected that immediately upon processing the REGISTER message, the ONU would program the local LTE function to accept all traffic sent on the unicast logical link, i.e., all frames with LLID value assigned to this ONU.

4. Following the transmission of REGISTER MPCPDU, the DBA agent allocates a normal grant to the newly registered ONU. This grant is needed to give the ONU an opportunity to transmit an acknowledgment back to the OLT. The final registration process issues a normal GATE MPCPDU, which has the group MAC control DA, but is transmitted on point-to-point logical link toward only one ONU.

In the ONU, the normal GATE MPCPDU is forwarded to the gate reception process, and again, the received grant is stored until the local MPCP clock reaches the grant start time value.

5. When the grant activates, the discovery process at the ONU transmits a REGISTER_ACK MPCPDU. This message is transmitted on the point-to-point logical link and has the group MAC control DA. The reception of the REGISTER_ACK MPCPDU at the OLT concludes the registration procedure.

8.5.1 Discovery gate generation at the OLT

When the discovery agent at the OLT decides to initiate a discovery round, it instructs the discovery gate generation process to send a discovery GATE message, advertising the start time of the discovery slot and its length. The state diagram of this process is shown in Fig. 8.23.

8.5.1.1 IDLE state.

Upon initialization, the discovery gate generation process enters the IDLE state, where it sets the insideDiscoveryWindow variable to false. This variable is shared between the discovery gate generation and the request reception processes, informing the latter about the beginning and the end of the discovery window.

The discovery gate generation process remains in the IDLE state until the discovery agent initiates a new discovery round by issuing a request to transmit the discovery GATE. Upon such request, the process transitions to the SEND DISCOVERY WINDOW state.

8.5.1.2 SEND DISCOVERY WINDOW state.

In the SEND DISCOVERY WINDOW state, a discovery GATE message is created by setting all

```
                          BEGIN
┌─────────────┐            │
│     ┌───────┴───┐        │
│     │           ▼        ▼
│  ╔══════════════════════════════════════════╗
│  ║                 IDLE                      ║
│  ╟──────────────────────────────────────────╢
│  ║  insideDiscoveryWindow = false            ║
│  ╚══════════════════════════════════════════╝
│                            │
│                            │   MACR( DA,
│                            │         GATE,
│                            │         discovery,
│                            │         start,
│                            │         length,
│                            │         discovery_length,
│                            │         sync_time)
│                            ▼
│  ╔══════════════════════════════════════════╗
│  ║         SEND DISCOVERY WINDOW             ║
│  ╟──────────────────────────────────────────╢
│  ║  data_tx[0:15]    = GATE                  ║
│  ║  data_tx[48:50]   = 1                     ║
│  ║  data_tx[51]      = 1                     ║
│  ║  data_tx[56:87]   = start                 ║
│  ║  data_tx[88:103]  = length                ║
│  ║  data_tx[104:119] = sync_time             ║
│  ║                                           ║
│  ║  TransmitFrame(DA, SA, MAC_Ctrl_type, data_tx) ║
│  ╚══════════════════════════════════════════╝
│                            │
│                            │   localTime = start
│                            ▼
│  ╔══════════════════════════════════════════╗
│  ║           DISCOVERY WINDOW                ║
│  ╟──────────────────────────────────────────╢
│  ║  insideDiscoveryWindow = true             ║
│  ║                                           ║
│  ║  [start discovery_window_size_timer, discovery_length] ║
│  ╚══════════════════════════════════════════╝
│            │
└────────────┘
 discovery_window_size_timer_done
```

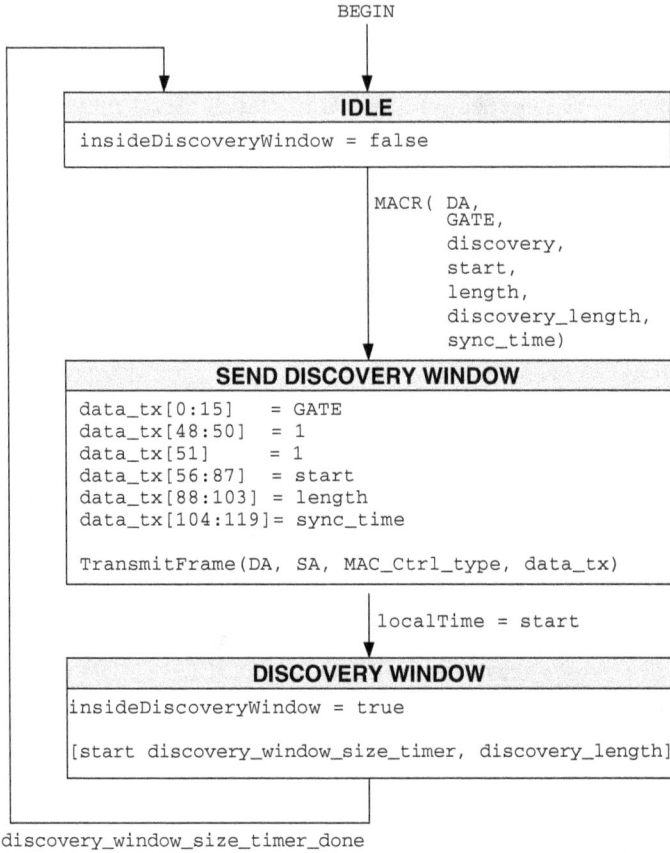

Figure 8.23   OLT discovery gate generation. (*Reprinted from IEEE Standard 802.3ah with permission from IEEE.*)

its fields to the values issued by the discovery agent. The fields of the discovery GATE message are shown in Fig. 8.5*a*. This message is then passed to the control multiplexer by calling the `TransmitFrame(...)` function.

The process remains in the SEND DISCOVERY WINDOW state until the beginning of the discovery window—a condition indicated by the local MPCP clock reaching the value specified as the grant start time. Upon this event, the process enters the DISCOVERY WINDOW state.

8.5.1.3 DISCOVERY WINDOW state. In this state, the variable `insideDiscoveryWindow` is set to true, and a timer representing the discovery window (`discovery_window_size_timer`) is started. When this timer expires, i.e., when the discovery window ends, the process returns to

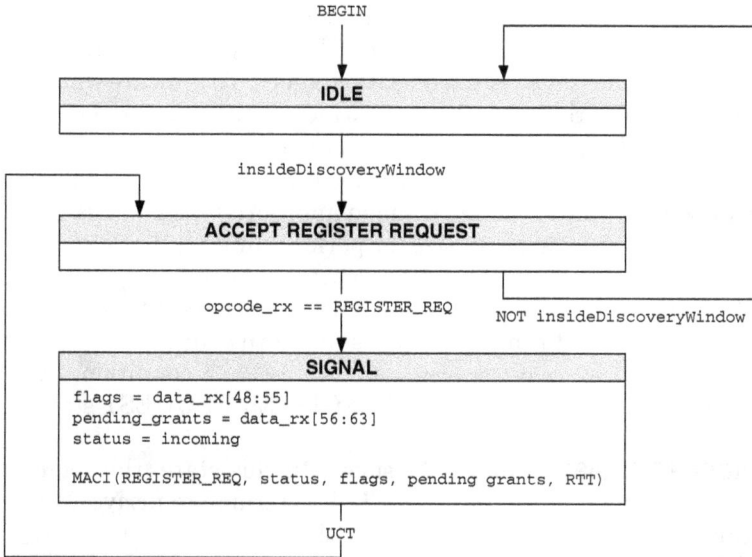

Figure 8.24   OLT request reception state diagram. (*Reprinted from IEEE Standard 802.3ah with permission from IEEE.*)

the IDLE state, where the `insideDiscoveryWindow` variable is immediately reset to false.

8.5.2 Request reception at the OLT

The request reception process is responsible for receiving the REGISTER_REQ messages and passing the received data to the DBA agent. This process only runs on the instance of MPCP associated with the broadcast LLID. The state diagram of the request reception process is shown in Fig. 8.24.

The request reception process only accepts REGISTER_REQ messages that arrive within the discovery window, as indicated by the asserted `insideDiscoveryWindow` variable. All messages that arrive outside the discovery window are discarded without informing the discovery agent. This behavior is not justified. The main reason to allocate the discovery window is to prevent REGISTER_REQ messages from colliding with normally scheduled data. A reception of REGISTER_REQ outside the allocated window indicates an abnormal situation: either missynchronized ONU or a misprovisioned discovery window length. But the fact that the REGISTER_REQ was received by the request reception process at the OLT means that, by lucky chance, this message has not collided with any data. Discarding this message at the OLT would force the ONU to repeat the discovery attempt again, possibly causing the collision a second time, if the problem remains. It would be a more robust behavior for the OLT to pass all the received REGISTER_REQ messages to the discovery agent. To indicate whether a message arrived within the allocated window, the `insideDiscovery-Window` value could be passed to the discovery agent as well. Passing this information to the discovery agent will let it know who the culprit is and will allow it to take an action deemed appropriate, e.g., alert operator, increase discovery window size, or perform unicast discovery.

8.5.2.1 IDLE state. Upon initialization, the request reception process enters the IDLE state. It remains in this state until a discovery window opens, as indicated by `insideDiscoveryWindow == true` condition, upon which the process enters the ACCEPT REGISTER REQUEST state.

8.5.2.2 ACCEPT REGISTER REQUEST state. In this state, the request reception process waits for a REGISTER_REQ frame to arrive. When such a frame arrives, which is signaled by the `opcode_rx == REGISTER_-REQ` condition being true, the process transitions to the SIGNAL state. When the discovery window ends, the process returns to the IDLE state, in which it will not accept REGISTER_- REQ messages anymore.

8.5.2.3 SIGNAL state. In the SIGNAL state, the incoming REGIS-TER_REQ frame is parsed, and the `flags` and `pending_grants` fields are extracted. These values, together with the `RTT`, which is calculated by the control parser (see Sec. 8.2.1), are passed to the discovery agent via the `MACI(REGISTER_REQ, status, flags, pending_grants, RTT)` service primitive.

From this state, the request reception process unconditionally transitions to the ACCEPT REGISTER REQUEST state, where it awaits for more REGISTER_REQ frames that may arrive in the same discovery window.

8.5.3 Register generation at the OLT

The register generation process is driven by the discovery agent. Upon receiving a request from the discovery agent, the register generation process forms a REGISTER message and transmits it to the OLT. The format of the REGISTER message was discussed in Sec. 8.1.4. The register generation state diagram is shown in Fig. 8.25.

8.5.3.1 WAIT FOR REGISTER state. Upon initialization, the register generation process enters the WAIT FOR REGISTER state, in which it remains until the discovery agent issues a request to transmit a

```
                              BEGIN

                                │
                                ▼                   ┌──────────┐
┌──────────────────────────────────────────────────┼─────┐    │
│              WAIT FOR REGISTER                     │     │    │
├────────────────────────────────────────────────────────┤    │
│                                                         │    │
└─────────────────────────────────────────────────────────┘   │
                                │                              │
                                │                              │
                         MACR( DA,                             │
                               REGISTER,                       │
                               LLID,                           │
                               status,                         │
                               pending_grants)                 │
                                │                              │
                                ▼                              │
┌─────────────────────────────────────────────────────────┐   │
│                     REGISTER                             │   │
├─────────────────────────────────────────────────────────┤   │
│ data_tx[0:15]  = REGISTER                               │   │
│ data_tx[48:63] = LLID                                   │   │
│ data_tx[64:71] = status                                 │   │
│ data_tx[72:87] = syncTime                               │   │
│ data_tx[88:96] = pending_grants                         │   │
│                                                         │   │
│ TransmitFrame(DA, SA, MAC_Ctrl_type, data_tx)           │   │
└─────────────────────────────────────────────────────────┘   │
                                │                              │
                               UCT ─────────────────────────────┘
```

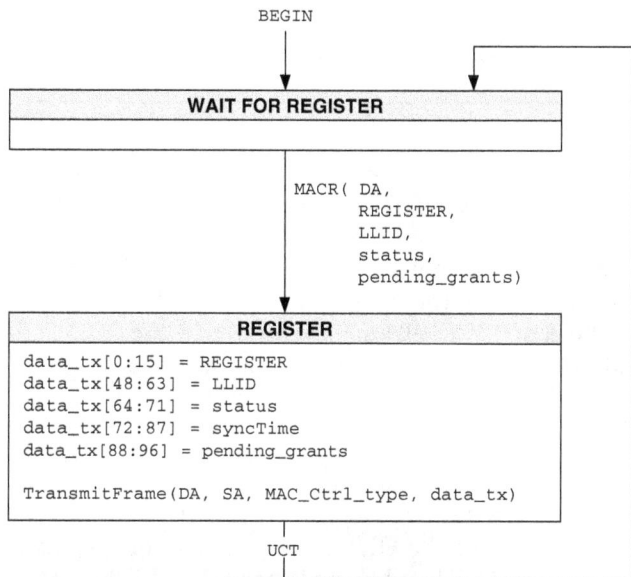

Figure 8.25 OLT register generation state diagram. (*Reprinted from IEEE Standard 802.3ah with permission from IEEE.*)

REGISTER message. This request is represented by the MACR (DA,REGISTER, LLID, status, pending_grants) service primitive, with the following parameters:

DA—MAC address of the ONU being registered. This address is learned from the received REGISTER_REQ message.

REGISTER—opcode identifying the REGISTER MPCPDU ($00\text{-}05_{16}$).

LLID—the logical link identification assigned by the discovery agent. It remains the agent's responsibility to ensure uniqueness of all assigned LLIDs.

status—this parameter represents the specific registration instructions to the ONU and is copied into the flags field of REGISTER MPCPDU (see Sec. 8.1.4.2).

pending_grants—this parameter echoes the value of pending grants received in the REGISTER_REQ message.

⚠️ Please note that in the IEEE 802.3ah, the description for MACR(DA,REGISTER, LLID, status, pending_grants) service primitive incorrectly states that the DA parameter has the value of "multicast MAC control address as defined in Annex 31B," i.e., the group address $01\text{-}80\text{-}C2\text{-}00\text{-}00\text{-}01_{16}$. REGISTER messages are

> addressed to individual ONUs, to which the point-to-point logical links are not established yet. Therefore, these messages are sent on the broadcast logical link, but with individual MAC addresses of destination ONUs.

Upon receiving a request from the discovery agent to send a REGISTER message, the process transitions to the REGISTER state.

8.5.3.2   REGISTER state.   In this state, a REGISTER MPCPDU is created, and its fields are set to the values issued by the discovery agent. The fields of the REGISTER frame are shown in Fig. 8.8. This frame is then passed to the control multiplexer by calling the Transmit-Frame(...) function, and the process unconditionally returns to the WAIT FOR REGISTER state.

8.5.4   Final registration at the OLT

The final registration process at the OLT is responsible for issuing a unicast GATE message to an ONU and receiving the REGISTER_ACK frame. This process is instantiated for each logical port at the OLT, except the port connected to the broadcast logical link. The state diagram of the final registration process is shown in Fig. 8.26.

8.5.4.1   WAIT FOR GATE state.   Upon initialization, the final registration process enters the WAIT FOR GATE state, where it initializes the registered variable to false. The process remains in this state until the DBA agent issues a request to transmit a GATE message. This request is represented by the MACR(DA, GATE, grant_number, start[4], length[4], force_report[4]) service primitive. It is easy to recognize that the same service primitive is expected by the gate generation process at the OLT (see Sec. 8.3.1). The gate generation process will only issue a GATE MPCPDU if the corresponding ONU is registered (i.e., if registered == true). Therefore, the first GATE message, which is transmitted as part of the registration procedure, is ignored by the gate generation process; instead, the corresponding GATE MPCPDU is generated by the final registration process in the state WAIT FOR REGISTER_ACK.

8.5.4.2   WAIT FOR REGISTER_ACK state.   In this state, the final registration process creates a GATE MPCPDU using the parameters issued by the DBA agent. The GATE frame is then passed to the control multiplexer by calling the TransmitFrame(...) function.

Additionally, the process calculates grantEndTime—the future time corresponding to the end of the first grant in the GATE MPCPDU. As was explained in Sec. 5.3.1.2, if the OLT expects the data from ONU to

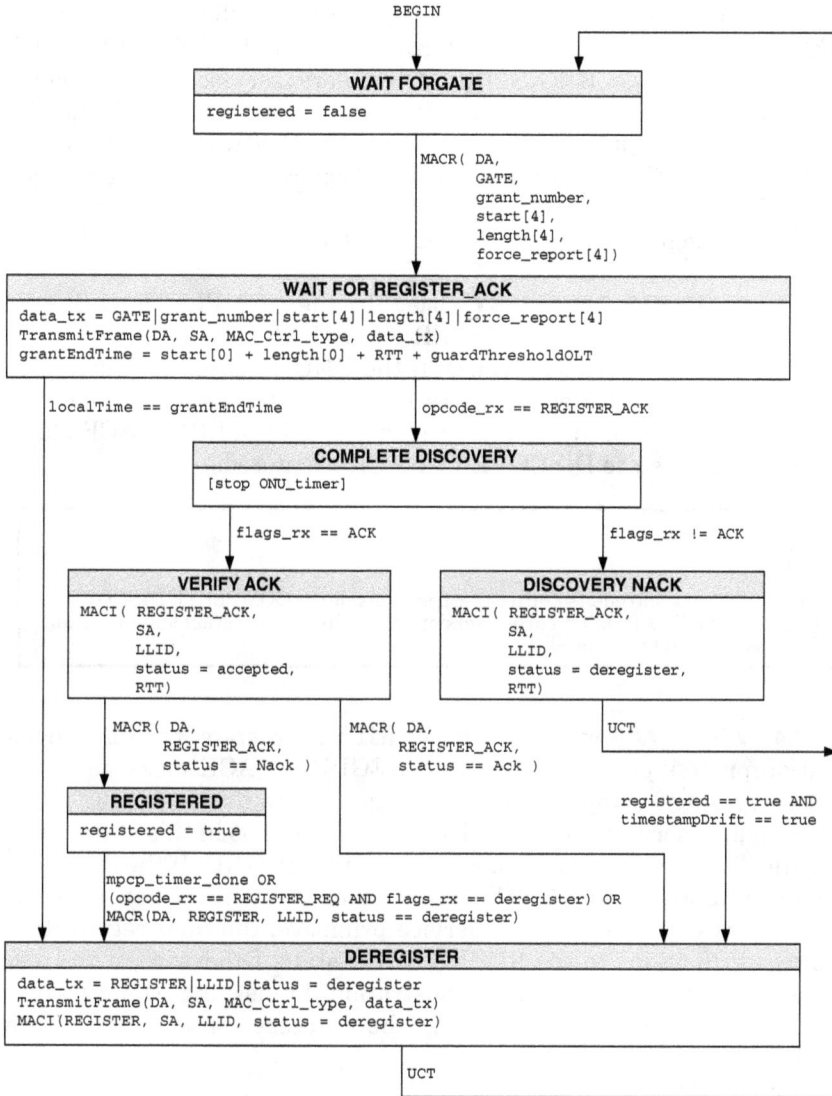

Figure 8.26 OLT final registration state diagram. (*Reprinted from IEEE Standard 802.3ah with permission from IEEE.*)

arrive at time T, it should set the grant start time *start* to $T - RTT$. Conversely, given *start*, we can find when the first bit of data is expected to arrive to the OLT: $T = start + RTT$. The last bit will arrive at time $start + RTT + length$. Finally, after adding the margin for delay variability, we get the latest possible time for data arrival to the OLT: `grantEndTime = start[0] + RTT + length[0] + guardThresholdOLT`.

The final registration process expects the ONU to transmit a REGISTER_ACK within the first grant given to the ONU. If the OLT does not receive a REGISTER_ACK before the local MPCP clock reaches the grantEndTime, it considers the registration failed and the final registration process transitions into the DEREGISTER state.

If a REGISTER_ACK frame is received before the local clock reaches the grantEndTime, as indicated by the opcode_rx == REGISTER_ACK condition, the process enters the COMPLETE DISCOVERY state.

8.5.4.3 COMPLETE DISCOVERY state. In the COMPLETE DISCOVERY state, the received REGISTER_ACK MPCPDU is further parsed to determine the ONU's response. If the ONU confirms the successful registration, as indicated by the received flags field having value Ack (or flags_rx == Ack), the process transitions to the VERIFY ACK state. Otherwise, the state DISCOVERY NACK is entered.

⚠️ Please note that the code [stop ONU_timer] inside the COMPLETE DISCOVERY state is an artifact from an earlier version of the final registration state diagram. This code should be removed.

8.5.4.4 VERIFY ACK state. In the VERIFY ACK state, the final registration process passes the received REGISTER_ACK message to the discovery agent using the MACI(REGISTER_ACK, SA, LLID, status = accepted, RTT) service primitive. The process remains in this state pending the discovery agent's decision to register the ONU. If the discovery agent authorizes the ONU's registration, i.e., it issues MACR(DA, REGISTER_ACK, status == Ack) service primitive, the final registration process transitions to the REGISTERED state. Otherwise, if the discovery agent denies ONU's registration by issuing the MACR(DA, REGISTER_ACK, status == Nack), the process enters the state DEREGISTER.

8.5.4.5 DISCOVERY NACK state. In the DISCOVERY NACK state, the final registration process informs the OLT's discovery agent of the ONU's refusal to register, by issuing the MACI(REGISTER_ACK, SA, LLID, status = accepted, RTT) service primitive.

From this state, the final registration process unconditionally returns to the WAIT FOR GATE state, where the registered variable is set to false.

8.5.4.6 REGISTERED state. Upon transition to the REGISTERED state, the process sets global variable `registered` to true, indicating to the gate generation process that it is allowed to issue GATE MPCPDUs. The final registration process will remain in the REGISTERED state for as long as the corresponding ONU remains registered. The ONU will remain registered until any of the following conditions becomes true:

1. `mpcp_timer_done == true`. This condition indicates that the OLT has not received a REPORT message from the corresponding ONU for the entire `mpcp_timeout` interval. The `mpcp_timer` is set by the report reception process described in Sec. 8.4.2.

2. `timestampDrift == true`.[2] This condition indicates that the difference between previously calculated RTT and the last calculated RTT is larger than the allowed margin `guardThresholdOLT`. The `guardThresholdOLT` is set by the OLT's control parser process, described in Sec. 8.2.1.

3. `opcode_rx == REGISTER_REQ AND flags_rx == deregister`. This condition indicates a situation in which the ONU initiates deregistration and transmits a REGISTER_REQ message with the flags field having the value deregister (see Sec. 8.1.3.1).

4. `MACR(DA, REGISTER, LLID, status == deregister)`. This is a situation in which the OLT's local discovery agent initiates deregistration by issuing the above service primitive with parameter `status == deregister`.

If any of the above conditions are satisfied, the final registration process enters the DEREGISTER state.

8.5.4.7 DEREGISTER state. In the DEREGISTER state, the final registration process creates a REGISTER frame with the Flags field having value 2 (deregister), as explained in Sec. 8.1.4.2, and passes the frame to the control multiplexer by calling the `TransmitFrame(...)` function. The process also indicates to the discovery agent that the ONU has been deregistered and unconditionally returns to the WAIT FOR GATE state.

8.5.5 Discovery process at the ONU

As shown in Fig. 8.22, the discovery process in the ONU is responsible for generating REGISTER_REQ MPCPDUs, processing the received

---

[2] In the final registration state diagram (Fig. 8.26), the transition labeled `registered == true AND timestampDrift == true` is shown as a separate global transition. However, it can occur only between states REGISTERED and DEREGISTER, so we consider it here along with other conditions for the transition between these two states.

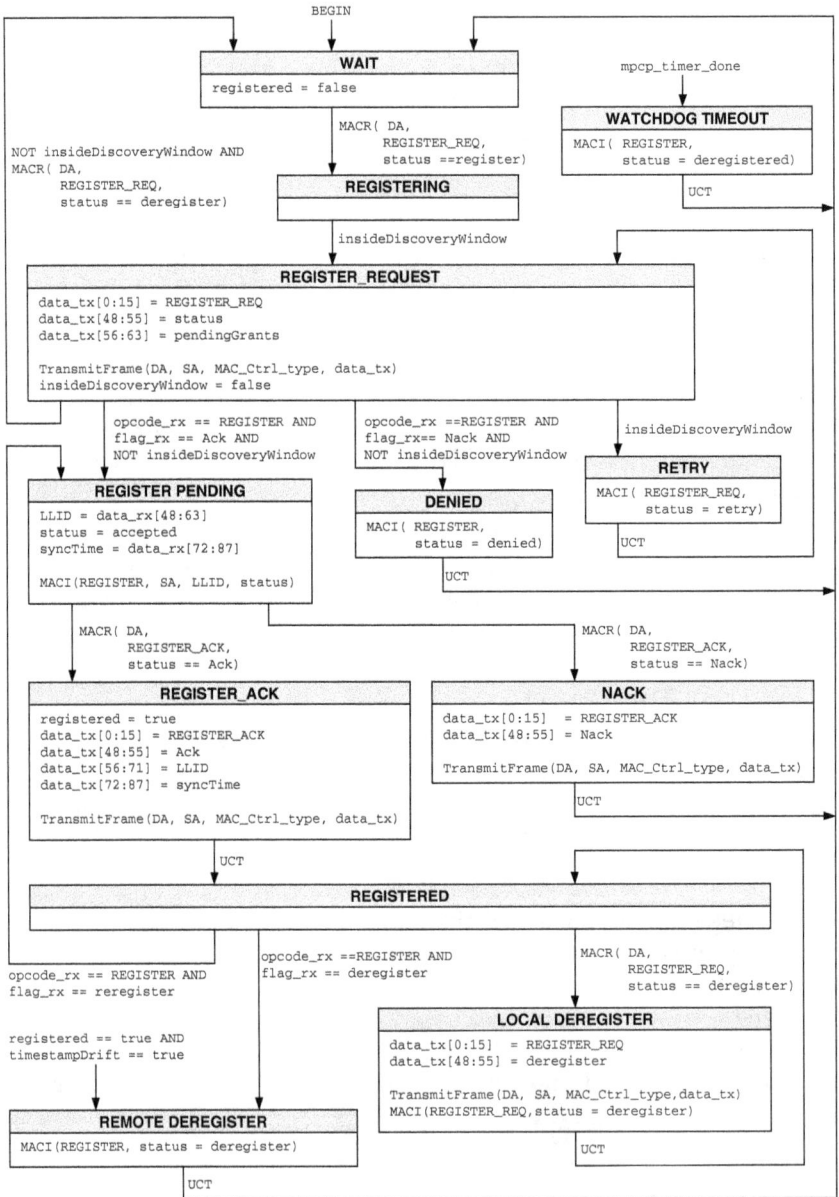

Figure 8.27   ONU discovery process state diagram. (*Reprinted from IEEE Standard 802.3ah with permission from IEEE.*)

REGISTER MPCPDUs, and issuing acknowledgments in the form of REGISTER_ACK MPCPDUs. The discovery process state diagram is shown in Fig. 8.27.

8.5.5.1 WAIT state. Upon initialization, the discovery process enters the WAIT state, where it initializes the `registered` variable to false. The process remains in this state until ONU's discovery agent issues a request to register. This request is represented by the `MACR(DA,-REGISTER_REQ, status == register)` service primitive. Upon receiving such request, the process transitions to the REGISTERING state.

8.5.5.2 REGISTERING state. The discovery process remains in the REGISTERING state until the next discovery window becomes available, as indicated by the `insideDiscoveryWindow == true` condition. The global variable `insideDiscoveryWindow` is set and cleared by the gate reception process, discussed in Sec. 8.3.2. When the discovery window becomes available, the process transitions to the REGISTER_REQUEST state.

8.5.5.3 REGISTER_REQUEST state. In this state, the discovery process forms a REGISTER_REQ MPCPDU. The fields of the REGISTER_REQ MPCPDU are shown in Fig. 8.7. This frame is then passed to the control multiplexer by calling the `TransmitFrame(...)` function. Upon transmission of the REGISTER_REQ frame, the variable `insideDiscoveryWindow` is reset to false.

The discovery process remains in the REGISTER_REQUEST state, waiting for REGISTER MPCPDU from the OLT. When REGISTER MPCPDU indicating a successful registration [i.e., with field flags = 3 (Ack)] arrives, the process transitions to the REGISTER PENDING state.

If the OLT denies the registration [i.e., the REGISTER MPCPDU has field flags = 4 (Nack)], the discovery process enters the DENIED state.

If, while waiting for the REGISTER MPCPDU, the discovery process receives a request from the local discovery agent to deregister, it will return to the WAIT state, where it will remain until the discovery agent decides to register again.

If an ONU does not receive the REGISTER message before the next discovery GATE, it will infer that a collision has occurred, and it will attempt to initialize again. The `insideDiscoveryWindow` variable becoming true while the discovery process remains in the REGISTER_REQUEST state indicates that a new discovery window became available to the ONU. In other words, that means that a new discovery GATE arrived and that the previous attempt to register did not succeed. In this situation, the discovery process will transition to the RETRY state, where it will inform the discovery agent of its intention to retry the registration.

8.5.5.4   DENIED state.   In the DENIED state, the discovery process informs the local discovery agent of the OLT's refusal to register the ONU and unconditionally returns to the WAIT state, where it will await for higher-layer request to register again.

8.5.5.5   RETRY state.   In the RETRY state, the discovery process informs the local discovery agent of the absence of OLT's response, possibly due to a collided REGISTER_REQ message, and unconditionally returns to the REGISTER_REQUEST state, where it will immediately attempt to register again.

8.5.5.6   REGISTER PENDING state.   The decision of whether to accept registration from the OLT is not made by the discovery process; rather it is delegated to the local discovery agent. In the REGISTER PENDING state, the received REGISTER MPCPDU is parsed and the relevant parameters are passed to the discovery agent via the MACI (REGISTER, SA, LLID, status) service primitive. If the discovery agent accepts the registration, which is signaled by its issuing the MACR(DA, REGISTER_ACK, status==Ack) service primitive, the process enters the REGISTER_ACK state. If the discovery agent denies registration by issuing MACR(DA, REGISTER_ACK, status==Nack), a transition to the NACK state occurs.

8.5.5.7   REGISTER_ACK state.   The REGISTER_ACK state is entered when both the discovery agent in the OLT and the local discovery agent in the ONU agree on registration parameters. In this state, the discovery process sets the global variable registered to true and transmits a REGISTER_ACK MPCPDU to the OLT. The process then unconditionally transitions to the REGISTERED state.

8.5.5.8   NACK state.   The NACK state is entered if the local discovery agent refused registration. In this case, the discovery process transmits a REGISTER_ACK with flags field value equal to Nack and unconditionally returns to the WAIT state.

8.5.5.9   REGISTERED state.   After the registration procedure is completed, the discovery process enters the REGISTERED state, in which it remains as long as the ONU remains registered. Below, we consider several conditions that will take the discovery process out of the REGISTERED state.

1. MACR(DA,   REGISTER_REQ,   status   ==   deregister).   This condition indicates that the local discovery agent initiates deregistration by issuing the above service primitive with parameter status ==

`deregister`. In this case, the discovery process transitions to the LOCAL DEREGISTER state.

2. `opcode_rx == REGISTER AND flag_rx == deregister`. This condition indicates a situation in which the OLT initiates deregistration and transmits a REGISTER message with flags field having the value "deregister" (see Sec. 8.1.4.2). Under this condition, the discovery process enters the REMOTE DEREGISTER state.

3. `opcode_rx == REGISTER AND flag_rx == reregister`. This condition indicates a situation in which the OLT directs the ONU to reregister, possibly with different parameters, such as a new LLID or a different syncTime. Following this request, the discovery process returns to the REGISTER PENDING state, where it further parses the received REGISTER MPCPDU and indicates new parameters to the local discovery agent.

8.5.5.10   LOCAL DEREGISTER state.   The discovery agent at the ONU is able to deny the registration after it receives a REGISTER MPCPDU. But if the previously registered discovery agent decides to deregister, it is not allowed to do so on its own. All it can do is to solicit deregistration from the OLT by issuing a REGISTER_REQ message with `status = deregister`.

In LOCAL DEREGISTER state, the discovery process creates a REGISTER_REQ frame and passes it to the control multiplexer by calling the `TransmitFrame(...)` function. The process then unconditionally returns to the REGISTERED state where it waits for an incoming REGISTER MPCPDU with `flags_rx == deregister`. Until such a message is received, the ONU remains registered and must continue participating in exchange of GATE and REPORT messages, etc.

8.5.5.11   REMOTE DEREGISTER state.   The REMOTE DEREGISTER is a transient state in which the discovery process informs the discovery agent that the ONU has been deregistered. The process then unconditionally returns to the WAIT state.

The REMOTE DEREGISTER state is also entered if `timestampDrift` error occurs. At the ONU, the condition `timestampDrift == true` indicates that the difference between the received timestamp and the local MPCP clock (`localTime` variable) is larger than the allowed margin `guardThresholdONU`. The global variable `timestampDrift` is set by the ONU's control parser process, described in Sec. 8.2.1.

8.5.5.12   WATCHDOG TIMEOUT state.   The WATCHDOG TIMEOUT is a transient state in which the discovery process informs the discovery agent that the registered ONU has not received GATE messages for the `mpcp_timeout` interval. The `mpcp_timer` is armed by the gate reception process described in Sec. 8.3.2.

⚠️

After an ONU becomes deregistered, it may not receive GATE MPCPDUs on a regular basis anymore. However, the `mpcp_timer` may remain armed since the last received GATE message. Expiration of this timer, indicated by the `mpcp_timer_done == true` condition, will cause a transition to WATCHDOG TIMEOUT state, possibly breaking an ongoing registration procedure.

This problem may be solved in several ways. One solution would allow transition to the WATCHDOG TIMEOUT state for only registered ONUs; i.e., the transition label would need to be changed to `registered == true AND mpcp_timer_done == true`.

Another solution would stop `mpcp_timer` immediately when the ONU became unregistered. In this case, the WAIT state would include the following code:

```
registered = false
[stop mpcp_timer]
```

Forward Error Correction

The IEEE 802.3ah standard specifies an optional *forward error correction* (FEC) mechanism. The FEC corrects errors that may occur during the transmission, thus reducing the *bit-error ratio* (BER). The gain provided by FEC can be used to increase the distance between the OLT and ONUs, or to increase the split ratio of EPON, or simply to improve the reliability of the digital channel.

9.1 Basics of FEC Coding

FEC is a method of error control in digital communications, which preprocesses data before the transmission. Such preprocessing involves adding redundancy to the original information, such that, using this redundant information, the receiving device is able to detect and correct some transmission errors. The main categories of FEC methods are block-coding, convolutional coding, and the relatively recently developed turbocoding.

One of the most widely used codes is the Reed-Solomon codes. *Reed-Solomon* (RS) codes are block-based error-correcting codes—encoding and decoding are done on one block at a time. A RS code is denoted as RS (n, k), where n is the length of the encoded block and k is the length of information block.

The RS encoding operates not over individual bits, but over m-bit symbols. The length of encoded block n is related to symbol size as $n = 2^m - 1$. Thus, if 8-bit symbols are used, n should be 255.

The RS encoder takes a block of k information symbols and adds $n - k$ redundant symbols to it. The IEEE 802.3ah standard refers to the redundant symbols as parity data; we will adhere to the same terminology here. The RS code is known as *systematic* code, which refers to the fact that parity data are added at the end of information symbols, leaving the information block unchanged. In Sec. 9.2 we will see that this property plays a crucial role in allowing non-FEC receivers (i.e., receivers without FEC decoders) to still be able to receive FEC-coded data.

The error-correcting capability of RS(n, k) code is determined by the number of parity symbols $n - k$: up to $(n - k)/2$ erroneous symbols per n-symbol block can be corrected. It is important to note that multiple bit errors in a single symbol are counted as a single error, a feature making RS coding especially efficient for correcting *burst errors.*

Sometimes, the position of an erroneous symbol in the block may be known. The symbol error with a known position is called an *erasure.* For example, not all 1024 possible 10-bit values are valid code-words. Thus, a received 10-bit value that does not represent a valid code-word can be marked as an erasure. RS coding schemes can correct up to $n - k$ erasures, twice the number of errors. A received block, which has a combination of r errors and s erasures, can be corrected as long as $2r + s \leq n - k$.

The IEEE 802.3ah standard has adopted the coding scheme RS(255, 239)—the same FEC scheme as specified in [G975]. The notation RS (255, 239) is enough to tell us that this scheme operates over 8-bit symbols, adds 16 parity symbols per block, and can correct up to 8 errors and up to 16 erasures.

This chapter will only focus on mechanisms and procedures necessary to enable FEC in EPON, such as FEC frame delineation and buffering. Interested readers are referred to [CC81] and [Wic95] for an in-depth treatment of the Reed-Solomon algorithm as well as error correction theory and techniques in general.

9.2   Stream-Based versus Frame-Based FEC

The choice of FEC framing structure has generated heated debates in the IEEE 802.3ah task force. The two main camps argued for *stream-based* versus *frame-based* FEC structure.

A stream-based mechanism treats the Ethernet frames and idles between them as just a stream of (uninterpreted) data symbols. This method is simpler to implement—after every block of $n \times 239$ symbols (octets or code-words), the FEC will insert $n \times 16$ parity symbols ($n = 1$, 2, 3 ...). The stream-based FEC adds a fixed overhead equal to $1 - k/n$, which, in the case of the RS(255, 239) scheme, is equal to 6.27 percent. Of course, stream-based encoding requires both the transmitting and the receiving devices to use this framing structure. A non-FEC-capable device will unavoidably become confused by added parity data and won't be able to recover any data. In EPON, this dependency translates to a situation such that if one ONU needs to use FEC, all ONUs must use FEC, and conversely, if one ONU is unable to use FEC, none of the ONUs may use FEC.

A frame-based FEC method seeks to encode only the useful data (i.e., Ethernet frames) and to leave the gaps between the frames unprotected. In this method, a frame is divided into 239-byte blocks, and 16 bytes of parity data is added for each block. Depending on the frame length, the last block may be shorter than 239 bytes. Such a block is padded with zeros to the length of 239, and the parity codes are calculated over the full-size block. However, the padding symbols are not transmitted. Similarly, the receiver reconstructs the shortened block to its full length by appending the necessary number of zeros, before applying the FEC decoder function to correct possible errors.

In the frame-based method, the parity symbols generated for each block are grouped together and are appended at the end of a frame, leaving the frame itself unchanged. The fact that the entire Ethernet frame is left unchanged is the major advantage of the frame-based FEC encoding. It allows a non-FEC-capable device to receive a FEC-encoded frame, albeit without any error correction. Thus, an EPON can contain a combination of FEC-capable and FEC-incapable ONUs. Only the FEC-capable ONUs will take advantage of the added FEC protection. FEC-incapable ONUs will not see any coding gain, but nevertheless will be able to receive frames.

Among the shortcomings of the frame-based FEC scheme, the main one is its variable overhead, which depends on a mix of packet sizes. In Chap. 12 we will find that, for an empirical packet size distribution, the average value for FEC overhead is 9.25 percent. This is significantly higher than the 6.27 percent overhead in the case of the stream-based FEC.

Despite its higher complexity and higher overhead, the frame-based FEC was adopted as a baseline proposal for EPON.

9.3 FEC Frame Delineation

To differentiate a FEC-encoded frame from a nonencoded frame, special frame markers are used. These frame markers need to be processed by the receiver before a frame can be delineated and the parity data can be accessed. Therefore, the frame markers are not protected by the FEC. To reduce the probability of false marker detection or misdetection under high BER, the markers use a longer sequence of symbols. Table 9.1 shows the frame markers and their corresponding symbol sequence.

TABLE 9.1   Frame Delimiters for FEC-Coded Frames

Notation	Description	Sequence
/S_FEC/	Start of FEC-coded packet	/K28.5/D6.4/K28.5/D6.4/S/
/T_FEC_E/	End of FEC-coded packet with even alignment	/T/R/I/T/R/
/T_FEC_O/	End of FEC-coded packet with odd alignment	/T/R/R/I/T/R/

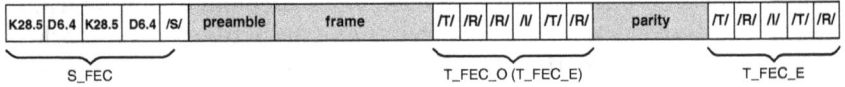

Figure 9.1   Structure of FEC-coded frame.

The symbols /T/, /R/, /S/, and /I/ are the same symbols that are used to delineate regular (non-FEC) frames and are described in Table 36-3 in [802.3]. The receiving device tries to correlate the received bit stream to the marker sequences, and declares a match if it finds a correlation with Hamming distance of less than 5 (i.e., no more than 4 bits different from the expected sequence).

Figure 9.1 presents the structure of a FEC-coded frame. A FEC-coded frame starts with the /S_FEC/ sequence. Following the FCS field, the first terminating marker is located. This marker may be either /T_FEC_O/ or /T_FEC_E/, depending on the alignment of the frame. The selection of the marker should ensure that the idle code-group /I/ located in the middle of this marker starts at the even position. After the first terminating marker, the parity data are appended. The length of the parity data depends on the frame length; for an n-byte frame (including preamble and FCS), the amount of parity data is equal to $\lceil n / 239 \rceil \times 16$. After the parity data, the second terminating delimiter is appended. This delimiter can only be /T_FEC_E/ because the length of parity data is always even.

9.3.1   Hamming distance between FEC delimiters

The FEC decoder scans the incoming bit stream for a possible match with a delimiter. The match is found if the incoming bit stream has less than 5 bit errors compared with the expected delimiter. However, there is a problem, in that the last 6 code-groups of /T_FEC_O/ delimiter and the entire /T_FEC_E/ delimiter only have the Hamming distance of 2 between them. Figure 9.2 illustrates this for the case when the starting running disparity is negative and when it is positive.

The fact that the Hamming distance between these two delimiters is only 2 leads to a peculiar situation, in which the FEC decoder may not

/T_FEC_E/

Code-group	Dx.y	/T/	/R/	K28.5	D16.2	/T/	/R/
10-bit value	xxx	2E8	3A8	0FA	245	2E8	3A8

2 bit difference   =   =   =   =   =

RD	−	−	−	−	−	+	−	−
10-bit value	2E8	3A8	3A8	0FA	245	2E8	3A8
Code-group	/T/	/R/	/R/	K28.5	D16.2	/T/	/R/

/T_FEC_O/

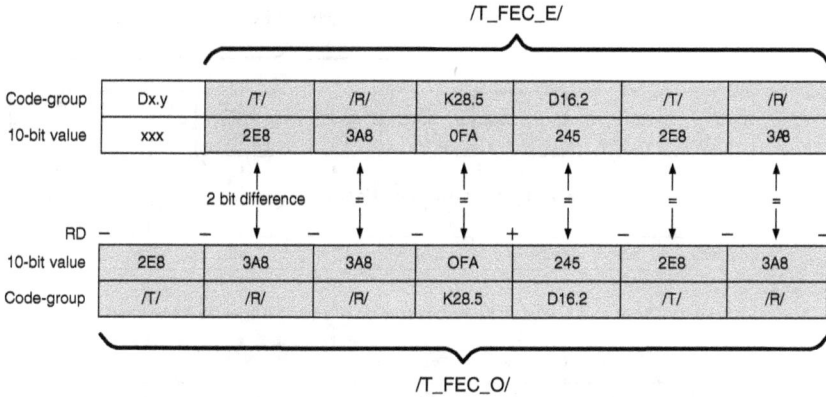

(a) Correlation between /T_FEC_E/ and /T_FEC_O/ when starting disparity is negative

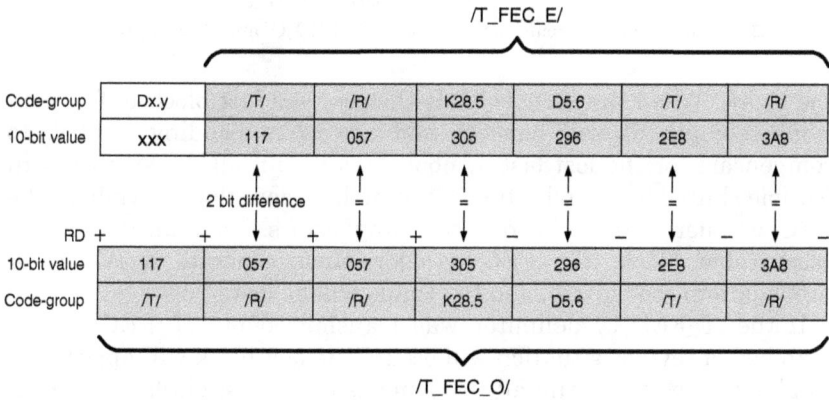

/T_FEC_E/

Code-group	Dx.y	/T/	/R/	K28.5	D5.6	/T/	/R/
10-bit value	xxx	117	057	305	296	2E8	3A8

2 bit difference   =   =   =   =   =

RD	+	+	+	+	−	−	−	−
10-bit value	117	057	057	305	296	2E8	3A8
Code-group	/T/	/R/	/R/	K28.5	D5.6	/T/	/R/

/T_FEC_O/

(b) Correlation between /T_FEC_E/ and /T_FEC_O/ when starting disparity is positive

Figure 9.2   Hamming distance between /T_FEC_O/ and /T_FEC_E/ delimiters.

be able to chose the correct delimiter. Consider, for example, a received sequence of code-groups as illustrated in Fig. 9.3. Both /T_FEC_O/ and /T_FEC_E/ delimiters can be matched with only two bit errors. At this moment the decoder has a 50 percent chance of guessing it right.

What happens if the decoder guesses it wrong? If the correct (transmitted) delimiter was /T_FEC_E/ but the decoder guessed /T_FEC_O/, then the last code-group of the FCS field (D28.1 in Fig. 9.3) will be consumed by the incorrectly matched delimiter. The last block in a frame, which most often is shortened as it is, will become one code-word shorter. Losing one symbol by itself would not be terribly dangerous, if not for the IEEE 802.3ah directive to pad a shortened block from the beginning of the block (on the left) instead of the end (on

Possible match /T_FEC_E/ (2 bit errors)

D9.3	D28.1	/T/	/R/	K28.5	D16.2	/T/	/R/
25C	0E9	2E8	3A8	0FA	245	2E8	3A8

No bit errors No bit errors 2 bit errors No bit errors No bit errors No bit errors No bit errors No bit errors

Received sequence

25C	0E9	3A8	3A8	0FA	245	2E8	3A8

No bit errors No bit errors 2 bit errors No bit errors No bit errors No bit errors No bit errors No bit errors

25C	2E8	3A8	3A8	0FA	245	2E8	3A8
D9.3	/T/	/R/	/R/	K28.5	D16.2	/T/	/R/

Possible match /T_FEC_O/ (2 bit errors)

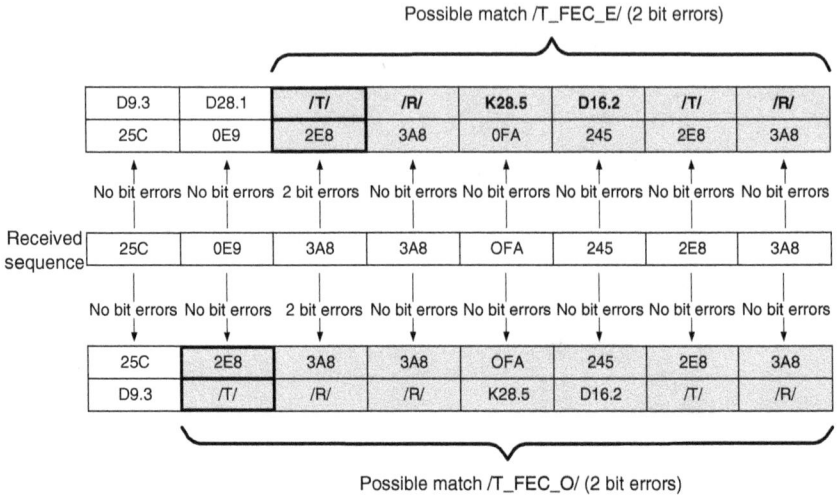

Figure 9.3   Example of bit stream matching both /T_FEC_O/ and /T_FEC_E/ delimiters.

the right). When the decoder pads the received last block to full 239-symbol length, it will have to add one extra padding symbol, to compensate for the lost last symbol. As a result, all the symbols will become shifted by 1, and if the last block has more than 8 symbols, the FEC decoder will choke on too many errors. To indicate an unrecoverable block, the FEC decoder must generate a /V/ (error propagation) code-group, and the frame will be lost.

If the /T_FEC_O/ delimiter was transmitted but /T_FEC_E/ was matched, a reverse situation will occur: The last block will appear one symbol longer, and again all the symbols in the last block will appear shifted by one position, in the opposite direction this time. The net result will be the same—the frame will be discarded.

> The IEEE 802.3 work group has recognized the problem with the /T_FEC_E/ and /T_FEC_O/ delimiter definitions only after approval of the standard. The work group is expected to correct the situation in the near future. The fix most likely will include a modification of the code-group sequence for one or both of the delimiters.

9.3.2 Backward compatibility

Let us now consider how a FEC-coded frame will be treated by a non-FEC-capable device. Such a device will look for an /I/S/ sequence to find a start of the frame. The IEEE 802.3 standard is strict on what can be transmitted as the idle ordered set (only /K28.5/D5.6/ and /K28.5/ D16.2/ are allowed), but at the same time it is very tolerant of the receiving data:

> A received ordered set which consists of two code-groups, the first of which is /K28.5/ and the second of which is a data code-group other than /D21.5/ or /D2.2/, is treated as an /I/ ordered set.

Therefore, a non-FEC-capable ONU, when it sees the /K28.5/D6.4/S/ sequence, will treat it as a *start-of-packet delimiter* (SPD). The code-groups that follow SPD up to a terminating sequence /T/R/ are considered a frame. The /T/R/ sequence, also called the *end-of-packet delimiter* (EPD), signals to the receiving device the packet boundary. It is easy to notice that the first two code-groups in the /T_FEC_x/ delimiter are exactly the same as in the EPD. The non-FEC-capable device will recognize the beginning of /T_FEC_x/ as the EPD delimiter and thus will delineate the frame correctly.

There is a slight complication, though. A non-FEC-capable device would expect idle ordered sets to continue up to the next SPD. However, in case of FEC-coded frame, the receiver will see the rest of the /T_FEC_x/ delimiter followed by parity data, followed by the second /T_FEC_E/ delimiter. These data will trigger a FALSE_CARRIER event at the receiving PHY sublayer. Such false FALSE_CARRIER events may not necessarily impede the operation of a non-FEC-capable device; however, they will increment a counter, called *aFalseCarriers*, after every FEC-coded frame. As a result, the true FALSE_CARRIER events will be obscured.

9.4 Encoding Procedure

The FEC encoder is located in the extended physical coding sublayer (see Sec. 4.1.4.2). There was a dilemma on the placement of the FEC encoder relative to the existing PCS transmit function. On one hand, the FEC encoder substitutes the default frame delimiters (SPD and EPD) with /S_FEC/ and /T_FEC/; therefore, the FEC encoder should be placed after the PCS transmit state machine. On the other hand, the FEC encoder should operate on 8-bit data, not on 10-bit data (otherwise, overhead increases significantly), so it should be placed before the PCS transmit state machine which does 8b/10b encoding. In the end, it seemed a simpler solution to locate the FEC function below the PCS transmit state machine and to require the FEC encoder to internally perform double conversion: 10-bit to 8-bit and 8-bit again to 10-bit. Of course, no reasonable implementation would actually do this double encoding; after all, the standard specifies the model whose externally observable behavior should be reproduced, so internal implementation does not matter.

Figure 9.4 shows the block diagram of the FEC encoding function. The FEC encoder has the *ten-bit interface* (TBI) on both ends, thus ensuring that it can be easily omitted without affecting its adjacent sublayers.

As a 10-bit coded frame enters the FEC encoder, it is passed to the 10b/8b decoder. The decoded frame, as a stream of 8-bit symbols, is shifted through a data buffer, and the parity symbols are calculated using the RS(255, 239) encoder. Note that RS encoding does not alter the contents of the data buffer. The unmodified frame passes through the selector and delimiter-detector (SDD) block, which performs the following tasks:

1. When the SDD detects sequence /I/I/S/, it replaces it with /S_FEC/.

2. When the SDD detects /T/R/I/ or /T/R/R/I/, it replaces this sequence with /T_FEC_O/ or /T_FEC_E/. (In fact, it does not replace this sequence, but rather inserts an additional /T/R/ pair, which complements the detected delimiters to /T_FEC_O/ or /T_FEC_E/.)

3. When FEC is enabled, the interframe gap is increased to allow transmission of parity data, courtesy of the control multiplexer process (see Sec. 8.2.2.7). Following the transmission of /T_FEC_O/ or /T_FEC_E/, the SDD replaces the idle ordered sets in this extended interframe gap with the data from the parity buffer.

4. When the parity buffer becomes empty, the SDD transmits /T_FEC_E/ and then transmits data (idles) emanating from the data buffer, until the /I/I/S/ sequence is detected, as in step 1.

Finally, the 8-bit data out of SDD passes through the 8b/10b encoder again, and the complete FEC-coded frame is passed to the data detector which is responsible for turning the laser on and off (see Chap. 7).

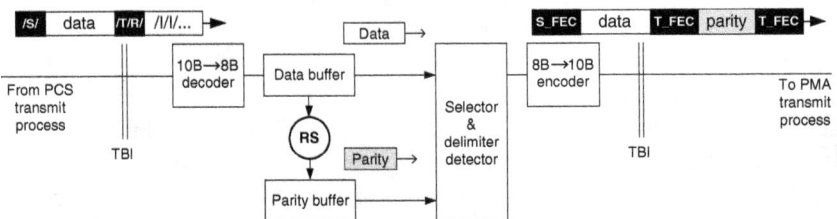

Figure 9.4   FEC encoder block diagram.

9.5 Decoding Procedure

Since FEC and non-FEC ONUs may be combined in the same EPON, the FEC decoding procedure should not only process and correct errors in FEC-coded frames, but also be able to transparently pass any non-FEC-coded frame to the regular PCS receive process.

As is the case with FEC encoder, the optional FEC decoder has a TBI on both ends (Fig. 9.5). Thus, implementations that do not require error correction capabilities may exclude the FEC decoder without affecting the adjacent functions.

The FEC parity data are appended at the end of the frame. Therefore, an entire frame should be buffered before the parity data can be accessed. The buffering delay is determined by the maximum frame size plus the necessary time to perform error correction. For normal MPCP operation, the delay between the control multiplexer and control parser should be constant. The bypass buffer provides a matching delay for non-FEC-coded frames (Fig. 9.5).

If an arriving frame is FEC-coded (a fact that is determined by a detection of the /S_FEC/ delimiter), it is passed to 8b/10b decoder. The decoded frame is stored in the data buffer until its parity data arrive.

When the look-ahead delimiter detector matches the first /T_FEC_x/, it directs the remaining portion of the incoming data, up to the second /T_FEC_E/, to the parity buffer. Once the first 16 bytes of the parity data is received, the error correction of the first block can begin. The corrected 8-bit data are encoded into 10-bit code-words again and passed to the existing PCS receive process. The FEC decoder replaces the /S_FEC/ delimiter with an /I/I/S/ sequence. The first /T_FEC_x/ delimiter is replaced with EPD followed by /I/I/. Finally, the parity data and the second terminating delimiter /T_FEC_E/ are replaced by idle ordered sets. Thus, to the regular PCS receive state machine, the received frame looks like a regular Ethernet frame with an extended run of idles (interframe gap) at the end.

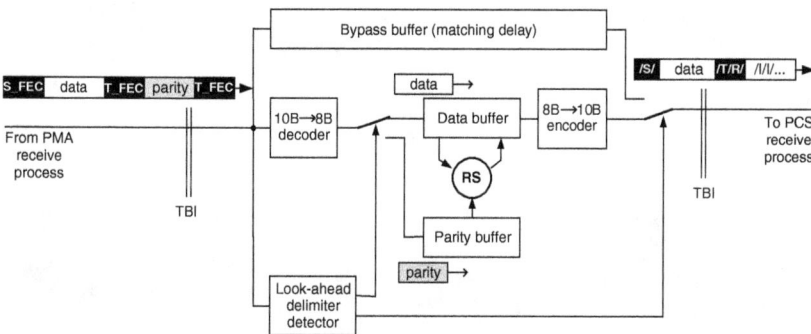

Figure 9.5 FEC decoder block diagram.

References

[802.3]   *IEEE Standard for Information technology - Telecommunications
             and information exchange between systems - Local and metro-
             politan area networks — Specific Requirements. Part 3: Carrier
             Sense Multiple Access with Collision Detection (CSMA/CD) Access
             Method and Physical Layer Specification,* ANSI/IEEE Standard
             802.3-2002, 2002 edition. Available at http://standards.ieee.org/
             getieee802/download/802.3-2002.pdf.

[CC81]   G. C. Clark and J. B. Cain, *Error-Correction Coding for Digital
             Communications,* Kluwer Academic Press/Plenum Publishers,
             1981.

[G975]   ITU-T Recommendation G.975, *Forward Error Correction for
             Submarine Systems,* in Series G: Transmission Systems and
             Media, Digital Systems and Networks, Telecommunication
             Standardization Sector of ITU, October 2000.

[Wic95]   S. B. Wicker, *Error Control Systems for Digital Communication
             and Storage,* Prentice-Hall, Inc., 1995.

System-Level Issues

EPON Encryption

Security requirements in EPON are based on the fact that EPON serves noncooperative, private users, but has a broadcasting downstream channel, potentially available to any interested party capable of operating an end station in promiscuous mode. In general, to ensure EPON security, network operators must be able to guarantee subscriber privacy, and must be provided mechanisms to control subscriber's access to the infrastructure. In a residential access environment, individual users expect their data to remain private. For the business access application, this requirement is fundamental. The two main problems associated with lack of privacy are subscriber's susceptibility to eavesdropping by neighbors (a subscriber issue) and susceptibility to theft of service (a service provider issue).

In EPON, eavesdropping is possible by operating an ONU in promiscuous mode: being exposed to all downstream traffic, such an ONU can listen to traffic intended to other ONUs. Point-to-point emulation adds logical link IDs (see Chap. 6) that allow an ONU to recognize frames intended for it and to filter out the rest. However, this mechanism does not offer the required security, as an ONU might disable this filtering and monitor all traffic.

The upstream transmission in an EPON is relatively secure. Due to directivity of a passive combiner, the upstream traffic is visible only to the OLT. Although reflections might occur in the passive combiner, sending some small fraction of upstream power downstream again, the downstream transmission is on a different wavelength than the upstream transmissions. Thus, the ONU is "blind" to reflected traffic that is not processed in the receive circuitry.

The upstream traffic can also be intercepted at the PON splitter/combiner, as splitters and combiners are most often manufactured as symmetric devices (see Sec. 2.2). That is, although only one coupler port is connected to the trunk fiber, more ports are available. A special device sensitive to the upstream wavelength can be connected facing downstream to one such unused port. This device will be able to intercept all upstream communications.

Theft of service occurs when a subscriber impersonates a neighbor and transmits frames that are not billed to the impersonator's account. An OLT obtains the identity of the subscriber through the logical link ID inserted by each ONU in the frame preamble. This link ID can be faked by the malicious ONU when transmitting in the upstream direction. To be able to transmit in the hijacked timeslot, the impersonating ONU must also be able to eavesdrop on the downstream to receive GATE messages addressed to a victim.

Encryption of downstream transmission prevents eavesdropping when the encryption key is not shared, providing privacy for subscriber data and making impersonation of another ONU difficult. Thus, a point-to-point tunnel is created that allows private communication between the OLT and the different ONUs.

Encryption of the upstream transmission prevents interception of the upstream traffic when a tap is added at the PON splitter. Upstream encryption also prevents impersonation, as the ONU generating a frame must possess a key presumably known only to one ONU.

10.1 Development of a Security Mechanism

Although EPON is vulnerable to eavesdropping and theft-of-service attacks, proposals to include security mechanisms in objectives of the IEEE 802.3ah task force did not find the necessary support. The main arguments against including a security mechanism in EPON specification were the lack of needed expertise among IEEE 802.3 participants and the belief that security is outside the traditional scope of IEEE 802.3. Nevertheless, the issue remained, and in November 2002, the 802 Eexecutive Committee formed a LinkSec (link security) study group to "evaluate link security architecture issues with the objective of identifying the broader scope that can be common to all MAC solutions, develop a security architecture for IEEE 802, and develop standards to address link security issues." Later, this study group was placed under auspice of the IEEE 802.1 work group (project 802.1ae).

Since the security specification is being developed by the IEEE 802.1 work group, it is not specific to Ethernet and even less so to EPON. The solution sought by the LinkSec would be MAC-independent and would allow establishment of multihop secure channels. As such, it encrypts message payloads, but must keep the MAC addresses in clear text. Also, implementing security above MAC and MAC control layers means that the MPCP control messages and OAM messages will have to be transmitted in clear text. Keeping the MAC addresses and the MPCP and OAM frames in clear text may facilitate traffic analysis as well as subscriber impersonation. In general, many experts believe that since specific privacy problems were created by the EPON architecture, the solution should also be within EPON [HPN02].

Presented below is an outline of an EPON-specific encryption mechanism; it addresses EPON issues, but also utilizes features available only in EPON, such as the MPCP clock. It should be emphasized that the mechanism described below is different from the approach taken by IEEE 802.1ae.

10.2   EPON-Specific Encryption

The proposed EPON encryption mechanism is based on the *Advanced Encryption Standard* (AES) algorithm, published by the National Institute of Standards and Technology (NIST) in the United States [FIPS197]. AES allows the use of 128-bit, 192-bit, or 256-bit keys. This chapter describes an encryption mechanism similar to the one used in GPON [G984.3], but with necessary modifications to adapt it to the EPON architecture. The Ethernet frame format, including the preamble and IPG, is not modified in order to remain compliant with IEEE standards and avoid potential issues from future extension of the IEEE 802.3 standard. To ensure a higher degree of privacy, this method encrypts a complete Ethernet frame, including the Ethernet header and FCS field. The MPCP and OAM control messages are also encrypted.

10.2.1   Block cipher mode

To encrypt messages that are larger than the 128-bit blocks, a block cipher mode is used—a mode of operation in which a long message is broken into fixed-size blocks and each block is encrypted separately. There are many different cipher modes available [Dwo01]. A mode called *counter* (CTR) *mode* is used in the proposed scheme.

In the CTR mode, the block cipher generates a stream of 128-bit output blocks which are produced by the cipher function applied to a stream of counters (input blocks). The cipher output blocks are then XORed (i.e., subjected to bitwise exclusive-OR operation) with the input plain text to produce the cipher text (see Fig. 10.1a).

Since a message length need not be equal to an integral number of code blocks, the last plain text block (1 to 16 bytes in length) is XORed

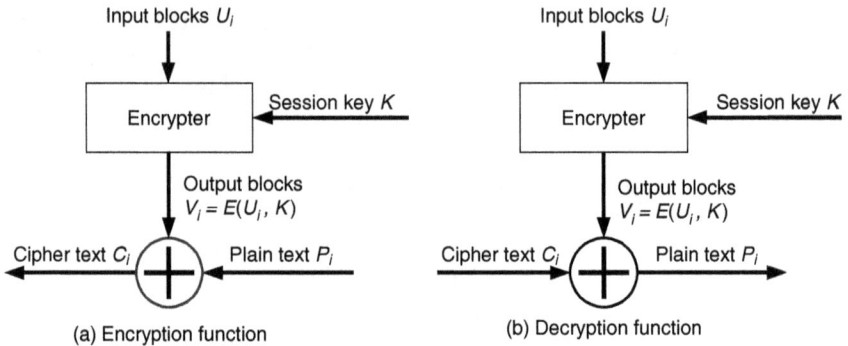

(a) Encryption function   (b) Decryption function

Figure 10.1   Implementation of counter mode.

with the most significant portion of the last output block. This approach does not require padding of the plain-text messages and provides an encryption function without overhead. The following is a formal representation of CTR block cipher mode:

$$V_i = E(K, U_i) \quad \text{for } i = 1, 2, ..., k \tag{10.1}$$

$$C_i = P_i \oplus V_i \tag{10.2}$$

where P_i is ith plain-text block, C_i is the ith cipher text block, K is the session key, V_i is ith cipher output block, and U_i is a series of input 128-bit values which are used only once during the lifetime of the given session key.

To decrypt a message, the cipher text is broken into 128-bit blocks and each block is XORed with the corresponding cipher output block to regenerate the plain text. To produce the output blocks, the receiving station uses the same stream of counters encrypted by the cipher function (Fig. 10.1b).

The advantage of using the counter mode is that both the transmitting and the receiving stations use the same cipher function, implementing only the encrypting (forward cipher) portion of the AES block cipher.

10.2.1.1   Cipher input values.   The counter mode requires a nonrepeating cipher input value (also called a counter value) to be associated with each 128-bit block of text. It is important that the cipher input values be synchronized with respect to the message being encrypted or decrypted. In EPON, some level of synchronization is provided by MPCP. Recall from Sec. 5.3.1.3 that MPCP clocks in the OLT and ONU are synchronized such that a frame transmitted by the OLT at time T will be received by an ONU when its local time is also equal to T.

Cipher counter

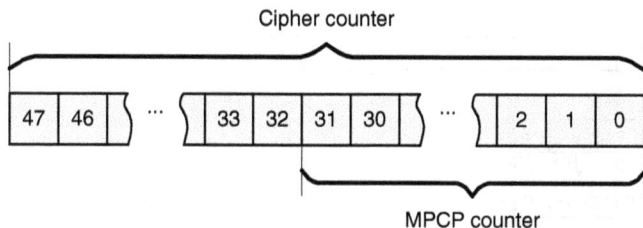

Figure 10.2   Relation of cipher counter to MPCP counter.

Therefore, at least in the downstream direction, the cipher input values can be derived from MPCP clock.

In CTR mode, there is a requirement that the cipher input values do not repeat for the lifetime of a given key. However, the MPCP counter, which is 32 bits long, wraps around approximately every 70 s, i.e., after 70 s the counter values begin to repeat. This would impose a shorter limit on the lifetime of a key, necessitating frequent key exchanges. Therefore, to prolong key lifetime, the MPCP counter is extended to 48 bits. To avoid confusion, let us call this extended counter a *cipher counter*. As shown in Fig. 10.2, the 32 least-significant bits of the cipher counter are aligned with the MPCP counter.

To produce the cipher input values, in addition to the cipher counter, a 7-bit counter called a *block counter* is used. The block counter counts 128-bit text blocks within a frame. This counter is reset to 0 at the beginning of each frame and is incremented by 1 for each 128-bit block.

Figure 10.3 illustrates the relationship between the cipher counter and block counter in producing the cipher input values.

The value of the cipher counter corresponding to the first byte of a frame is noted. As will be explained in Sec. 10.2.1.2, the MPCP counters may become slightly misaligned due to variability in propagation and processing delays at the OLT and an ONU. To remove this misalignment, only the 43 most-significant bits of the cipher counter are considered. The 43 bits of cipher counter are combined with 7 bits of block counter to produce 50-bit values that do not repeat for any text blocks for the lifetime of a session key.

Similarly to the approach taken in the recommendation ITU-T G.984.3, the 128-bit cipher block input values are produced by concatenating the 50-bit values three times and discarding the most-significant 22 bits of the resulting 150-bit block (see Fig. 10.4).

10.2.1.2 Cipher counter alignment. For proper operation of the counter mode, it is critical that the cipher input values (or cipher counters) be aligned with respect to the data being encrypted or decrypted. If the propagation and processing delays between the OLT and an ONU are

Cipher counter

* || denotes concatenation

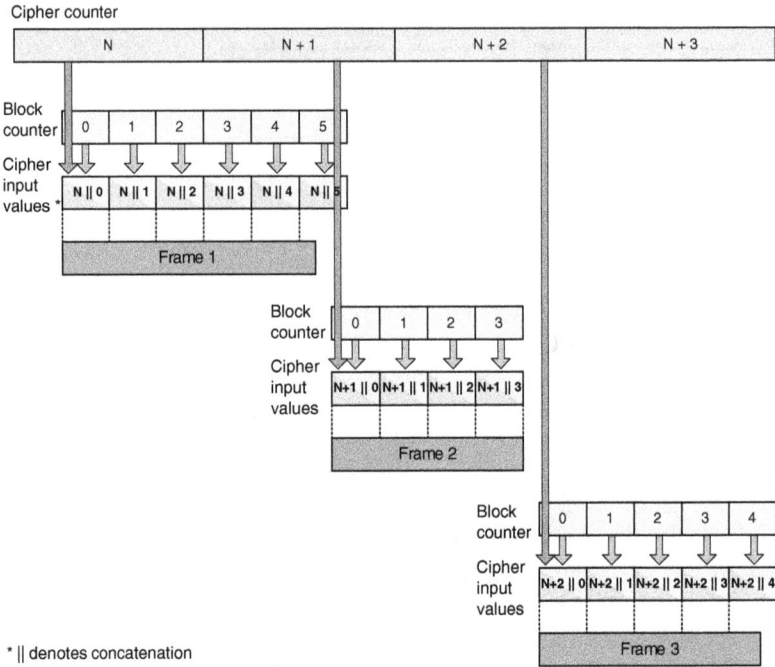

Figure 10.3   Relationship between cipher counter sequence, block counter sequence, and cipher input values.

128-bit cipher input values

Figure 10.4   The 128-bit cipher input block is produced by concatenation of cipher counter and block counter.

constant, the MPCP clocks and therefore the cipher counters remain aligned such that if a frame is transmitted by the OLT when its cipher counter equals N, this frame will be received by the ONU when the ONU's local cipher counter also has value N.

However, as we have discussed in Sec. 8.2.1.5 the IEEE 802.3ah standard allows up to 8 TQ (128 ns) of MPCP clock misalignment in the downstream direction (OLT to ONU) and up to 12 TQ (192 ns) for the round-trip (OLT to ONU to OLT) transmission.

To keep the cipher input values synchronized with respect to the user data, only the 43 most-significant bits of the cipher counter are taken to create cipher input values. However, it is still possible that even the 43 most significant bits of the cipher counters in the transmitting and receiving stations can become misaligned, if, e.g., a carryover from bit location 4 to bit location 5 occurs as a result of the propagation delay variability. To ensure that the counters remain synchronized, the value of bit 5 of the cipher counter corresponding to the first block of an Ethernet frame is carried with this frame.

A simple solution would be to carry bit 5 of the cipher counter in one of the reserved fields of the preamble. However, to remain compliant with IEEE standards and avoid potential issues from future extension of the IEEE 802.3 standard, the format of the frame preamble should not be modified. Below we consider an alternative solution, in which the value of bit 5 is conveyed by inverting the bits of the CRC-8 located in the frame preamble as follows:

```
if(cipher_counter [5] == 1)
      preamble[56:63] = CRC8 // CRC8 is not modified
else
      preamble[56:63] = ~CRC8 // CRC8 is inverted
```

At the receiving station, the value of bit 5 of the cipher counter will be recovered by matching the received preamble CRC against its calculated and inversed values as shown below:

```
if (preamble[56:63] == CRC8)
      cipher_bit5 = 1
else if (preamble[56:63] == ~CRC8)
      cipher_bit5 = 0
else
      ... // Preamble is invalid - discard the frame
```

The receiving station should note the value of its local cipher counter corresponding to the first byte of the frame. After the `cipher_bit5` is recovered, the cipher counter can be reliably adjusted by either incrementing or decrementing the counter value until bit 5 of the counter becomes equal to the recovered value of the `cipher_bit5`. The decision of whether to increment or decrement the counter is based on which direction would require the smaller absolute change. Since the maximum delay variability allowed by the IEEE 802.3ah standard specification is less than one-half of the period for bit 5 ($guardThresholdOLT < 2^5/2$), only one direction of adjustment is possible (i.e., the absolute value of increment or decrement would remain within the maximum-allowed bounds). The following pseudocode illustrates the counter adjustment procedure, which can be done in constant time:

```
if (cipher_counter[5] ^ cipher_bit5)
{
    if (cipher_counter[0:4] >= 16)
        // increment the cipher_counter to the nearest future value
        // which has the cipher_counter[5] == cipher_bit5
        cipher_counter = cipher_counter + ~cipher_counter[0:4] + 1;
    else
        // decrement the cipher_counter to the nearest past value
        // which has the cipher_counter[5] == cipher_bit5
        cipher_counter = cipher_counter - cipher_counter[0:4] - 1;
}
```

The above procedure is performed at the beginning of each received frame.

10.2.1.3 Lifetime of a key. The value of the cipher counter must not repeat during the lifetime of a session key. The cipher counter runs synchronously with the MPCP counter and wraps around after 2^{48} increments, which corresponds to approximately 1250 h or 52 days. This imposes the maximum key lifetime of 1250 h.

10.2.2 Downstream encryption

In the proposed scheme, the OLT needs only one cipher counter for encrypting downstream frames, even though different logical links will use different keys. A straightforward modification may enable different cipher counters for each logical link. The initial value of the OLT's cipher counter is conveyed to an ONU during the initial key exchange.

The chosen approach does not result in undesirable frame chaining in which a lost or corrupted frame would inhibit proper decryption of all subsequent frames, as would be the case when, for example, cipher counter counts frames. In the current scheme, each frame is independently encrypted and decrypted based on the time of the frame's departure and arrival. The departure and arrival times correspond to the time of transmission of the first octet of the frame (first octet of the destination address field).

10.2.3 Upstream encryption

In the upstream direction, the cipher counters are not aligned as they are in the downstream direction. For encrypting upstream frames, the ONU uses the value of its cipher counter corresponding to the grant start time. The first block of the first frame in a slot would be associated with the cipher counter value corresponding to the start time of the given grant. Starting at the time the first byte of the first frame is transmitted, the crypto counter will continue running with 16-ns increments.

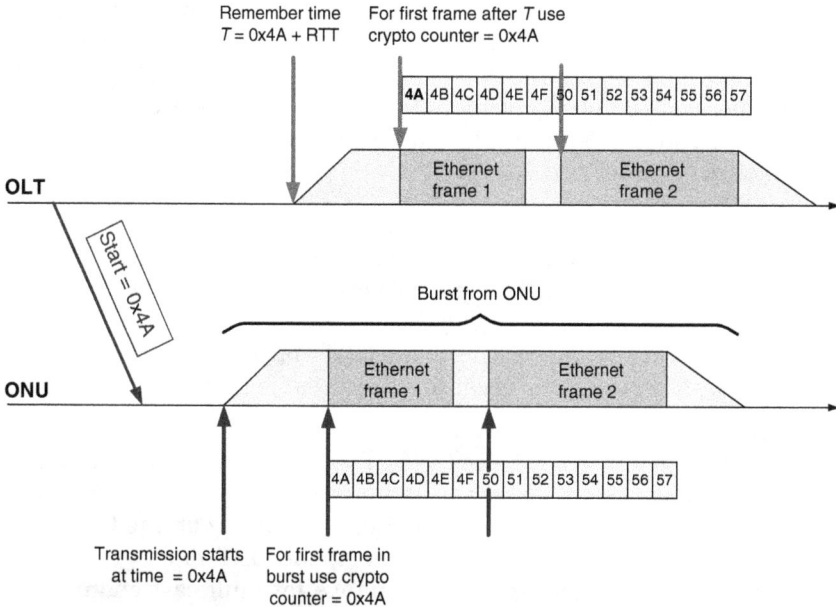

Figure 10.5   Alignment of cipher counter with an upstream burst.

To decrypt the received frames, the OLT remembers the future time when the grant will arrive. The first frame arriving after this time will be decrypted using the cipher counter value associated with the remembered future time[1] (see Fig. 10.5). The actual procedure for the upstream encryption is identical to that for the downstream encryption.

10.2.4   Key exchange and switch-over scheme

Since the upstream channel is relatively more secure than the downstream channel, it is reasonable to generate a new key at the ONU and transmit it to the OLT. This would allow operators to implement only downstream encryption, if this were deemed sufficient. Therefore, for unicast logical links, the key exchange request is initiated by the OLT, and a new key is generated by the ONU.

However, in the case of multicast logical links, multiple ONUs must use the same key. In this case, the OLT initiates key exchange and generates the new key value.

---

[1] Note that in the upstream direction, if the first frame in a burst is lost (i.e., the start-of-packet delimiter could not be detected), the entire burst may not be decrypted properly. However, such a chaining effect is limited only to one burst.

10.2.4.1 Message format. The IEEE 802.3ah standard defines the *operation, administration, and maintenance* (OAM) sublayer, which allows the exchange of organization-specific messages (OAMPDUs). Such custom messages can be conveniently used to perform key exchange. To allow both ONU-based and OLT-based key generation, the following messages may be defined:[2]

KEY_REQUEST(switch_counter). This message is issued by the OLT to request a new key from an ONU. This message also conveys to an ONU a future value of the cipher counter at which a new key is to become active (we call this value a switch_counter). It is important that the new key became active in the OLT and an ONU synchronously with transmission and reception of the same message.

KEY_ASSIGN(switch_counter, key). This message is issued by the OLT to assign a new key to an ONU. This message carries a new key as well as a switch_counter value at which the key switch-over should take place. Typically, an ONU generates a new key on the OLT's request. However, in some circumstances, the OLT must generate a key. One such example is a key exchange for multicast channel, in which case, all receiving devices should use the same key.

KEY_RESPONSE(switch_counter, key). This message is issued by an ONU in response to KEY_REQUEST, but can also be used as an acknowledgment for the KEY_ASSIGN message. In this message, the ONU conveys to the OLT the new key value to be used after the switch-over and also confirms the key switch-over counter value.

10.2.4.2 Key exchange protocol using ONU-generated key. We start with the assumption that the cipher counters in the OLT and ONUs are synchronized. The bottom 32 bits is synchronized as part of the MPCP synchronization during the autodiscovery. To synchronize the higher 16 bits, the OLT needs to convey the initial value of the cipher counter to the ONU. This can be done once, following the autodiscovery. Alternatively, it is also possible to include the value of the cipher counter with each KEY_REQUEST or KEY_ASSIGN message, similar to the timestamping mechanisms used by MPCP. However, the MPCP timestamping is done by the hardware (control multiplexer), exactly at the moment when a frame is ready to be transmitted. To allow the key exchange messages to be generated by software and to avoid dependency of the message content on the transmission time, it is reasonable to deliver

---

[2] Here, we will not discuss the precise format of OAM organization-specific messages. Rather, we will focus on the information related to key exchange and key switch-over.

only the higher 16 bits of the cipher counter `cipher_counter[32:47]` with each message.

The protocol to exchange an ONU-generated key performs the following steps:

1. The OLT initiates key exchange by generating a KEY_REQUEST (`switch_counter`, `cipher_counter[32:47]`) message. The `switch_counter` field carries the value of the cipher counter at which a new key is to be activated. The value of the `cipher_counter[32:47]` field is set to the 16 most-significant bits of the OLT's own cipher counter corresponding to the first block of the KEY_REQUEST message. Simultaneously with sending the KEY_REQUEST message, the OLT starts a key exchange timer.

2. Upon receiving the KEY_REQUEST message, the ONU loads its cipher counter with a combination of the received 16 most-significant bits of the OLT's cipher counter (`cipher_counter` field) and its local MPCP counter. The ONU stores the received `switch_counter` value. It then generates and stores a new 128-bit key and responds by transmitting a KEY_RESPONSE(`switch_counter`, `key`) message in which it conveys the new key to the OLT. The `key` field contains a new 128-bit value to be used as a new key.

Successful reception of the KEY_RESPONSE message concludes the key exchange procedure. If no KEY_RESPONSE message arrives before the key exchange timer expires, or if the returned value of the `switch_counter` does not correspond to the value transmitted in the KEY_REQUEST message, the OLT will initiate another key exchange by issuing a new KEY_REQUEST message. The ONU should always respond to the KEY_REQUEST message. When a new KEY_REQUEST message arrives, the ONU should discard any stored key it may have generated previously and generate a new key. The ONU shall always use the most recently received `switch_counter` value and the most recently generated key.

The ONU should be able to process the KEY_REQUEST message and generate the KEY_RESPONSE message in time significantly less than the timeout interval set by the OLT. The OLT should initiate the key exchange procedure well in advance of the intended key switch-over time, such that, if necessary, the key exchange may be repeated several times.

The above protocol may be used for exchanging the keys for both upstream and downstream channels. However, if the upstream and downstream keys can be exchanged independently and concurrently, the protocol messages should have an additional field identifying the channel.

10.2.4.3 Key exchange protocol using OLT-generated key. The OLT-generated key exchange protocol is similar to the ONU-generated key exchange, except that the key is generated by the OLT. In all cases, the key exchange is initiated by the OLT. The key exchange protocol performs the following steps:

1. The OLT initiates key exchange by generating a KEY_ASSIGN (switch_counter, cipher_counter[32:47]) message. The switch_counter field carries the value of the cipher counter at which a new key is to be activated. The value of cipher_counter[32:47] is set to 16 most-significant bits of the OLT's own cipher counter corresponding to the first block of the KEY_ASSIGN message. The key field contains a new 128-bit value to be used as a new key. Simultaneously with sending the KEY_ASSIGN message, the OLT starts a key exchange timer.

2. Upon receiving the KEY_REQUEST message, the ONU loads its cipher counter with a combination of received 16 most-significant bits of the OLT's cipher counter (cipher_counter[32:47] field) and its local MPCP counter. The ONU generates a KEY_RESPONSE message in which it echoes the new key and the switch_counter values back to the OLT.

The OLT-generated key is typically used when more than one ONU should have the same key. Similarly to the ONU-generated key exchange protocol, the OLT repeats the procedure if not all KEY_RESPONSE messages arrive at the OLT before the key exchange timer expires, or if the returned value of the key or switch_time field does not correspond to the value set in the KEY_ASSIGN message.

10.3 Summary

Expansion of Ethernet into public networks serving subscriber markets brings with it a new slew of challenges, including requirements for security and privacy. This chapter provided an overview of a possible EPON encryption method—a method that is tailored for EPON needs and uses specific features available in EPON.

As important as they are, such details as initial key exchange and authentication were not discussed here. These mechanisms are usually specified by network operators and are determined by existing provisioning systems and the ease of integration with other parts of the network.

References

[Dwo01]   M. Dworkin, *Recommendation for Block Cipher Modes of Operation—Methods and Techniques*, National Institute of Standards and Technology, December 2001.

[FIPS197]   Federal Information Processing Standard 197, *Advanced Encryption Standard*, National Institute of Standards and Technology, U.S. Department of Commerce, Nov. 26, 2001.

[G984.3]   ITU-T Recommendation G.984.3, *Transmission Convergence Layer for Gigabit Passive Optical Networks*, October 2003.

[HPN02]   O-P. Hiironen, A. Pietiläinen, and A. Nylund, "Privacy in EPON," presented at IEEE 802.3ah meeting in Edinburgh, UK, May 2002. Available at http://www.ieee802.org/3/efm/public/may02/hiironen_1_0502.pdf.

11

Path Protection in EPON

In some critical deployments, the access network may require fast protection switching. To achieve this, a certain path redundancy should be added to a PON by providing several alternative, diversely routed paths. Dissimilar access network environments may require different protection schemes. Redundancy may be added to an entire PON's topology, or to only a part of the PON, say, the trunk or the branches of the tree. See Fig. 11.1.

Let us denote by u_c the *probability of unavailability* (or simply the unavailability) of a component c, and by U_g the unavailability of a group (a serial chain) g of components. Both u_c and U_g are the expected fractions of time during which the corresponding component or group of components is unavailable. It is convenient to define a trunk group consisting of trunk fiber, OLT transmitter, and OLT receiver and a branch group consisting of the branch fiber, ONU transmitter, and ONU receiver. We keep the splitter as a separate group consisting of only one component. The unavailability of the trunk group can be calculated as

$$U_{\text{trunk}} = 1 - \left(1 - u_{\text{OLT\_Tx}}\right)\left(1 - u_{\text{OLT\_Rx}}\right)\left(1 - u_{\text{trunk}}\right)$$

where $u_{\text{OLT\_Tx}}$ = unavailability of the OLT transmitter, $u_{\text{OLT\_Rx}}$ = unavailability of the OLT receiver, and u_{trunk} = unavailability of the trunk fiber.

Similarly, we obtain the unavailability of the branch group as

$$U_{\text{branch}} = 1 - \left(1 - u_{\text{ONU\_Tx}}\right)\left(1 - u_{\text{ONU\_Rx}}\right)\left(1 - u_{\text{branch}}\right)$$

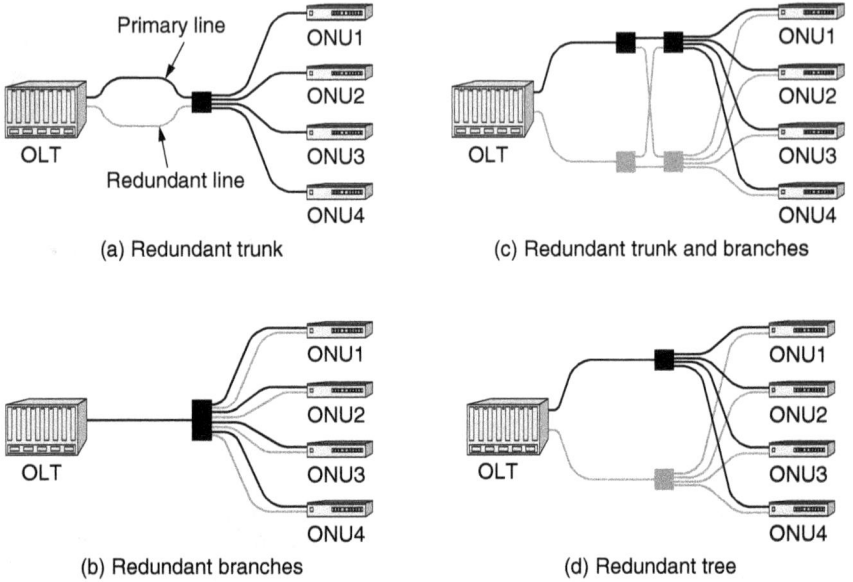

(a) Redundant trunk

(c) Redundant trunk and branches

(b) Redundant branches

(d) Redundant tree

Figure 11.1 Redundant PON topologies.

where $u_{\text{ONU\_Tx}}$ = unavailability of the ONU transmitter, $u_{\text{ONU\_Rx}}$ = unavailability of the ONU receiver, and u_{branch} = unavailability of the branch fiber. Since we are concerned with bidirectional transmissions, we consider both the transceiver and the receiver in each group.

Finally, for the splitter group we just have $U_{\text{split}} = u_{\text{split}}$, where u_{split} = unavailability of the splitter.

11.1 Unprotected Tree

We first consider the unavailability of an unprotected PON, as shown in Fig. 2.4a. Since the trunk, the splitter, and the branch groups are connected serially, it is easy to see that the overall service unavailability in unprotected PON is

$$U = 1 - \left(1 - U_{\text{trunk}}\right)\left(1 - U_{\text{split}}\right)\left(1 - U_{\text{branch}}\right)$$

11.2 Protected Trunk

Figure 11.1a illustrates a PON with a protected trunk. The trunk fiber, OLT receiver, and OLT transmitter are the most critical elements in a PON; a failure of any of them will result in all PON users losing the service. The protected trunk configuration aims at protecting the critical PON elements with the least amount of redundancy. In a protected trunk configuration, the OLT is equipped with primary and secondary transceivers (a transmitter-receiver combination) and uses two diversely routed trunk fibers to reach $2 \times N$ splitter. Under normal conditions, the OLT uses the primary transceiver. If a link failure is detected, the OLT switches to the secondary transceiver. The protected trunk configuration shown in Fig. 11.1*a* has a service unavailability value equal to

$$U = 1 - \left(1 - U^2_{\text{trunk}}\right)\left(1 - U_{\text{split}}\right)\left(1 - U_{\text{branch}}\right)$$

Numerical values for the unavailability parameter of various redundant topology schemes will be presented below.

The major cost-increase component for this scheme, compared to the unprotected PON, is providing the diversely routed (sheath-disjoint) trunk fiber from the CO to the splitter.

11.3   Protected Branches

Figure 11.1*b* presents a PON with protected branches. The assumption behind this protection scheme is that even though a branch failure would affect only one ONU, the final fiber drop represents the most hazardous environment, and hence, the branch fiber has a much higher failure rate compared to the trunk. Indeed, a field data analysis suggests that the drop cable failure rates are about an order of magnitude higher than those of the trunk cable [CFL99].

The protected branch scheme uses one trunk fiber connected to a $1 \times 2N$ splitter. The splitter, in turn, has two diversely routed branches connected to each ONU. When an ONU detects loss of signal, it will switch to the secondary transceiver. This scheme has the unavailability parameter equal to

$$U = 1 - \left(1 - U_{\text{trunk}}\right)\left(1 - U_{\text{split}}\right)\left(1 - U^2_{\text{branch}}\right)$$

A shortcoming of this scheme is that it requires a twice-larger split ratio, which introduces additional 3-dB splitting loss. This may affect the maximum reach or the maximum fan-out of such PON. The major cost component of this scheme is associated with the cost of N additional redundant transceivers, one for each ONU.

11.4   Protected Trunk and Branches

A PON with protected trunk and branches is shown in Fig. 11.1*c*. This topology has two trunk fibers connected to $2N$ branches, 2 branches per

ONU. To allow better protection flexibility, it is desirable to use a cascade of splitters instead of a single $2 \times 2N$ splitter device, which would remain a single point of failure. As Fig. 11.1c illustrates, the signal in the downstream direction passes through a 1×2 splitter first and then through a $2 \times N$ splitter. If these splitters are separate devices, the failure of one of them will not affect the rest of them. This configuration is also affected by an increased split ratio. However, it provides lower unavailability. To simplify the unavailability equation, we include the first splitter with the trunk group and call it the *trunk+splitter* group. We similarly include the second splitter with the branch group and call it *branch+splitter* group. The unavailability for these two new groups may be obtained as

$$U_{\text{trunk+splitter}} = 1 - \left(1 - U_{\text{trunk}}\right)\left(1 - U_{\text{split}}\right)$$

$$U_{\text{branch+splitter}} = 1 - \left(1 - U_{\text{branch}}\right)\left(1 - U_{\text{split}}\right)$$

Then the resulting service unavailability for the redundant trunk and branch topology can be expressed as

$$U = 1 - \left(1 - U_{\text{trunk+splitter}}^{2}\right)\left(1 - U_{\text{branch+splitter}}^{2}\right)$$

The cost of implementing this protection scheme is higher than the cost of the above schemes and is dominated by the cost of providing sheath-disjoint trunk and dual transceivers per ONU.

11.5   Protected Tree

Finally, Fig. 11.1d presents a redundant tree PON, which has a complete duplication of the entire fiber plant. The OLT and all ONUs are equipped with two transceivers. This topology does not suffer from an increased splitting ratio, as do the protected branch and protected trunk and branch schemes. An important advantage of this topology is that, in the absence of failure, both the primary and the secondary trees can carry data traffic; i.e., this PON has twice the capacity of other configurations. The protected tree configuration uses the same amount of fiber and number of transceivers as the protected trunk and branch scheme. However, it has a higher unavailability value compared to that of the protected trunk and branch scheme. The unavailability value can be found as

$$U = \left[1 - \left(1 - U_{\text{trunk}}\right)\left(1 - U_{\text{split}}\right)\left(1 - U_{\text{branch}}\right)\right]^{2}$$

TABLE 11.1   Failure Rates and Repair Times for Various Components

Component	Failure in time (FIT)	Mean time to repair (MTTR), h	u_c	Unavailability = FIT × MTTR/10^9 h
Cable cut (trunk = 20 km)	2283	12	u_{trunk}	2.74×10^{-5}
Cable cut (branch)	1712	12	u_{branch}	2.05×10^{-5}
Splitter failure	114	12	u_{split}	1.37×10^{-6}
OLT transmitter failure	3424	2	u_{OLT_Tx}	6.85×10^{-6}
OLT receiver failure	1142	2	u_{OLT_Rx}	2.28×10^{-6}
ONU transmitter failure	1712	6	u_{ONU_Tx}	1.03×10^{-5}
ONU receiver failure	1142	6	u_{ONU_Rx}	6.85×10^{-6}

We obtain numerical values for component unavailability from the failure rates reported in [CFL99] and summarized in Table 11.1.[1] The *failure-in-time* (FIT) parameter represents the average number of failures in 10^9 h. The unavailability of a component is simply its FIT per unit of time multiplied by its *mean time to repair* (MTTR). The typical value for fiber repair is 12 h. A component replacement in CO takes 2 h, while a component replacement at the customer premises typically takes 6 h.

Table 11.2 compares the service unavailability per ONU for an unprotected passive tree, shown in Fig. 2.4a, and the four redundant topologies, shown in Fig. 11.1. The downtime is calculated by multiplying the unavailability by number of minutes in a year.

An unprotected PON (i.e., PON without redundant topology and transceivers) is expected to have approximately 40 min/yr of downtime, which corresponds to an availability of 99.992 percent.

The data presented in Table 11.2 suggest approximately equal downtimes for trunk protection and branch protection schemes. Both of these schemes reduce the downtime by one-half compared to the unprotected PON. This reduction of downtime from 40 to 20 min/yr is equivalent to an increase in availability from 99.992 to 99.996 percent, a very marginal improvement.

The schemes that provide full redundancy (protected tree and protected trunk and branch) are expected to have a significantly lower

---

[1] [CFL99] reports the components failure rate, which is a probability of component failure in a 1-yr interval. The FIT parameter can be obtained from the failure rate r as FIT = $r \times 10^9$ h/(365 days/yr × 24 h/day).

TABLE 11.2   Unavailability and Expected Downtime for
Various PON Topologies

Configuration	Service unavailability	Expected downtime, min/yr
Unprotected tree	7.56×10^{-5}	39.72
Protected trunk	3.90×10^{-5}	20.52
Protected branch	3.79×10^{-5}	19.92
Protected trunk and branch	1.31×10^{-9}	0.00069
Protected tree	5.71×10^{-9}	0.0030

downtime (an order of several milliseconds per year) due to fiber cuts or equipment failures.

In the above examples, we considered only the downtime caused by fiber cuts, splitter failures, and transceiver failures. The overall downtime is also affected by such factors as power failure and operator errors. In addition, failures outside the access portion of the network can contribute to the downtime experienced by subscribers.

Reference

[CFL99]   W. Circiora, J. Farmer, and D. Large, *Modern Cable Television Technology: Video, Voice, and Data*, Morgan Kaufmann Publishers, 1999.

EPON Performance

12

Baseline Efficiency

EPON efficiency depends on many parameters, such as packet size distribution, configuration of the scheduler, and the speed of the laser driver and clock recovery circuits. Making unrealistic assumptions about any of these parameters can result in efficiency numbers being far off from the true value. It is, therefore, clear that to answer the question of EPON efficiency, one has to come up with a realistic and unambiguous set of EPON operational parameters and traffic characteristics. In this chapter, we attempt to identify all the parameters affecting the efficiency and to justify the chosen values for these parameters.

By network efficiency we usually mean the *throughput efficiency*, also called *utilization*. Throughput is a measure of how much user data (application-level data) the network can carry through in a unit of time. Throughput efficiency is the ratio of maximum throughput to the network bit rate.

Perhaps, the easiest way to calculate the efficiency is to find the overhead components associated with frame encapsulation, scheduling, and optional FEC.

12.1 Encapsulation Overhead

The encapsulation (framing) overhead was briefly considered in Sec. 3.2.1. This overhead is a property of all Ethernet architectures (not specific to EPONs); it is a result of adding an 8-byte frame preamble, 14-byte Ethernet header, and 4-byte FCS field to MAC service data units (m_sdu) comprised of user's data. Additionally, at least a 12-byte

(96-ns) minimum interframe gap (IFG) should be left between two adjacent frames. Thus, the absolute overhead per frame is constant and equal to 38 bytes. Short payloads are padded to a minimum length of 46 bytes. This also contributes to the Ethernet encapsulation overhead and is counted in our calculations.

The average value of the encapsulation overhead depends on the distribution of packet (m_sdu) sizes. The distributions of packet sizes were reported in the literature and typically have a trimodal shape, which is similar for backbone networks [CMT98] and access networks [SG01].

The value of average overhead can be obtained from the following formula:

$$avgOverhead = 1 - \frac{\displaystyle\sum_{s=S^{min}}^{S^{max}} s \times f(s)}{\displaystyle\sum_{s=S^{min}}^{S^{max}} encap(s) \times f(s)} \tag{12.1}$$

where s = size of the payload (m_sdu), $f(\cdot)$ = probability distribution function for packet sizes, and $encap(s)$ is the size of an encapsulated payload s. For the Ethernet encapsulation, the $encap(s)$ function is equal to $\max\{s, minPayload\} + |DA| + |SA| + |Length/Type| + |preamble| + |IFG| = \max\{s, 46\} + 38$.

Using the IP packet size distribution obtained in a headend of a cable network [SG01], we get the Ethernet encapsulation overhead to be 7.42 percent.

12.2   Scheduling Overhead

The scheduling overhead in EPON consists of *control channel overhead, guard band overhead, discovery overhead*, and *frame delineation overhead*. Some of the parameters affecting the overhead, such as cycle time or frequency of the discovery attempts, are outside the scope of IEEE 802.3ah. Therefore, in some cases we will present multiple values of overhead for the different choices of the configuration parameters.

12.2.1   Control channel overhead

The control channel overhead represents bandwidth lost due to use of in-band control messages, such as GATE and REPORT messages. The amount of overhead depends on the number of ONUs and the cycle time, i.e., an interval of time in which each ONU should receive a GATE message and send a REPORT message. We make an assumption here that the scheduling algorithm requires only one GATE message and one REPORT message to be exchanged between each ONU and the OLT in one cycle time. ITU-T Recommendation G.114, *One-Way Transmission Time*, specifies the delay for voice traffic in an access network at 1.5 ms [G.114]. To achieve the average delay of 1.5 ms for frames carrying voice data, the cycle time should be about 1 ms. If the maximum delay is to not exceed the 1.5-ms limit, the cycle time should be fixed at 750 µs. We therefore present the control message overhead values for both of these cycle times.

The control message overhead is calculated as

$$controlOverhead = 1 - \frac{|MPCPDU| \times N_{\text{ONU}}}{T_{\text{cycle}} \times R_{\text{EPON}}} \qquad (12.2)$$

where $|MPCPDU|$ is the size of the GATE and REPORT messages (including the preamble and the IFG), N_{ONU} = number of ONUs (i.e., number of messages sent in one cycle time), T_{cycle} is a cycle time, and R_{EPON} is EPON data rate (1 Gbps).

For a 1-ms cycle, we get 1.08 percent overhead for the system with 16 ONUs and 2.15 percent overhead for a 32-ONU system. If the cycle time is reduced to 750 µs, the overhead reaches 1.43 percent and 2.87 percent, respectively, for 16-ONU and 32-ONU systems. Assuming that the scheduling algorithm uses as many REPORT as it uses GATE messages, the equal control channel overhead is present in both upstream and downstream directions.

12.2.2   Guard band overhead

The guard bands are intervals of time during which ONUs turn the lasers on and off and the OLT performs automatic gain control and clock synchronization (see Sec. 2.5). The guard band overhead is present only in the upstream direction and depends on PMD and PMA parameters, such as laser-on/laser-off (T_{on} and T_{off}) times, OLT's automatic gain control (T_{AGC}) and clock and data recovery (T_{CDR}) times, and code-group alignment $T_{\text{code\_group\_align}}$ time. We consider a worst-case scenario where both T_{AGC} and T_{CDR} are equal to 400 ns, which is the maximum value allowed by the IEEE 802.3ah standard. The T_{on} and T_{off} times are fixed at 512 ns each. The $T_{\text{code\_group\_align}}$ is fixed at 32 ns.

The laser-off time may partially overlap the laser-on time of the next ONU. However, as shown in Fig. 12.1, the guard band should include at least a 24-TQ dead zone to allow for timing variability of the multi-point control protocol. This dead zone should be twice the value of the `guardThresholdOLT` constant, which was discussed in Sec. 8.2.1.5.

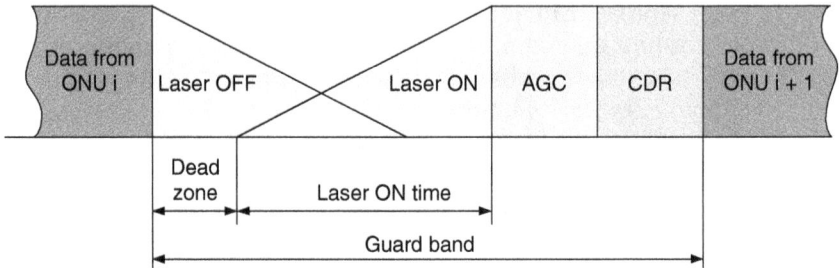

Figure 12.1 Guard band components.

The guard band overhead can be calculated from the following formula:

$$guardOverhead$$
$$= \frac{\left(T_{\text{ON}} + T_{\text{dead\_zone}} + T_{\text{AGC}} + T_{\text{CDR}} + T_{\text{code\_group\_align}}\right) \times N_{\text{ONU}}}{T_{\text{cycle}}} \quad (12.3)$$

Here again, we consider two cycle times: 1-ms cycle and 750-µs cycle. The guard band overhead in a system with 1-ms cycle and 16 ONUs is 2.76 percent. A system with the same cycle time and 32 ONUs would have this overhead increased to 5.53 percent. If the cycle time is reduced to 750 µs, the overhead increases to 3.69 and 7.37 percent, respectively, for 16-ONU and 32-ONU systems.

12.2.3 Discovery overhead

The discovery overhead represents the bandwidth lost due to periodic allocation of a discovery window. The discovery window should be larger than the maximum round-trip time of 200 µs. In our calculations we assume the discovery window of size 300 µs. Frequency of the discovery attempts is not specified in the IEEE 802.3ah draft. Intelligent algorithms may detect a situation when all ONUs are operational and cease all discovery attempts. We, however, will assume a simpler algorithm that performs periodic discovery regardless of the number of registered ONUs. The discovery period can be very large, for example, 500 ms or more.

With a 500-ms discovery period, the discovery overhead is 300 µs/500 ms = 0.06 percent. This overhead affects only the upstream channel, since the discovery windows are allocated only for the upstream transmission.

12.2.4   Frame delineation overhead

The variable-size Ethernet frames may not be able to completely occupy the fixed-size timeslot. The nonfragmentability of Ethernet frames was the main reason for introducing multiple queue sets in the REPORT messages—the scheduler would always try to select one of the reported queue lengths in order for the granted timeslot to be filled completely. Grants to ONUs are based on their reported queue lengths. However, multiple grants, which are based on the reported queue lengths, with their associated guard bands may not fill the fixed cycle time exactly. At least one grant should have a length that is not based on the reported queue length, but is dependent on the available transmission time left in the current cycle. The frame delineation overhead represents the expected unused timeslot remainder when the grant length is independent of the reported queue lengths.

What follows is a derivation of the expected size of such an unused remainder. Let us first introduce the following notation:

W = timeslot size

R = random variable representing unused remainder

S = random variable representing packet sizes

S^{min}, S^{max} = range for packet sizes ($S^{min} \leq S \leq S^{max}$). In Ethernet, S^{min} = 64 bytes and S^{max} = 1518 bytes.

By definition, the expected remainder is

$$E(R) = \sum_{r=1}^{S^{max}-1} r \times P(R = r) \tag{12.4}$$

Obviously, the remainder can only be in the range from 0 to $S^{max} - 1$. If the remainder is more than $S^{max} - 1$ bytes long, then we are guaranteed that the next packet will fit into the current timeslot, thus reducing the remainder. Here, we assume that we always have packets waiting, i.e., the load is heavy.

Given that we have placed k packets in the timeslot, what is the probability of getting a remainder of size r? By taking $X_k = S_1 + S_2 + \cdots + S_k$, this probability is

$$P(R = r \mid K = k) = P(X_k = W - r \cap X_{k+1} > W \mid K = k)$$

$$= P(X_k = W - r \cap S_{k+1} > r \mid K = k) \tag{12.5}$$

$$= P(X_k = W - r \mid K = k) \times P(S_{k+1} > r \mid K = k)$$

Assuming that all S are independent and identically distributed, we get

$$P(S_{k+1} > r \mid K = k) = P(S > r) \tag{12.6}$$

To get the probability of $R = r$, we sum Eq. (12.5) for all k.

$$P(R = r) = \sum_{k=1}^{\infty} P(R = r \mid K = k) \times P(K = k) \tag{12.7}$$

$$= P(S > r) \times \sum_{k=1}^{\infty} P(X_k = W - r \mid K = k) \times P(K = k)$$

We sum it for all k because we don't care how many packets fit in a timeslot. All we care about is that after we have added some number of packets, we still have the unused remainder of size r. Strictly speaking, we don't need to sum for all k. Timeslot of a specific size W can only accommodate m packets, where

$$\left\lfloor \frac{W}{S^{\max}} \right\rfloor \leq m \leq \left\lfloor \frac{W}{S^{\min}} \right\rfloor$$

Now, the summation in Eq. (12.7) denotes the probability that the sum of several packet sizes equals $W - r$, without any references to the number of packets used in summation. In other words, this is the probability that any number of packet sizes sums to $W - r$. Thus, we have

$$P(R = r) = P(S > r) \times P(X = W - r) \tag{12.8}$$

We can view X as a renewal process with interrenewal times S. Thus, we expect to have one renewal every $E(S)$ bytes. The probability that some renewal will occur exactly at epoch $W - r$ is, therefore, $1/E(S)$; that is,

$$P(X = W - r) = \frac{1}{E(S)} \tag{12.9}$$

After substituting Eqs. (12.8) and (12.9) into Eq. (12.4), we get

$$E(R) = \sum_{r=1}^{S^{\max}-1} r \times \frac{P(S > r)}{E(S)} = \frac{1}{E(S)} \sum_{r=1}^{S^{\max}-1} r \times [1 - F_S(r)] \tag{12.10}$$

where $F_S(\cdot)$ is a cumulative distribution function of S.

Finally, for the remainder probability density function, we have

$$f_R(r) = \begin{cases} \dfrac{1 - F_S(r)}{E(S)} & 0 \le r \le S^{\max} - 1 \\ \\ 0 & \text{otherwise} \end{cases} \tag{12.11}$$

The interesting result here is that $E(R)$ does not depend on the timeslot size W. It only depends on the distribution of packet sizes. This agrees very well with simulations.

Using the empirical packet size distribution from [SG01], we get the average remainder approximately equal to 595 bytes. That means that we should expect, on average, 595 bytes wasted due to variable-size frames not packing the fixed cycle completely.

The frame delineation overhead is calculated as

$$delineationOverhead = \frac{E(R)}{T_{\text{cycle}} \times R_{\text{EPON}}} \tag{12.12}$$

where R_{EPON} is the data rate in EPON. With a 1-ms cycle time, this overhead is equal to 0.48 percent. With the reduced cycle time of 750 µs, the overhead is 0.63 percent.

12.3 FEC Overhead

The EPON standard specifies an optional frame-based forward-error correction scheme. This method appends FEC parity data at the end of each frame. Sixteen bytes of parity data are added for each 239-byte block of data being FEC-protected. In addition, the frame-based FEC scheme uses extended delimiters to delineate protected data from parity code. The extended delimiters are described in Chap. 9.

Figure 12.2 illustrates the format of a FEC-encoded frame. The shaded areas represent added fields, which are responsible for FEC overhead.

Given a frame payload of size s, the FEC overhead can be calculated as

$$fecOverhead(s) = \left\lceil \frac{encap(s)}{239} \right\rceil \times 16 + |/\text{I/T/R}/| + |\text{T\_FEC\_E}| \tag{12.13}$$

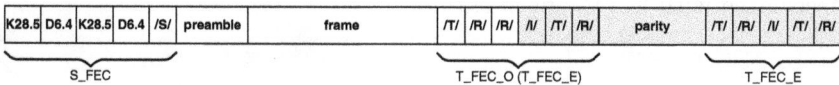

K28.5	D6.4	K28.5	D6.4	/S/	preamble	frame	/T/	/R/	/R/	/I/	/T/	/R/	parity	/T/	/R/	/I/	/T/	/R/

S_FEC   T_FEC_O (T_FEC_E)   T_FEC_E

Figure 12.2 Structure of a FEC-coded frame.

where encap(s) represents the Ethernet frame encapsulation and is equal to max{s, $minPayload$} + |DA| + |SA| + |Length/Type| + |preamble| + |IFG| = max{s, 46} + 38. The term |/I/T/R/| represents the length of the tail (shaded) portion of the first T_FEC delimiter (see Fig. 12.2) and is 4 bytes long. The term |T_FEC_E| represents the length of the second delimiter and is equal to 6 bytes.

The average value of the FEC overhead depends on the distribution of packet (m_sdu) sizes and can be obtained from

$$avgFecOverhead = \frac{\displaystyle\sum_{s=S^{\min}}^{S^{\max}} fecOverhead(s) \times f(s)}{\displaystyle\sum_{s=S^{\min}}^{S^{\max}} [encap(s) + fecOverhead(s)] \times f(s)} \qquad (12.14)$$

where s = size of the payload (m_sdu) and $f(\cdot)$ = probability distribution function for packet sizes.

Using IP packet size distribution obtained in a headend of a cable network [SG01], we get the average FEC overhead of 9.25 percent. The worst-case overhead occurs if the traffic consists of the minimum-size packets, in which case the overhead reaches 23.6 percent.

12.4   Summary

Table 12.1 summarizes the overhead components and shows the efficiency and net (application-level) throughput for upstream and downstream channels. The values in the table represent the worst-case situation—a small cycle time (750 µs) and the maximum-allowed values of the optical overhead components, such as T_{AGC} and T_{CDR}.

In the above calculations we considered the overhead and efficiency at the GMII. One may reasonably argue that Ethernet's 8b/10b line coding contributes an additional 20 percent of overhead if each transition on the line is considered a bit. Of course, this would result in a lower percent value of efficiency; however, considering the actual line rate of 1.25 Gbps, the net throughput will remain as shown in Table 12.1.

It is quite possible that a particular scheduling algorithm or implementation will have lower efficiency; however, that would only be a result of particular design decisions and not an intrinsic overhead associated with EPON specification.

TABLE 12.1   EPON Overhead Components and Efficiency for 750-μs Fixed Cycle Time

	Downstream		Upstream	
	16 ONUs	32 ONUs	16 ONUs	32 ONUs
Encapsulation overhead, %	7.42	7.42	7.42	7.42
Control channel overhead, %	1.43	2.87	1.43	2.87
Guard band overhead, %	—	—	3.69	7.37
Discovery overhead, %	—	—	0.06	0.06
FEC overhead (optional), %	9.25	9.25	9.25	9.25
Maximum total overhead with FEC, %	17.19	18.39	20.29	24.45
Minimum net throughput with FEC, Mbps	828.15	816.05	797.11	755.45
Minimum efficiency relative to 1 GbE point-to-point link,* %	98.57	97.13	94.88	89.92

* The relative efficiency includes only the scheduling overhead. The encapsulation overhead is present in both 1 Gbps point-to-point links and EPON. The FEC overhead is also ignored, because (1) it is optional and (2) point-to-point links may use the same FEC method to increase their reach (although, at present, the FEC method is only defined for P2MP media).

References

[CMT98]   K. C. Claffy, G. Miller, and K. Thompson, "The nature of the beast: Recent traffic measurements from an Internet backbone," in *Proceedings INET '98*, Geneva, Switzerland, July 1998.

[SG01]   D. Sala and A. Gummalla, "PON functional requirements: Services and performance," presented at IEEE 802.3ah meeting in Portland, OR, July 2001. Available at http://grouper.ieee.org/groups/802/3/efm/public/jul01/presentations/sala_1_0701.pdf.

[G.114]   ITU-T Recommendation G.114, *One-Way Transmission Time*, in Series G: Transmission Systems and Media, Digital Systems and Networks, Telecommunication Standardization Sector of ITU, May 2000.

13

Discovery Slot Allocation

According to the IEEE 802.3ah standard, the size and periodicity of the discovery windows are left to implementation. A simple implementation may settle on periodic discovery windows of fixed size, as was the case with examples considered in Chap. 12. But more sophisticated schemes can also be considered. For example, since the normal traffic is suspended during the discovery, one may desire to minimize the impact on the traffic. Such a scheme would require each individual discovery window to be as small as possible, probably at the expense of having a large number of collisions and repeating attempts. Alternatively, the goal of optimization may be to reduce the overall channel unavailability time due to the discovery process. Such a scheme would try to minimize the combined size of all discovery windows, while not being particularly concerned with the size of each individual window.

Below we consider a method to adjust the discovery slot size in order to minimize the overall channel unavailability. The basic premise of this method is the assumption that the OLT can always estimate the maximum number of ONUs still requiring registration. Indeed, it is typically known to a network operator how many ONUs are provisioned on each PON. Also, at any time, it is known how many ONUs are registered. The difference between the two is the number of ONUs that may attempt initialization at the next discovery opportunity. (We say *may* because, some ONUs may be turned off or disconnected.) The OLT may safely suspend all discovery attempts if all the provisioned ONUs are already registered. If not all ONUs are registered, a discovery window should be allocated. Below we will derive the optimal size of the discovery window as a function of the number of ONUs attempting registration. But first let us introduce some notation:

W Discovery slot size (time)

n Number of contending ONUs

E Extended guard band around the discovery slot. This guard band should be larger than the largest round-trip time. (Then the discovery window is equal to $W + E$, as explained in Sec. 5.3.2.1.)

M ONU's transmission size (time) during registration: $M = T_{\mathrm{on}} + T_{\mathrm{AGC}} + T_{\mathrm{CDR}} + |MPCPDU| + |preamble| + |IFG| + T_{\mathrm{off}}$.

13.1   Pairwise Collision Probability

In this section, we will find the probability that REGISTER_REQ messages, transmitted by two ONUs, will collide.

If the round-trip times to the two ONUs are different by more than the discovery slot size W, the messages from these ONUs will never collide, because their discovery slots will not overlap at the OLT. Here we consider a worst-case scenario, in which all ONUs are located at the same distance from the OLT; thus their REGISTER_REQ messages will all arrive at the OLT within the interval of size W. In other words, their discovery slots will completely overlap at the OLT.

Recall from Sec. 5.3.2.2 that to avoid persistent collisions, each ONU applies a random delay to the transmission of the REGISTER_REQ message. The ONU's transmission should never extend beyond the granted slot, so the random delay is chosen from the range $[0, W - M]$ and, according to the IEEE 802.3ah standard, should have a uniform distribution. Let's denote this random delay as x for the first REGISTER_REQ message and y for the second REGISTER_REQ message.

Even if each ONU starts its transmission on the local MPCP clock boundary, due to processing and propagation delay variability (Sec. 8.2.1.5), the messages will not align on clock boundaries when they arrive at the OLT. Therefore, we consider x and y to have a continuous uniform distribution in the range $[0, W - M]$.

A collision between two transmissions will occur if the two messages arrive closer than interval M to each other, that is, $P_{\mathrm{coll}} = P(|x - y| < M)$.

We find P_{coll} by using a graphical method, as follows. Let each point on a plane represent random delays (offsets) of two REGISTER_REQ messages, such that the x coordinate represents the random delay of the first message and the y coordinate represents the random delay of the second message. Then all possible combinations of two random delays will form a square with one corner at the origin and the diagonal corner having coordinates $(W - M, W - M)$, as shown in Fig. 13.1a. The points below the diagonal represent all combinations where the first random delay was longer than the second random delay, and all points above the diagonal represent situations where the first delay is shorter.

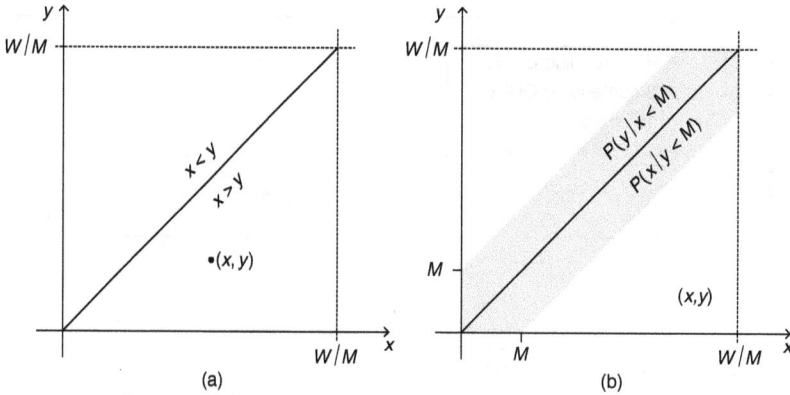

Figure 13.1   Graphical representation of pairwise collision probability.

The shaded region in Fig. 13.1*b* shows the area where the difference between the two random delays is less than M, that is, $|x - y| < M$. Therefore, the probability of a collision is equal to the ratio of the shaded area to the total sample space, which is the area of a square with the side length $= W - M$:

$$P_{coll} = P(|x - y| < M) = \frac{(W - M)^2 - (W - 2M)^2}{(W - M)^2} = 1 - \left(1 - \frac{M}{W - M}\right)^2 \quad (13.1)$$

13.2 Average Success Rate

When n ONUs are attempting registration, the probability of success for each ONU is $P_{success}$:

$$P_{success} = (1 - P_{coll})^{n-1} = \left(1 - \frac{M}{W - M}\right)^{2n-2} \quad (13.2)$$

From Eq. (13.2), the average number of successful registrations in one attempt is $V(n, W)$:

$$V(n, W) = n \times P_{success} = n \times \left(1 - \frac{M}{W - M}\right)^{2n-2} \quad (13.3)$$

Figure 13.2 shows the average number of successful registrations as a function of discovery slot size.

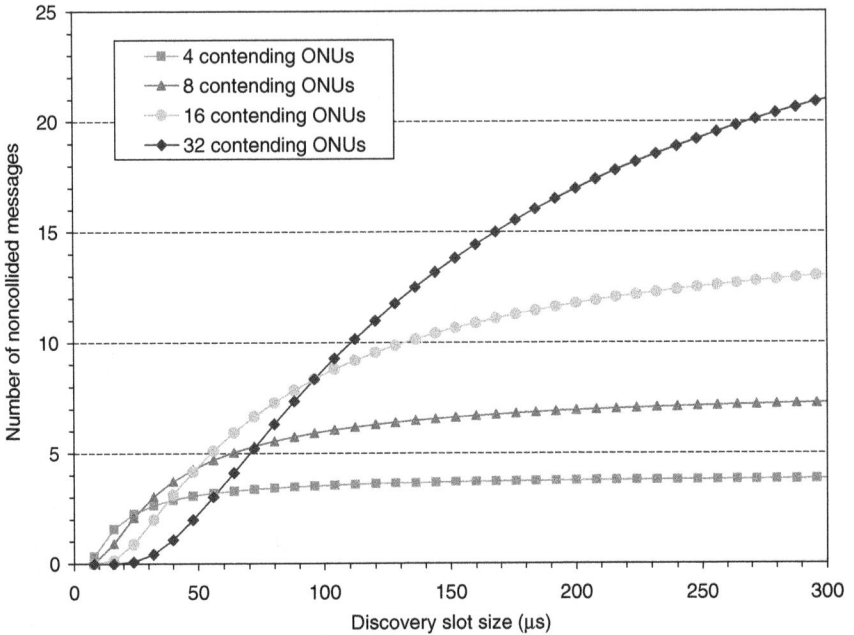

Figure 13.2   Average number of successful registrations in one attempt.

13.3   Efficiency of Discovery Slot Size

Efficiency (utilization) of the discovery slot can be measured as a ratio of the number of successful registrations to the discovery slot size:

$$U(n, W) = \frac{V(n, W)}{W} \tag{13.4}$$

To account for the extended guard band needed during discovery, Eq. (13.4) is amended as follows:

$$U(n, W) = \frac{V(n, W)}{W + E} = \frac{n}{W + E} \times \left(1 - \frac{M}{W - M}\right)^{2n - 2} \tag{13.5}$$

The coordinate of the maximum of the $U(n, W)$ function represents the optimal discovery slot size for a given number n of contending ONUs (see Fig. 13.3).

13.4   Optimal Discovery Slot Size

To find the best slot size, we solve $dU(n, W) / dW = 0$ for W.

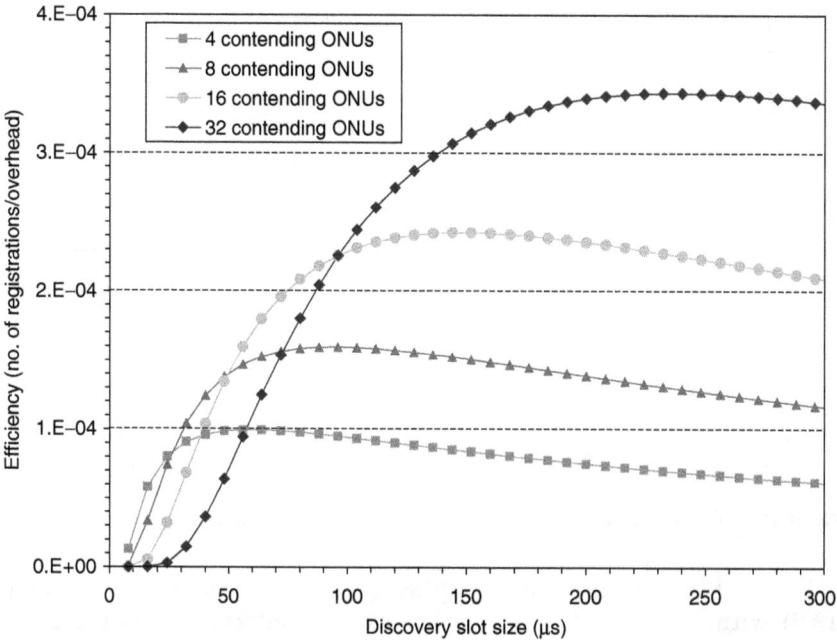

Figure 13.3   Efficiency of the discovery slot size.

$$\frac{d}{ds}\left[\frac{n}{W + E} \times \left(1 - \frac{M}{W - M}\right)^{2n - 2}\right] = 0$$

$$\frac{n(2n - 2)M}{(W + E)(W - M)^2}\left(1 - \frac{M}{W - M}\right)^{2n - 3} - \frac{n}{(W + E)^2}\left(1 - \frac{M}{W - M}\right)^{2n - 2} = 0$$

(13.6)

Solving Eqs. (13.6) for W, we find the expression for the best slot size for a given number n of contending ONUs:

$$W = M\left(n + \tfrac{1}{2}\right) + \sqrt{M^2\left(n^2 + n + \tfrac{9}{4}\right) + 2ME(n - 1)}$$
(13.7)

Figure 13.4 shows the best slot size when the number of contending ONUs varied from 2 to 32. The calculations were done for the worst-case PMD and PMA overhead, that is, $T_{AGC} = 400$ ns and $T_{CDR} = 400$ ns. The values of the rest of the components of transmission size M are fixed per the standard as follows: $T_{on} = 512$ ns, $T_{off} = 512$ ns, $T_{code\_group\_align} = 32$ ns, $|MPCPDU| + |preamble| + |IFG| = 672$ ns.

Therefore, $M = 2.528$ μs or 158 TQ. The extended guard band value, which corresponds to the maximum RTT value, was 200 μs, or 12,500 TQ.

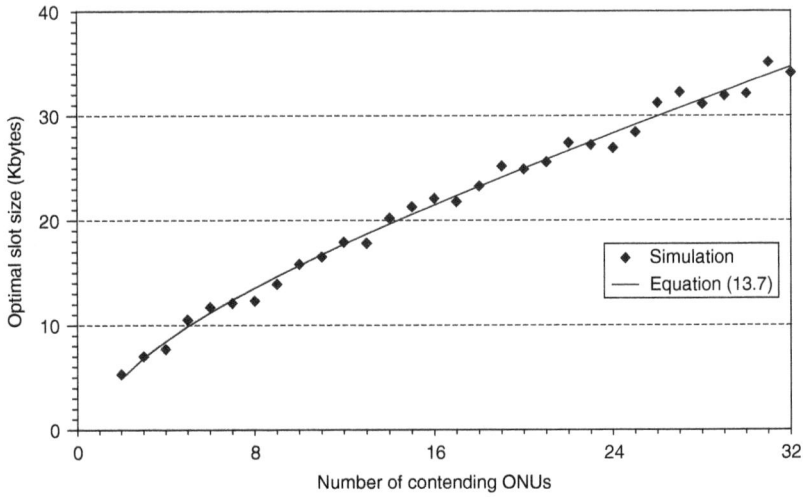

Figure 13.4   Optimal discovery slot size as a function of the number of contending ONUs.

Figure 13.4 compares the analytically derived optimal slot size (Eq. 13.7) with a result of a simulation experiment that finds the most efficient slot size averaged over 10,000 iterations.

14

EPON with Static Slot Assignment

14.1 Introduction

In this chapter, we investigate the performance of EPON with a *static slot assignment* (SSA). Specifically, we investigate how the static (fixed) slot structure can be combined with highly bursty data traffic consisting of variable-length Ethernet frames. We shall try to answer these questions in this study: What is the average delay the packets will experience in the ONU buffer? How big should this buffer be? And what link utilization can we achieve?

To obtain an accurate and realistic performance analysis, it is important to simulate the system behavior with appropriate traffic injected into the system. There are studies showing that most network traffic flows, such as http, ftp, and video streams, can be characterized by *self-similarity* and *long-range dependence* (LRD) (see App. A).

The simulation analysis was performed using synthetic self-similar traffic. Self-similar (or fractal) traffic has the same or similar degree of burstiness observed at a wide range of time scales. The method to generate self-similar synthetic traffic is described in App. B.

14.2 System Architecture

Let us consider an EPON segment consisting of an OLT and N ONUs (Fig. 14.1). The distances between the OLT and each ONU are randomly (uniformly) distributed over the interval [0.5 km, 20 km], which corresponds to round-trip times ranging from 5 to 200 μs.

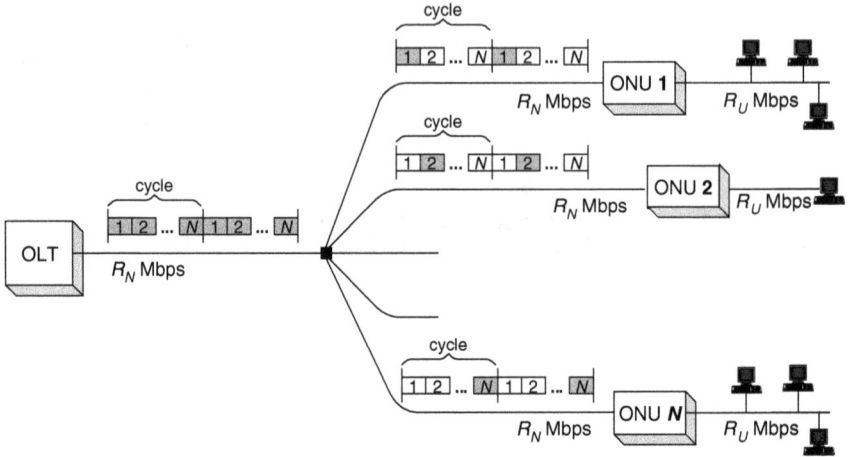

Figure 14.1   Simulation model of EPON segment with SSA.

The transmission speed of the EPON and that of the user access link need not necessarily be the same. In our model, we consider R_U Mbps to be the data rate of the access link from a user to an ONU, and R_N Mbps to be the rate of the upstream link from an ONU to the OLT. Line rates for each link are the same in the upstream and downstream directions.

As explained in Sec. 12.2.2, any two adjacent timeslots must have a guard interval G between them. A set of N timeslots together with their associated guard intervals is called a *cycle*. In other words, a cycle is a time interval between two successive timeslots assigned to one ONU (see Fig. 14.1). We denote cycle time by T. Making T too large will result in increased delay for all the packets, including high-priority (real-time) packets. Making T too small will result in more bandwidth being wasted by guard intervals.

Table 14.1 summarizes the system parameters used in the simulation experiment reported here.

14.3   Traffic Model

To illustrate the effects of traffic burstiness on the efficiency of the SSA, we performed simulations with the following traffic profiles:

LRD   This is self-similar and long-range dependent traffic, which presents, with nonnegligible probability, extremely large bursts of data (packet trains) and extremely long periods of silence (interburst gaps). This traffic remains bursty at many timescales and most closely models the empirical traffic profiles observed both in LANs and the backbone

TABLE 14.1   System Parameters

Parameter	Description	Value
N	Number of ONUs	16
R_U	Line rate of user-to-ONU link	100 Mbps
R_N	EPON line rate	1000 Mbps
Q	Buffer size in ONU	1 Mbyte
G	Guard interval between timeslots	1 µs
T	Cycle time	2 ms
W	Timeslot size $$W = R_N \left(\frac{T}{N} - G \right)$$	15,500 bytes

networks ([LT+94], [PF95], [CB96]). Appendix B describes the method we used to generate the self-similar traffic. This traffic has the same packet size distribution as was measured for upstream traffic in [SG01].

SRD   This is short-range dependent (SRD) traffic, and it can be characterized by the bursts and gaps described by the Poisson process. It is also highly bursty, but without the long-range dependence. As shown in Fig. A.1 in App. A, aggregating the SRD traffic over a large interval of time reduces the burstiness significantly faster than it does for the LRD traffic. This traffic has the same packet size distribution as the LRD traffic.

CBR   The constant-bit-rate (CBR) traffic profile is created by aggregating multiple constant packet rate substreams. Depending on the frequency (period) and phase of each substream, the aggregated traffic may still form some bursts, due to packets from multiple substreams aggregating into a single packet train. However, if we look at the profile of this traffic at a scale equal to the common multiple of all the substream periods, we will see a constant traffic rate.

We need to emphasize that SRD and CBR profiles do not bear any resemblance to the real network traffic. However, comparing packet delays in simulations using LRD, SRD, and CBR traffic profiles will let us visualize how much traffic burstiness contributes to the delay.

14.4 Performance Analysis

In this section, we investigate the EPON performance under varying offered loads. The performance matrix in our experiments comprises packet delay, average queue size, packet loss ratio, and bandwidth or timeslot utilization.

Each ONU receives ingress traffic of load φ, which is called ONU's *offered load*; φ is normalized to the user link capacity R_U. *Offered network load* Φ is the sum of the loads offered to each ONU and scaled based on R_U and R_N rates. Clearly, since the network capacity is less than the aggregated ingress bandwidth from all ONUs ($R_N < NR_U$), the offered network load can exceed 1:

$$\Phi = \frac{R_U}{R_N} \sum_{j=1}^{N} \varphi_j \qquad (14.1)$$

It is important to differentiate between the offered load and *carried load*. The carried load is determined by packets that were transmitted by the ONUs. Thus, the carried load is equal to the offered load only if the packet loss rate is zero; otherwise the carried load is less than the offered load. The carried load can never exceed 1. The carried load multiplied by the link capacity is also called *throughput*.

14.4.1 Average packet delay

In this section, we investigate the dependence of packet delay on the network load. We consider ONUs with a simple FIFO queue. In such a queue, if the next packet to be sent is larger than the remaining timeslot, then this packet and all packets that arrived after it will wait for the next timeslot.

Before we present our results, let us consider what the constituents of the packet delay are. Packets arrive to the ONU at random times. Every packet has to wait for the next timeslot to be transmitted upstream. This delay is termed *TDM delay*. The TDM delay is the time interval between packet arrival and the beginning of the next timeslot. In [HOR87], this delay is called the *slot synchronization delay*.

Due to the bursty nature of network traffic, even at light or moderate network load, some timeslots may be filled completely and still more packets may be waiting in the queue. These packets will have to wait for later timeslots to be transmitted. This additional delay is called the *burst delay*. Burst delay may span multiple cycles (recall that a cycle consists of N timeslots, where N is the number of ONUs).

Figure 14.2 shows the results of our first simulation. Here, we can see that CBR traffic experiences the shortest delay. In fact, only the TDM delay is present up to about 55 percent load. The reason for such nice behavior is that every timeslot is getting approximately the same number of bytes. Then when one timeslot finally overflows, all overflow, and we have an avalanchelike increase in the packet delay.

The SRD traffic shows a very slow increase in delay with the load up to 40 percent. At this load, the number of timeslots that overflow starts increasing exponentially fast; i.e., the burst delay begins to dominate.

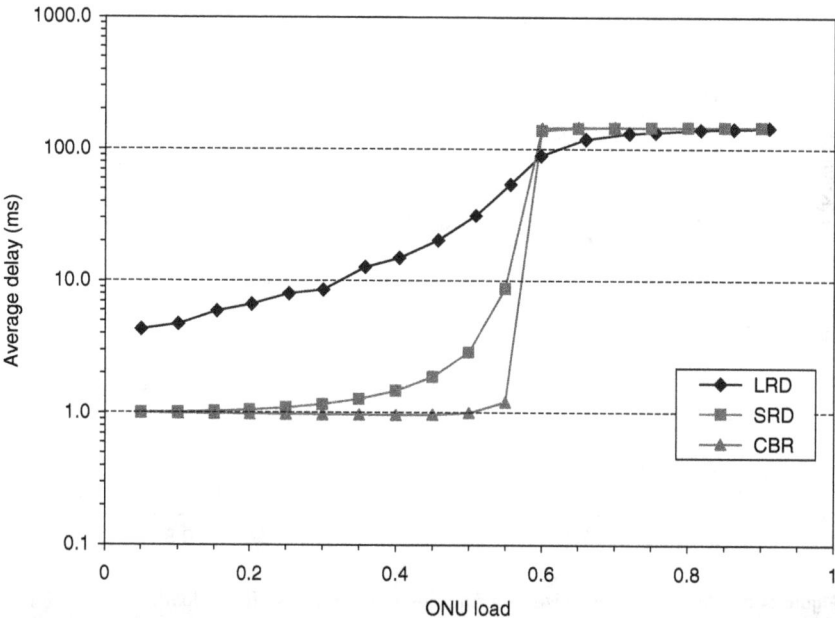

Figure 14.2   Average packet delay as a function of the ONU's offered load.

At 60 percent ONU load, the CBR and SRD queues are completely saturated and the profile of input traffic has no effect on the packet delay.

The LRD traffic shows a significant burst delay even at a very light load of 5 percent. The fact that some timeslots overflow means that there were some bursts of traffic that delivered more than 15,500 bytes (timeslot size) in 2 ms (cycle time). This means that while the average ONU load was only 5 percent or 5 Mbps, there were periods when the short-term data rate achieved at least 15,500 bytes × (8 bits/byte)/2 ms = 60 Mbps.

14.4.2   Average queue size

Figure 14.3 represents the average queue size. The queue behavior is very similar to that of the average packet delay. Queue size for CBR traffic grows linearly up to the 55 percent load. For loads above 55 percent, every timeslot sees the queue of size larger than the timeslot can accommodate, and the queue saturates very quickly.

The SRD traffic shows a queue size increase associated with burst delay—more packets waiting for several cycles in the queue contributes to the average queue size.

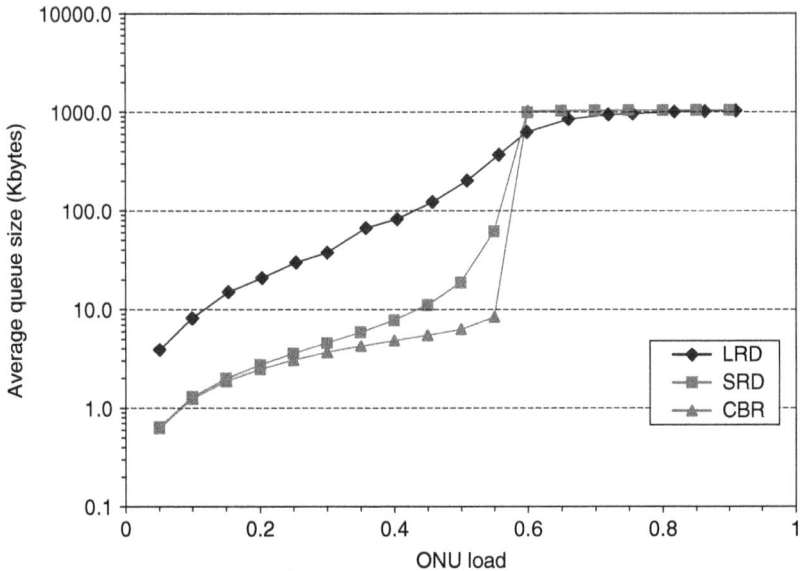

Figure 14.3   Average queue size as a function of the ONU's offered load.

However, it is the high burstiness of the LRD traffic that makes buffering less efficient. Even at very low load, it introduces a fairly large packet train, such that the average queue length is about 6 times larger than with CBR or SRD traffic. With the LRD traffic, significantly larger buffers are needed.

It is interesting to note that the point of queue saturation with LRD traffic comes later than that for CBR or SRD traffic. As the large frame bursts present even at the very light load, so the large interburst gaps still remain even at the very high load. These large gaps allow queues to occasionally drain, resulting in lower average delay.

14.4.3   Frame loss

The frame loss measurement (Fig. 14.4) provides yet another insight into the behavior of LRD traffic. From the frame delay analysis we know that the burst delay is present at a very light load. This means the queue backlog often exceeds the timeslot size, so the queued frames have to wait multiple cycles before being transmitted.

The loss chart shows a significantly worse situation: often the queues become completely saturated, and packets are lost even at a very light load. Figure 14.4 shows 0.5 percent packet loss at 5 percent load. Of course, to observe this effect, the simulation should be run for a longer time, as the large enough bursts may not be encountered otherwise.

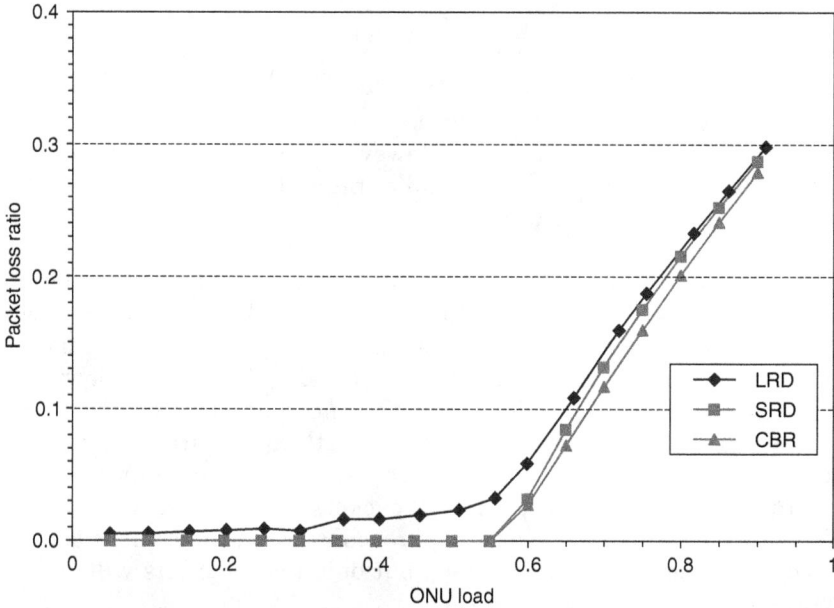

Figure 14.4   Packet loss ratio as a function of the ONU's offered load.

14.4.4 Bandwidth utilization

This section will evaluate the effective ONU's egress bandwidth. In Sec. 12.2.4, we showed that when variable-size frames are packed into a fixed-size slot, the expected unused slot remainder $E(R)$ can be calculated as

$$E(R) = \sum_{r=1}^{S^{\max}-1} r \times \frac{P(S > r)}{E(S)} \qquad (14.2)$$

where S is a random variable representing packet (Ethernet frame) sizes and $S^{\max}$ is the maximum size of Ethernet frame. The $E(R)$ is independent of the slot size and only depends on the distribution of packet sizes.

It follows that, for the timeslot of size W, the expected timeslot utilization is

$$E(U) = \frac{W - E(R)}{W} \qquad (14.3)$$

With the packet size distribution reported in [SG01], the average remainder is approximately equal to 595 bytes. Given the slot size of 15,500 bytes, we get a utilization of 96 percent.

The egress bandwidth available to an ONU can be obtained as

$$B = \frac{W - E(R)}{T_{\text{cycle}}} \approx 59.62 \text{ Mbps}$$

Obviously, increasing the timeslot size should result in increased utilization and egress bandwidth. However, increasing the timeslot size also leads to a larger cycle time and to proportionally increased TDM delay (or medium access delay).

14.4.4.1 Improving utilization using packet scheduling. In the simulation experiment described above, which used a simple FIFO queue in an ONU, if a packet that is currently at the head of the queue does not fit in a partially occupied timeslot, this packet and all packets following it will wait for the next timeslot, resulting in *head-of-line* (HOL) blocking.

However, if some later-arrived packets in the queue are small enough to fit into the current timeslot, then why wait? A smarter approach may be to attempt to reorder some packets waiting in the buffer. Figure 14.5 compares these two approaches. Here, three timeslots are needed without packet reordering, but only two timeslots will suffice with reordering.

This is a variation of the bin-packing problem. Different flavors of the algorithm may be used: first fit, best fit, prediction, etc. Figure 14.6 presents link utilization plots for the "no reordering" scheme (FIFO) and reordering using first fit. These results were obtained using empirical traffic traces [LT+94] to ensure that not only the packet size distribution, but also the relative order of the packets, remains realistic.

This reordering can be easily implemented in hardware as well as software. It is easy to see that $E(R)$ will depend not only on the packet size distribution, but also on the network load. Indeed, the higher the load, the more packets will be waiting in the queue, and the higher is the possibility of finding one packet that fits in the remaining timeslot.

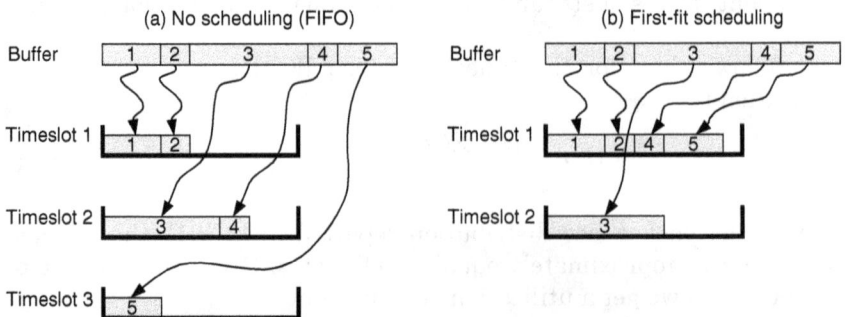

Figure 14.5 Illustration of packet scheduling.

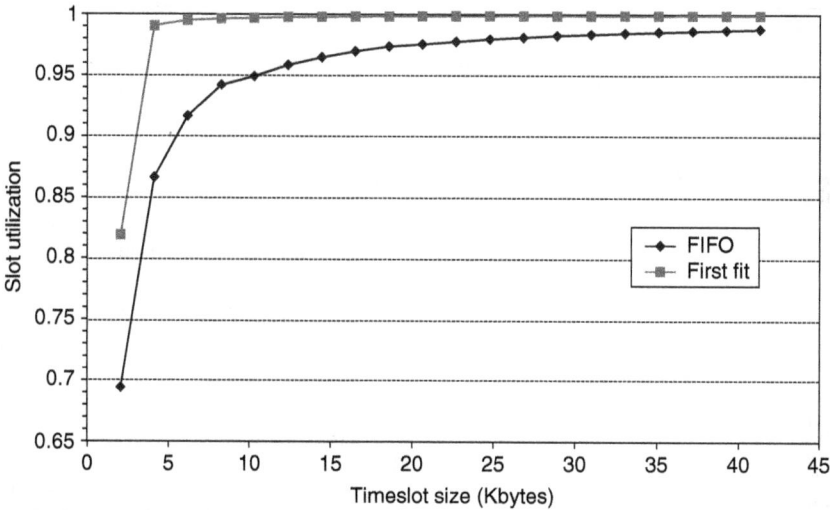

Figure 14.6   Average link utilization (FIFO versus first fit).

However, as it turns out, first-fit scheduling is not such a good approach. To understand the problem, we need to look at the effects of packet reordering from the perspective of *transmission control protocol/Internet protocol* (TCP/IP) payloads carried by Ethernet packets. Even though the transmission control protocol will restore the proper sequence of packets, excessive reordering may have the following consequences:

1. According to the fast retransmission protocol, the TCP receiver will send an immediate ACK for any out-of-order packet, whereas for in-order packets, it may generate a cumulative acknowledgment (typically for every other packet) [Ste94]. This will lead to more unnecessary packets being placed in the network.

2. Even more important, packet reordering in an ONU may result in a situation where n later packets are being transmitted before an earlier packet. This would generate n ACKs ($n - 1$ duplicate ACKs) for the earlier packet. If n exceeds a predefined threshold, it will trigger packet retransmission and reduction of the TCP's congestion window size (the *cwnd* parameter). Currently, the threshold value in most TCP/IP protocol stacks is set to 3 (e.g., refer to the *fast retransmission protocol* (FRP) in [Ste94]).

Even if special care is taken at the ONU to limit out-of-order packets to only 1 or 2, the network core may contribute additional reordering. While true reordering typically generates less than 3 duplicate ACKs

```
Let      Q be the queue of packets q₁,q₂, ..., qₙ waiting in an ONU
         C(qᵢ) - connection ID of packet qᵢ
         P - set containing IDs of packets that were postponed
         R - slot remainder

1        Repeat for every timeslot
2        {
3                 i = 1
4                 P ∈ ∅                              (Clear the set P)
5                 R = |timeslot|
6                 while i ≤ n and R ≥ Sᵐⁱⁿ
7                 {
8                         if qᵢ ≤ R then             (packet fits into timeslot)
9                         {
10                                if C(qᵢ) ∉ P       (i.e., packets from this connection
11                                                   were not postponed yet)
12                                {
13                                        send qᵢ
14                                        R = R - |qᵢ|
15                                }
16                        }
17                        else                       (packet doesn't fit into timeslot)
18                                P = P ∪ C(qᵢ)      (add connection ID to set P)
19                        i = i + 1
20                }
```

Figure 14.7   Algorithm for connection-based packet reordering.

and is ignored by the TCP sender, together with reordering introduced by the ONU, the number of duplicate ACKs may often exceed 3, thus forcing the sender to retransmit a packet. As a result, the overall throughput of user's data may decrease.

So, what is the solution? Let us assume that the traffic entering the ONU is an aggregate of multiple flows. In the case of business users, it would be the aggregated flows from multiple workstations. In the case of a residential network, we still may expect multiple connections at the same time. This is so because, as a converged access network, EPON will carry not only data, but also *voice-over-IP* (VOIP) and video traffic. Also, home appliances are becoming network plug-and-play devices. The conclusion is that if we have multiple flows, we can reorder packets that belong to different flows, and never reorder them if they belong to the same flow.

The outline of the algorithm is given in Fig. 14.7. This algorithm preserves the order of packets within a flow (connection) by keeping track (in set P) of all the connection identifiers of packets that were postponed. Obviously, the finer the granularity of connection identifiers, the more reordering possibilities the ONU will have, but more memory would need to be allocated for set P (which probably should be implemented as a hash table). So, if a connection is identified only by source address, then, in the case of a single user with multiple connections, the ONU will not be able to reorder any packets.

Looking at the destination address instead of the source address may improve the situation for ONUs with a single user, but this has a potential drawback when multiple users send packets to the same destination. In this situation, even though packets originate from different senders, the ONU will not reorder them.

A reasonable solution may be to look simultaneously at a *source-destination* pair, plus include the source and destination port numbers. Then the ONU will have maximum flexibility. More studies need to be done to determine the statistical properties of the connections to estimate the advantages of fine-granularity connection identifiers.

However, an important point is that the improvement achieved by this scheduling will be between the FIFO case and first-fit case. Clearly, the above algorithm will reorder some packets, which will make its utilization better than in FIFO. It is also true that some packets that belong to the same connection (and that first fit will reorder) will not be reordered in the given algorithm; thus, its performance will be lower than that of first fit.

14.5 Summary

This chapter illustrated the severe impact of traffic burstiness on the SSA performance and that analytic models or simulations employing traditional negative exponential distributions for burst sizes may often provide overly optimistic estimates for the network performance.

The packets in self-similar long-range dependent traffic streams experienced a significantly larger delay and nonnegligible packet loss even at very light loads. This performance of the system with self-similar traffic provides a startling contrast to models employing the Poisson arrival process. Under Poisson models, it is possible to increase the buffer and timeslot size just enough that traffic averaged over the cycle time would appear smooth and would fit entirely in the buffer, so that negligible packet loss will be observed. In a real network, the traffic bursts have a heavy-tail distribution. The tail of the distribution function for such distributions decreases subexponentially, unlike the Poisson, where the decrease is exponential. This leads to the fact that the probability of extremely large bursts in real traffic is greater than that in the Poisson model. This also means that no efficient traffic smoothing is possible in real networks and that packet loss cannot be prevented. It can only be mitigated (linearly reduced) at the expense of an exponential increase of buffer space and packet delay.

In this chapter we also witnessed that under the static slot assignment discipline, an ONU does not use all the bandwidth available to it; some fraction of bandwidth is wasted due to the unused remainder at the end of a timeslot. We considered a method to improve the utilization by using packet scheduling. We showed that the first-fit algorithm improves the utilization, but may have a negative impact on the TCP/IP connection behavior. We then suggested a connection-oriented first-fit algorithm.

References

[CB96]   M. Crovella and A. Bestavros, "Self-similarity in World Wide Web traffic: Evidence and possible causes," in *Proceedings of ACM SIGMETRICS International Conference on Measurement and Modeling of Computer Systems,* Philadelphia PA, May 1996.

[HOR87]   J. L. Hammond and P. J. P. O'Reilly, *Performance Analysis of Local Computer Networks*, Addison-Wesley, 1987.

[LT+94]   W. Leland, M. Taqqu, W. Willinger, and D. Wilson, "On the self-similar nature of Ethernet traffic (extended version)," *IEEE/ACM Transactions on Networking,* vol. 2, no. 1, pp. 1–15, February 1994.

[PF95]   V. Paxson and S. Floyd, "Wide-area traffic: The failure of Poisson modeling," *IEEE/ACM Transactions on Networking,* vol. 3, no. 3, pp. 226–244, June 1995.

[SG01]   D. Sala and A. Gummalla, "PON functional requirements: Services and performance," presented at IEEE 802.3ah meeting in Portland, OR, July 2001. Available at http://grouper.ieee.org/groups/802/3/efm/public/jul01/presentations/sala_1_0701.pdf.

[Ste94]   W. R. Stevens, *TCP/IP Illustrated*, vol. 1, Addison-Wesley, 1994.

15

EPON with Dynamic Slot Assignment

In Chap. 14, we considered a static TDM scheme, in which every ONU receives a fixed timeslot. While this scheme is very simple, it has the drawback that no statistical multiplexing between the ONUs is possible.

In has been shown that network traffic exhibits a high degree of burstiness [LT+94]. Traffic aggregation does not solve the problem as the variance of aggregated traffic decreases more slowly than the variance of a conventional Poisson process (see App. A). The long-range dependence (heavy-tailness) of the arrival process creates a situation in which some timeslots overflow even under very light load, resulting in packets being delayed for several timeslot periods. It is also true that some timeslots remain underutilized (not filled completely) even if the traffic load is very high. This leads to the PON bandwidth being underutilized. A dynamic scheme which reduces the timeslot size when there is no data would allow the excess bandwidth to be used by busy ONUs.

In this chapter we consider a simple DBA algorithm that provides statistical multiplexing for the ONUs. We investigate the improvements in performance (multiplexing gain) compared to the SSA scheme. We also consider and analyze several flavors of statistical multiplexing.

15.1 DBA Algorithm

The DBA algorithm considered here is based on the *interleaved polling with adaptive cycle time* (IPACT) method [KMP02]. In its original form,

```
Denote
t_scheduled   - time up to which the upstream channel has been scheduled
RTT_i         - round-trip time of the i^th ONU
T_guard       - the guard-band interval (constant)
T_REPORT      - time interval needed to transmit a report message (constant)
T_ON          - laser-on time (constant)
T_OFF         - laser-off time (constant)
T_process     - message processing delay (constant)
syncTime      - synchronization interval including T_AGC, T_CDR, and T_code_group_align
maxLength     - maximum limit on timeslot size (constant per ONU).
localTime     - value of local MPCP clock

1        For every received REPORT i
2        {
3                startTime = t_scheduled + T_guard
4                if startTime < localTime + T_process then
5                        startTime = localTime + T_process

6                length = REPORT.length + T_REPORT + T_ON + syncTime + T_OFF
7                if length > maxLength then
8                        length = maxLength

9                GATE = {startTime-RTT_i, length, forceReport = true}
10               send GATE

11               t_scheduled = startTime + length
12       }
```

Figure 15.1   Operation of DBA agent at the OLT.

the IPACT method used control messages that were embedded in data frames or in interframe gaps using an escape code. Such control messages were transmitted by the OLT just-in-time (see Sec. 5.3.1.2), eliminating the need for the ONU's synchronization and significantly reducing the control message overhead. However, for reasons discussed in Sec. 5.3, the IEEE 802.3ah task force adopted the MAC control frame format for control messages. Correspondingly, for this study, we modified IPACT to use MPCP control frames, as specified in the IEEE 802.3ah standard.

The MPCP behavior related to processing of GATE and REPORT messages was described in Chap. 8. Here we outline the behavior of the DBA agent at the OLT (see Fig. 15.1). Upon receiving a REPORT message from an ONU, the DBA agent calculates the new GATE parameters, such as the grant start time and grant length.

The OLT maintains a variable $t_{scheduled}$ which represents a future time up to which the upstream channel has been scheduled. To maintain a high utilization of the upstream channel, the DBA agent allocates the next timeslot immediately adjacent to the already allocated timeslots, with only the guard time interval (T_{guard}) left between the end of the previously allocated timeslot and the start of the timeslot currently being allocated.[1] This calculation is shown in line 3 of Fig. 15.1.

---

[1] Recall from Sec. 12.2.2 that the laser-on time may overlap the laser-off time of a previous timeslot, therefore T_{guard} may have a negative value.

The DBA agent needs to make sure that the ONU will have enough time to process the received GATE message before the granted timeslot is scheduled to begin. From the GATE reception process shown in Fig. 8.17, we know that an ONU will reject all grants for which the interval between the reception of the GATE frame and the start time of the grant is less than min_processing_time. We also know that ONU adjusts its local time to the timestamp value received with the GATE message. Therefore, to ensure that enough processing time is left for an ONU, it is enough for the DBA agent to ensure that the difference between the timestamp and grant start time is at least as large as min_processing_time. Of course, there will be an additional delay between the creation of a GATE message by the DBA agent and the timestamping of this message by the control multiplexer; this delay will be due to processing time, as well as possible blocking of the GATE message behind a long data frame, that started its transmission just before the DBA agent issued the GATE. This OLT delay together with ONU's min_processing_delay is taken into account in lines 4 and 5 which ensure that the grant start time is not closer than $T_{process}$ to the current time. The $T_{process}$ constant represents the combined maximum delay at the OLT and min_processing_time at the ONU. Recall that the IEEE standard specified the min_processing_time as 1024 TQ, or 16.384 µs.

Lines 6 through 8 determine a particular timeslot allocation discipline. In the example presented in Fig. 15.1, the granted timeslot size is set to the size requested by an ONU plus some additional size needed for physical layer overhead (T_{on}, T_{off}, and $syncTime$) and for transmission of a REPORT message (T_{REPORT}).

If the DBA agent allows each ONU to send the entire buffer contents in one transmission, ONUs with larger buffers and high data volume could monopolize the entire bandwidth. To avoid this, the OLT will limit the maximum transmission size. Thus, the DBA agent imposes a maximum on the timeslot size that can be assigned to an ONU ($maxLength$). We call such scheduling discipline a $limited\ service$, since the maximum timeslot size is limited to a predefined constant. In Sec. 15.1.3 we consider additional disciplines: $fixed$, $gated$, $constant$ $credit$, $linear\ credit$, and $elastic$ services.

Upon determining the start time and length parameters, the DBA agent forms a GATE message, consisting of a single grant (line 9), and passes this message to the GATE generation process (line 10) for transmission.

Finally, in line 11, the DBA agent updates its future channel allocation time $t_{scheduled}$, which now points to the end of the allocated timeslot.

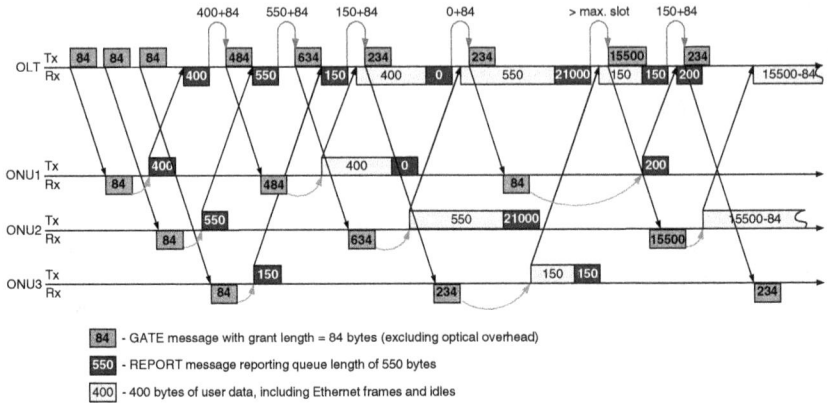

Figure 15.2   Time diagram of the limited service in IPACT algorithm.

Figure 15.2 presents a time diagram for the limited service. For simplicity of illustration, only three ONUs are shown, and the optical overhead, consisting of T_{on}, T_{off}, and *syncTime,* is ignored.

Upon completion of the discovery procedure, the OLT issues individual GATE messages to all discovered ONUs. These GATE messages contain only one grant and require the ONUs to send REPORT messages in the corresponding timeslots (i.e., the *forceReport* flags are set). The timeslot length in each GATE message is sufficient to send only one MPCPDU.

When a REPORT from the first ONU arrives, the OLT allocates a timeslot, as described above. Note that when ONU 2 requests a timeslot of size 21,000 bytes, it is granted only 15,500 bytes, which is the specified maximum.

If an ONU emptied its buffer completely, it will report 0 bytes back to the OLT. Correspondingly, in the next cycle, this ONU will be granted a small timeslot sufficient to send only a REPORT message, but no data.

Note that, after the first cycle, the OLT's receive channel is almost 100 percent utilized (REPORT messages and guard times consume some bandwidth). Idle ONUs (without data to send) are given very short transmission windows. This leads to a shortened cycle time, which, in turn, results in more frequent polling of active ONUs.

15.1.1 Maximum transmission window

To prevent the upstream channel monopolization by one ONU with high data volume, there should be a maximum transmission window size limit assigned to every ONU. We denote an ONU-specific maximum transmission window size as W_i^{max} (in bits). The choice of specific values of W_i^{max} determines the maximum polling cycle time T^{max} under heavy-load conditions:

$$T^{\text{max}} = \sum_{i=1}^{N} \left(G + \frac{W_i^{\text{max}}}{R_N} \right) \tag{15.1}$$

where W_i^{max} = maximum window size for ith ONU, bits; G = guard interval, s; N = number of ONUs; and R_N = EPON line rate, bps.

In addition to the maximum cycle time, the W_i^{max} value determines the guaranteed bandwidth available to ONU$_i$. Let B_i^{min} denote the guaranteed bandwidth of ONU$_i$. Obviously,

$$B_i^{\text{min}} = \frac{W_i^{\text{max}} - W^{\text{REPORT}}}{T^{\text{max}}} \tag{15.2}$$

i.e., ONU i is guaranteed to be able to send W_i^{max} bits, less the window size reserved for the REPORT message (W^{REPORT}), in at most T^{max} time. Of course, an ONU's bandwidth will be limited to its guaranteed bandwidth only if all other ONUs in the system also use all their available bandwidth. If at least one ONU has less data, it will be granted a shorter transmission window, thus making the cycle time shorter, and therefore the available bandwidth to all other ONUs will increase proportionally to their W_i^{max}. This is the mechanism behind dynamic bandwidth distribution: by adapting the cycle time to the instantaneous network load (i.e., queue occupancy), the bandwidth is automatically distributed to ONUs based on their W_i^{max} values. In the extreme case, when only one ONU has data to send, the bandwidth available to that ONU will be

$$B_i^{\text{max}} = \frac{W_i^{\text{max}} - W^{\text{REPORT}}}{N \times G + \dfrac{W_i^{\text{max}}}{R_N}} \tag{15.3}$$

In our simulations, we assume that all ONUs have the same SLA, that is, $W_i^{\text{max}} = W^{\text{max}}$, $\forall i$. This results in

$$T^{\text{max}} = N \left(G + \frac{W^{\text{max}}}{R_N} \right) \tag{15.4}$$

Using the configuration parameters shown in Table 14.1, we get W^{max} = 15,500 bytes. With this choice of parameters, every ONU will get a guaranteed bandwidth of ~61.66 Mbps and a maximum (best-effort) bandwidth of ~881 Mbps [see Eqs. (15.2) and (15.3)].[2]

---

[2] These numbers refer to row bandwidth values. In reality, the ONU may not be able to get these values due to packet delineation overhead (i.e., unused timeslot remainders).

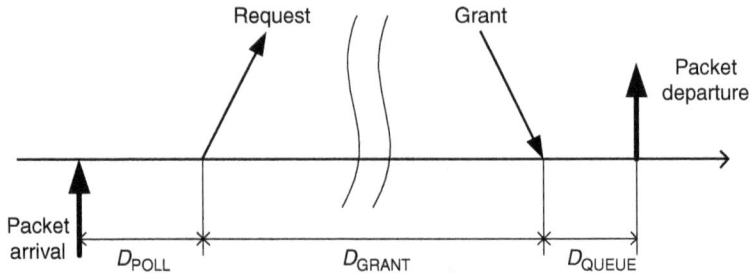

Figure 15.3   Components of packet delay at the ONU.

15.1.2   Components of packet delay

First, let us take a look at the components of the packet delay (Fig. 15.3). The packet delay D is equal to

$$D = D_{\text{POLL}} + D_{\text{GRANT}} + D_{\text{QUEUE}} \qquad (15.5)$$

where

D_{POLL} = time between packet arrival and next request sent by that ONU. On average, this delay equals one-half of the cycle time.

D_{GRANT} = time interval from an ONU's request for a transmission window until the beginning of the timeslot in which this frame is to be transmitted. This delay may span multiple cycles (i.e., a frame may have to skip several timeslots before it reaches the head of the queue), depending on how many frames there were in the queue at the time of the new arrival.

D_{QUEUE} = delay from the beginning of the timeslot till the beginning of frame transmission. On average, this delay is equal to half of a slot time and is insignificant compared to the previous two components.

15.1.3   IPACT allocation disciplines

How should the OLT determine the timeslot size $w_{i,k}$ which it is to grant to ONU i in cycle k, given the requested window size $v_{i,k}$? Table 15.1 summarizes a few approaches (services) the OLT may take in making its decision.

15.2   Results from Simulation Experiments

In this simulation experiment, all ONUs have identical load. Unless mentioned explicitly, all simulation parameters remain the same as in

Chap. 14 (refer to Table 14.1). After all the ONUs were discovered, the discovery process was suspended; i.e., performance parameters were not affected by discovery windows.

In Fig. 15.4, we present the mean packet delay for different scheduling services as a function of an ONU's offered load. First, we note a dramatic reduction of the average packet delay for all DSA services compared to SSA (fixed) service. The fixed service plot is interesting as an illustration of the traffic long-range dependence. Even at the very light load of 5 percent, the average packet delay is already very high (~4.4 ms). This is so because most packets arrive in very large packet trains. In fact, the packet trains were so large in our experiments that the 1-Mbyte buffers overflowed and some packets were dropped. Why do we observe this anomalous behavior only with fixed service? The reason is that, in fixed service, the cycle is large (fixed) under any load; several bursts that arrive close to one another can easily deliver more than W^{max} bits of data, so that more than one timeslot is needed to clear the queue. By the time the next timeslot arrives, even more data are waiting in the queue. Thus, after several cycles, the buffer will overflow. It can be shown that, in order to have buffer overflow, a time interval of length Δt should have an average arrival rate λ such that

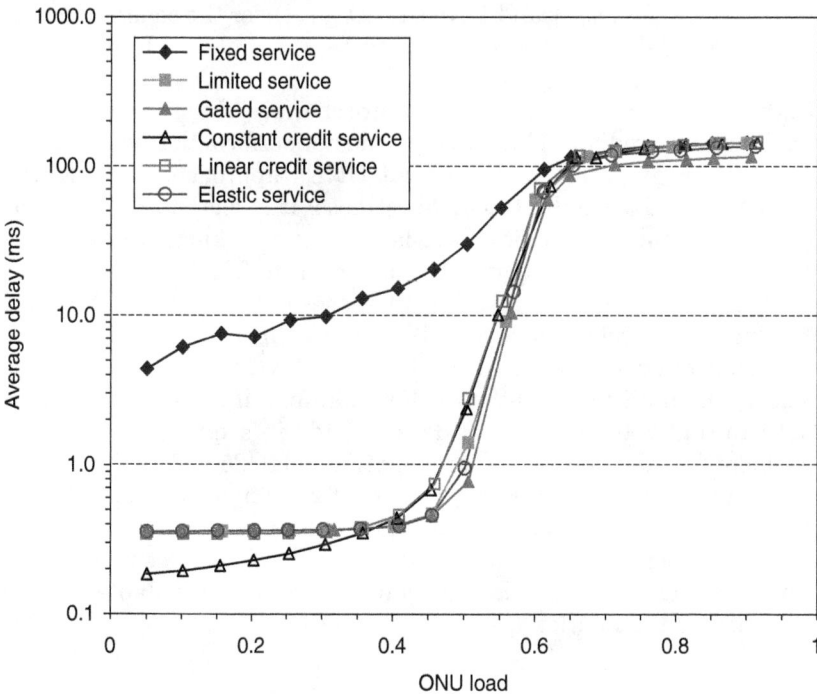

Figure 15.4   Average packet delay.

$$\lambda \geq \frac{Q}{\Delta t} + \frac{W^{\max} - W^{\text{REPORT}}}{T^{\max}} \tag{15.6}$$

Given our numerical values for Q, $W^{\max}$, and $T^{\max}$ (see Table 14.1), buffer overflow will occur if, e.g., we have a 1-s interval with an average arrival rate of 70 Mbps. And because the traffic is heavy-tailed, we do encounter such intervals.

This behavior is only characteristic of fixed service. All other services have a much shorter cycle time; there is just not enough time in a cycle to receive more bytes than $W^{\max}$, thus the queues never build up. We want to note here that the reduced cycle time which adapts exactly to the amount of data available in the ONUs is the main advantage of the IPACT algorithm.

We analyze the average delay for the remaining DSA services separately for three regions: light-load region (ONU offered load of less than 40 percent), moderate load (from 40 percent to 60 percent), and heavy load (above 60 percent).

In the light-load region, the D_{GRANT} delay component seldom exceeds 1 cycle time. However, if constant credit is given to an ONU, this component can be eliminated completely. Thus, while for limited, gated, linear credit, and elastic services the average delay approaches 1.5 cycle times (0.5 cycle for D_{POLL} and 1 cycle for D_{GRANT}), for the constant credit service, this delay approaches 0.5 cycle and only includes D_{POLL} component.[3]

In the moderate-load region, the picture changes; the credit-based schemes (constant credit and linear credit) do not perform that well. In this region, the aggregated load from all ONUs approaches the EPON capacity. Considering the traffic burstiness, it is easy to predict that some ONU in this region may have large queue backlogs, while others may have almost empty queues. Giving credit to ONUs with empty or almost empty queues unnecessarily increases the cycle time. The effect of the increased cycle time is amplified many times for the ONUs with backlogged queues, because the D_{GRANT} delay component for these ONUs spans multiple cycles now. For example, if a frame arrives at ONU i to find 250,000 bytes of data already in a queue, the D_{GRANT} component for this frame will approximately equal $\lceil 250,000 / 15,500 \rceil = 17$ cycle times. Giving unnecessary credits of 2000 bytes to only 4 ONUs will increase D_{GRANT} by as much as 1.088 ms.

In the high-load region, the average delay for all services reaches the saturation delay, which is determined by the available egress bandwidth and the buffer space Q.

---

[3] The D_{QUEUE} component is also present but is negligible at light load.

TABLE 15.1   Grant Scheduling Services Used in Simulation Experiments

Service	Formula	Description
Fixed $w_{i,k} = W^{\max}$		This scheduling discipline ignores the requested window size and always grants the maximum window. As a result, it has a constant cycle time $T^{\max}$. Essentially, this approach corresponds to the SSA PON system described in Chap. 14. It is shown here only for comparison.
Limited $w_{i,k} = \min \begin{cases} v_{i,k} \\ W^{\max} \end{cases}$ $v_{i,k} =$ requested window size[*]		This discipline was discussed in detail in the beginning of Sec. 15.1. It grants the requested number of bytes, but no more than $W^{\max}$. It is the most conservative scheme and has the shortest cycle of all the schemes.
Gated $w_{i,k} = v_{i,k}$		This service discipline does not impose the $W^{\max}$ limit on the granted window size; i.e., it will always authorize an ONU to send as much data as it has requested. Of course, without any limiting parameter, the cycle time may increase unboundedly if the offered load exceeds the network capacity. In this discipline, such a limiting factor is the buffer size Q; that is, an ONU cannot store more than Q bytes, and thus it will never request more than Q bytes.
Constant credit $w_{i,k} = \min \begin{cases} v_{i,k} + \text{const} \\ W^{\max} \end{cases}$		This scheme adds a constant credit to the requested window size. The idea behind adding the credit is the following: assume that x bytes arrived between the time when an ONU sent a REPORT and received the grant. If the granted window size equals the requested window + x (i.e., it has a credit of size x), then the D_{GRANT} delay component will be zero for these x bytes and the total delay will be shorter.
Linear credit $w_{i,k} = \min \begin{cases} v_{i,k} \times \text{const} \\ W^{\max} \end{cases}$		This scheme uses a similar approach to the constant credit scheme. However, the size of the credit is proportional to the requested window. The reasoning here is the following: LRD traffic possesses a certain degree of predictability (see [PW00]); i.e., if we observe a long burst of data, then this burst is likely to continue for some time into the future. Correspondingly, the arrival of more data during the last cycle time may signal that we are observing a burst of packets.
Elastic $w_{i,k} = \min \begin{cases} v_{i,k} \\ NW^{\max} - \sum\limits_{n=i-N}^{i-1} w_{n,k} \end{cases}$		Elastic service is an attempt to get rid of a fixed maximum window limit. The only limiting factor is the maximum cycle time $T^{\max}$. The maximum window is granted in such a way that the accumulated size of last N grants (including the one being granted) does not exceed $NW^{\max}$ bytes ($N =$ number of ONUs). Thus, if only one ONU has data to send, it may get a grant of size up to $NW^{\max}$.

[*]In these calculations, in adddition to a queue length received in a REPORT message the $v_{i,k}$ includes the required optical overhead T_{on}, T_{off}, and $syncTime$, and an additional space for an ONU to send the next REPORT message.

$$D_{\text{saturation}} \approx \left\lceil \frac{Q}{W^{\max} - W^{\text{REPORT}} - E(R)} \right\rceil \times T^{\max} \qquad (15.7)$$

where $E(R)$ is the expected unused timeslot remainder (see Sec. 12.2.4). Given the simulation parameters in Table 14.1, we should expect $D_{\text{saturation}} \approx 141.5$ ms, as seen in Fig. 15.4.

The gated service performs slightly better than the rest of the services. It transmits the entire buffer contents in one jumbo timeslot, eliminating multiple guard bands, needed when a timeslot is limited by $W^{\max}$. Detailed analysis of this service revealed oscillations in granted timeslot sizes: after ONU transmitted the entire queue, it reports an empty or a very small queue size (only the frames that arrived during the current transmission). Correspondingly, in the next cycle, it will be granted a small timeslot, but will report a full queue again. In the following cycle, it will be given a large timeslot and will report a short remaining queue again, and so on. Removing such oscillations can further improve the average packet delay. However, we will show that, even though the gated service has lower delay under high load, it is not a suitable service for an access network under consideration. The problem lies in the much longer cycle times (see Fig. 15.5). As a result, the D_{POLL} delay will be much larger, and therefore the packet latency will be much higher. Clearly, large D_{GRANT} and D_{QUEUE} delay components can be avoided for high-priority packets by using priority queuing. But D_{POLL} is a fundamental delay, which cannot be avoided, in general. This makes gated service not feasible for a multiservice access network.

Thus, we conclude that neither of the discussed service disciplines is better than the limited service. As such, for the remainder of this chapter, we shall focus our attention on the limited service discipline. In the next section, we will analyze the QoS characteristics of the limited service.

15.2.1 Performance of limited service

In this section, we analyze the performance of one selected ONU (referred to as a *tagged ONU*) as a function of its offered load and the effective (carried) load of the entire network. In Fig. 15.6, we present the average packet delay.

When the effective network load is low, all packets from the tagged source experience very little delay, no matter what the ONU's offered load is. This is a manifestation of dynamic bandwidth allocation—when the network load is low, the tagged source gets more bandwidth.

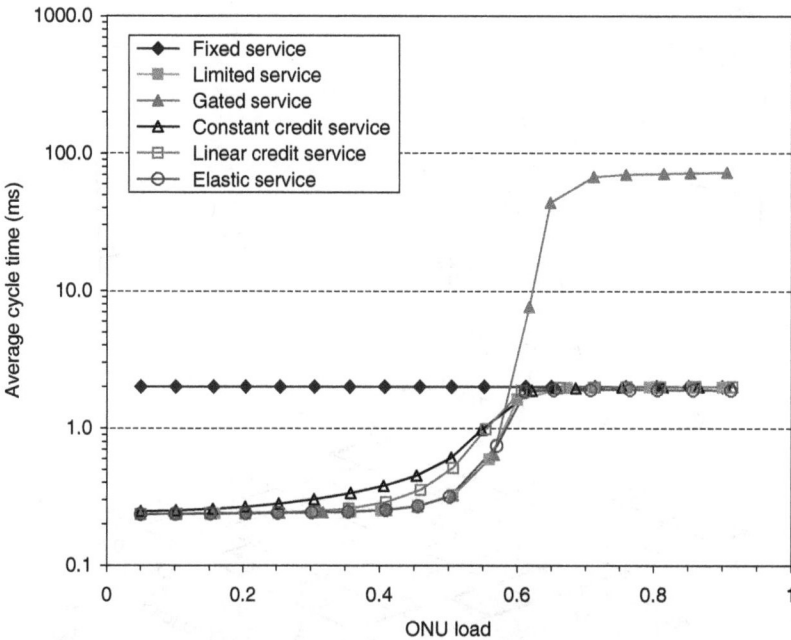

Figure 15.5 Average cycle times for various service disciplines.

The opposite situation—low offered load at the ONU and high effective network load—results in a higher delay. The only reason for this is the burstiness (i.e., long-range dependence) of the traffic. This is the same phenomenon observed with fixed service: high network load results in increased cycle time. This cycle time is large enough to receive more than W^{max} bytes of data during a burst. Hence, the D_{GRANT} delay for some packets will increase beyond one cycle time. We will discuss a way to combat this phenomenon by using priority queuing in Chap. 16.

Figure 15.7 shows the probability of a packet loss in a tagged ONU_i as a function of its offered load and the effective (carried) load of the entire network. Once again, we observe that packet loss is zero or negligible if the effective network load is less than 80 percent. When the network load is above 80 percent and the tagged ONU offered load is above 50 percent, we observe considerable packet loss due to buffer overflow.

15.3 Summary

In this chapter, we analyzed a simple algorithm for dynamic bandwidth allocation based on an IPACT algorithm. We found that the dynamic slot assignment provides a tremendous improvement in system

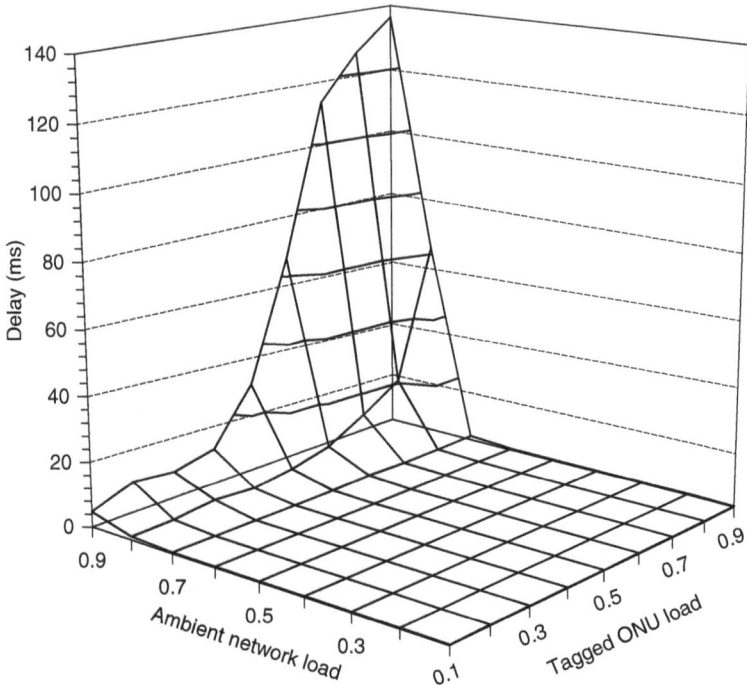

Figure 15.6   Average packet delay as a function of effective network load and ONU offered load.

performance, compared to the performance afforded by a static allocation scheme.

In IPACT, because each ONU uses the window size that is required at the moment, the polling cycle time adapts to the instantaneous queue loads, leading to an adaptive cycle time. This is the basic idea behind the fair unused bandwidth redistribution: reduced cycle time leads to an increase in the amount of best-effort bandwidth available to busy ONUs. This increase is proportional to their guaranteed bandwidth values; i.e., IPACT provides rate-proportional fair service. We also showed that the guaranteed bandwidth available to a user could easily be reprovisioned by simply changing a single parameter (W^{max}).

References

[KMP02]   G. Kramer, B. Mukherjee, and G. Pesavento, "Interleaved Polling with Adaptive Cycle Time (IPACT): A dynamic bandwidth distribution scheme in an optical access network," *Photonic Network Communications*, vol. 4, no. 1 pp. 89–107, January 2002.

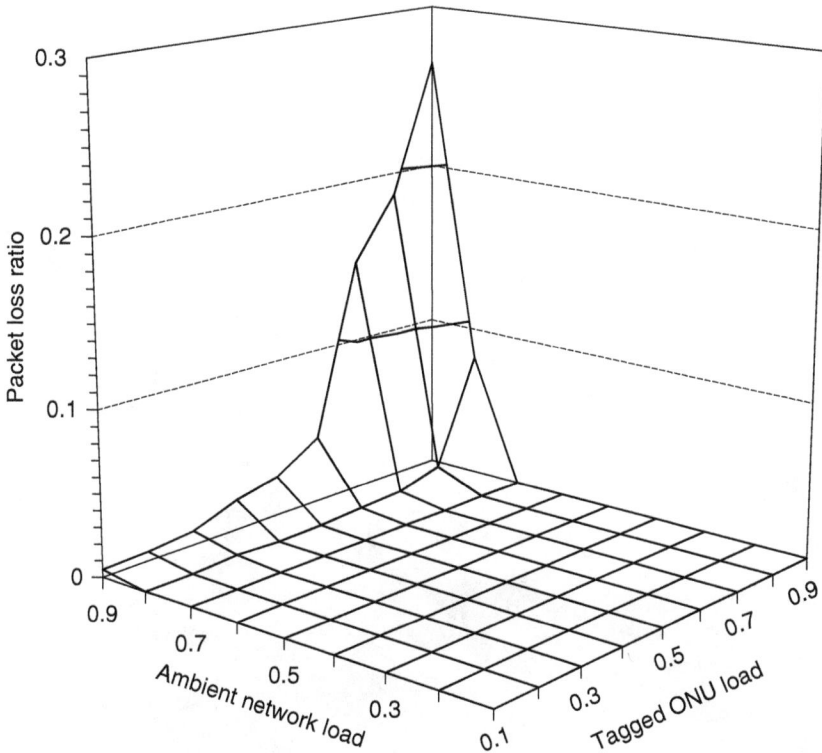

Figure 15.7   Packet-loss ratio as a function of effective network load and ONU offered load.

[LT+94]   W. Leland, M. Taqqu, W. Willinger, and D. Wilson, "On the self-similar nature of Ethernet traffic (extended version)," *IEEE/ACM Transactions on Networking*, vol. 2, no. 1, pp. 1–15, February 1994.

[PW00]   K. Park and W. Willinger, "Self-similar network traffic: An overview," in K. Park and W. Willinger, eds., *Self-Similar Network Traffic and Performance Evaluation*, Wiley Interscience, 2000.

16

Support for Differentiated Classes of Service

16.1 Introduction

Not being constrained by distance and bandwidth limitations of copper outside plant, EPON is able to deliver tens to hundreds of megabytes per second to and from users. A giant step forward compared to cable modems and DSL technologies, EPON is expected to be a truly converged network, supporting voice communications, standard and high-definition television (STV and HDTV), videoconferencing (interactive video), real-time and near-real-time transactions, and data traffic. To support this multitude of applications, EPON must guarantee appropriate performance for each such application.

Performance of a packet-based network (and EPON in particular) can be conveniently characterized by several parameters: bandwidth, packet delay (latency), delay variation (jitter), and packet-loss ratio. *Quality of service* (QoS) refers to a network's ability to provide bounds on some of or all these parameters on a per-connection (flow, session) basis. Not all networks, however, can maintain a per-connection state or even identify connections. To support diverse application requirements, such networks segregate all the traffic into a limited number of classes and provide differentiated service for each class. Such networks are said to maintain *classes of service* (CoS).

In this chapter, we focus on how to provide CoS differentiation mechanisms in EPON. These mechanisms include *intra-ONU scheduling* and *inter-ONU scheduling,* as shown in Fig. 16.1.

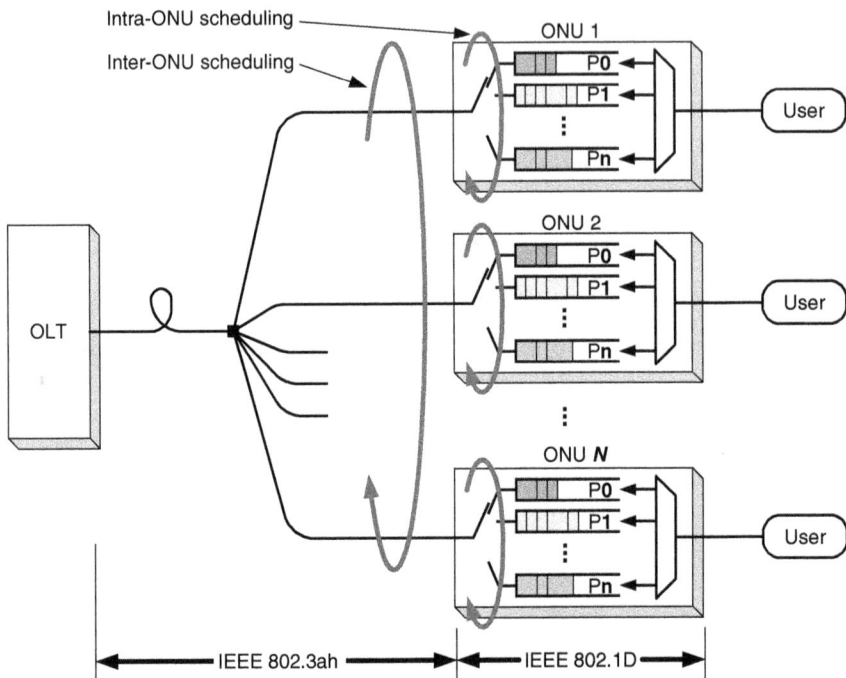

Figure 16.1   Intra-ONU and inter-ONU scheduling.

Being part of the IEEE 802 family of standards, EPON must be compliant with bridging defined in IEEE 802.1D [802.1D], including compliance with CoS mechanisms in this standard. Specifically, IEEE 802.1D, clause 7.7.4, states that the default *per-hop behavior* (PHB) of bridges (intra-ONU scheduling, in our terminology) is a *strict priority scheduling*.

In Chap. 15, we introduced a dynamic slot assignment scheme based on IPACT and compared several flavors of this algorithm. We found that the limited service scheme has the best performance among all considered variations of the modified IPACT algorithm.

The focus of this chapter is to investigate how an inter-ONU scheduler, which uses MPCP and the limited service DBA discipline, can be combined with a strict priority-based intra-ONU scheduler. We demonstrate that this combination results in quite an unexpected network behavior, where the queuing delay for some traffic classes increases when the network load decreases (a phenomenon we call *light-load penalty*). Since the light-load penalty affects only some traffic classes, it violates the fairness property among the traffic classes (i.e., performance for some classes degrades, while it improves for other classes as the load is increased).

Further, we discuss possible optimization schemes that improve the performance and eliminate the light-load penalty (either partially or completely).

16.1.1   Overview of IEEE 802.1D support for CoS

To support CoS, Ethernet networks must be able to classify traffic into classes of service and provide differentiated treatment to each class. The task of classification and differentiation of Ethernet frames was enabled by the introduction of two new standard extensions: P802.1p, *Supplement to MAC Bridges: Traffic Class Expediting and Dynamic Multicast Filtering* (later merged with IEEE 802.1D) and IEEE 802.1Q, *Virtual Bridged Local Area Networks*. IEEE 802.1Q defines a frame-format extension allowing Ethernet frames to carry priority information field in their header. The standard distinguishes the following traffic classes:

Network control—characterized by a "must get there" requirement to maintain and support the network infrastructure.

Voice—characterized by less than 10-ms delay, and hence maximum jitter (one-way transmission through the LAN infrastructure of a single campus).

Video—characterized by less than 100-ms delay.

Controlled load—important business applications subject to some form of "admission control," be that preplanning of the network requirement at one extreme to bandwidth reservation per flow at the time the flow is started, at the other extreme.

Excellent effort—or "CEO's best effort," the best-effort type of services that an information services organization would deliver to its most important customers.

Best effort—LAN traffic as we know it today.

Background—bulk transfers and other activities that are permitted on the network but that should not impact the use of the network by other users and applications.

If a bridge or a switch has less than seven queues, some of the traffic classes are grouped together. Table 16.1 illustrates the standard-recommended grouping of traffic classes.

IEEE 802.1D standard in clause 7.7.4 specifies the default bridge (switch) scheduling algorithm for multiple queues:

7.7.4 Selecting frames for transmission
The following algorithm shall be supported by all Bridges as the default algorithm for selecting frames for transmission:

TABLE 16.1   Mapping of Traffic Classes into Priority Queues (IEEE 802.1p)

Number of queues	Traffic type assignments						
1	Network control	Voice	Video	Controlled load	Excellent effort	Best effort	Background
2	Network control	Voice	Video	Controlled load	Excellent effort	Best effort	Background
3	Network control	Voice	Video	Controlled load	Excellent effort	Best effort	Background
4	Network control	Voice	Video	Controlled load	Excellent effort	Best effort	Background
5	Network control	Voice	Video	Controlled load	Excellent effort	Best effort	Background
6	Network control	Voice	Video	Controlled load	Excellent effort	Best effort	Background
7	Network control	Voice	Video	Controlled load	Excellent effort	Best effort	Background

a. For each Port, frames are selected for transmission on the basis of the traffic classes that the Port supports. For a given supported value of traffic class, frames are selected from the corresponding queue for transmission only if all queues corresponding to numerically higher values of traffic class supported by the Port are empty at the time of selection;

b. For a given queue, the order in which frames are selected for transmission shall maintain the ordering requirement specified in 7.7.3.

Additional algorithms, selectable by management means, may be supported as an implementation option, so long as the requirements of 7.7.3 are met.

16.2   System Architecture: Integrating Priority Queuing in EPON

In this study, we consider an EPON access network consisting of an OLT and N ONUs. Each ONU is equipped with n queues serving n priority classes (denoted $P0$, $P1$, ... Pn), with $P0$ being the highest priority and Pn being the lowest. When a packet is received from a user, the ONU classifies its type and places it in the corresponding queue. The queues in each ONU share common memory of size Q bytes. If an arriving packet with priority Pi finds the full buffer in the ONU, it can preempt one or more lower-priority packets Pk $(k > i)$ from their queues, such that the Pi packet can itself be placed into the Pi queue. Between timeslots, an ONU stores all the packets received from the user in their respective queues. When a transmission timeslot opens, the ONU serves a higher-priority queue to exhaustion before serving a lower-priority queue. With the exception of multiple queues per ONU, all the settings in our simulation experiments remain the same as in Chap. 14 (refer to Table 14.1).

In our model, we used the *limited* service discipline described in Chap. 15. Under this discipline, the OLT assigns to an ONU a timeslot of size equal to what the ONU had requested (through a previous REPORT message), but not greater than some predefined maximum W^{max}. The limit W^{max} is needed to guarantee a maximum interval between timeslots (cycle time) T^{max} and to avoid bandwidth hogging by a "hungry" ONU. This scheme was shown to efficiently share the bandwidth while maintaining fairness among ONUs.

In this study, we use simulation experiments rather than analytical methods. Our objective was to build a realistic model and evaluate the system behavior with a specific set of parameters; analytical modeling for such a system becomes extremely complex.

To obtain an accurate and realistic performance analysis, we generated synthetic self-similar traffic, using the method described in App. B. We used a trimodal packet size distribution similar to that observed in backbone networks [CMT98] and in a cable network headend [SG01]. The three main modes correspond to most-frequent packet sizes: 64, 582/594, and 1518 bytes (including Ethernet headers). The weights of the modes slightly differ in the backbone measurements and in the access network, so we used the distribution for the upstream traffic in a CATV network [SG01].

An important characteristic of a self-similar process is its heavy-tailed behavior (with tail decay coefficient a, $1 < a < 2$). This leads to a property of infinite variance (i.e., no moments $m > a$ exist). Therefore, most analytical methods only model approximate behavior. Having built a simulation model, we could concentrate on the transient system behavior and evaluate the behavior when the offered load exceeds the network capacity.

16.2.1 Traffic modeling

As a multiservice access network, the proposed architecture should support a multitude of services, i.e., best-effort data, variable-bit-rate (VBR) video stream, constant-bit-rate (CBR) stream for legacy equipment such as *plain old telephone service* (POTS) lines and *private branch exchange* (PBX) boxes.

In our simulation experiments, we classify our data into three priority classes: *P0*, *P1*, and *P2*. The three classes may be used for delivering voice, video, and data traffic. Using three classes also allows easy mapping of DiffServ's *expedited forwarding* (EF), *assured forwarding* (AF), and *best-effort* (BE) classes into 802.1D classes.

Class P0 is used to emulate a circuit-over-packet connection. *P0* traffic has constant bit rate (CBR). In our model, we chose to emulate a T1 connection. The T1 data arriving from the user are packetized in the ONU by placing 24 bytes of data in a packet. Including Ethernet and UDP/IP headers results in a 70-byte frame generated every 125 µs. Hence, the *P0* data consume 4.48 Mbps of bandwidth.

Class P1 in our experiment consists of VBR video streams that exhibit properties of self-similarity and long-range dependence (as shown in [GW94] for real MPEG-coded video streams). Since the *P1* traffic is highly bursty, it is possible that some packets in long bursts will be lost. This will happen if the entire buffer is occupied by *P1* and *P0* packets. Packet sizes in *P1* streams range from 64 to 1518 bytes.

Class P2 has the lowest priority. This priority level is used for non-real-time data transfers. There are no delivery or delay guarantees for this service. This is also self-similar and long-range-dependant traffic with variable-size packets ranging from 64 to 1518 bytes.

When we vary the load in our simulation experiments, we always keep the *P0* load constant (4.48 Mbps) and split the remaining load between *P1* and *P2* equally. For example, an ONU's offered load of 40 Mbps means that *P1* and *P2* classes generated $(40 - 4.48)/2 = 17.76$ Mbps each. In all the performance diagrams, the ONU's offered load values are normalized to the ONU's ingress link capacity ($R_U = 100$ Mbps).

16.3   Packet Delay Analysis

We start our performance analysis by investigating the combination of a limited service inter-ONU scheduler with strict priority queuing as the intra-ONU scheduler (we call this combination a *limited/priority* service discipline). This discipline grants the requested number of bytes, but no more than W^{max}. As our performance measures, we consider the average and maximum packet delay (Fig. 16.2). The horizontal axis represents load of an individual ONU. At each load, we collected statistics per 500 million packets. In our simulation experiments, all ONUs have uniform load. Thus, a network load of 1.0 corresponds to an ONU's load of 0.625 (62.5 Mbps). Understandably, the delay plots show clear knees at a load of around 0.625. At this point, the network begins to exhibit signs of saturation: buffers are full and a large number of packets are dropped.

One can immediately observe that combining the default priority-queuing per-hop behavior (PHB) with a simple polling mechanism in

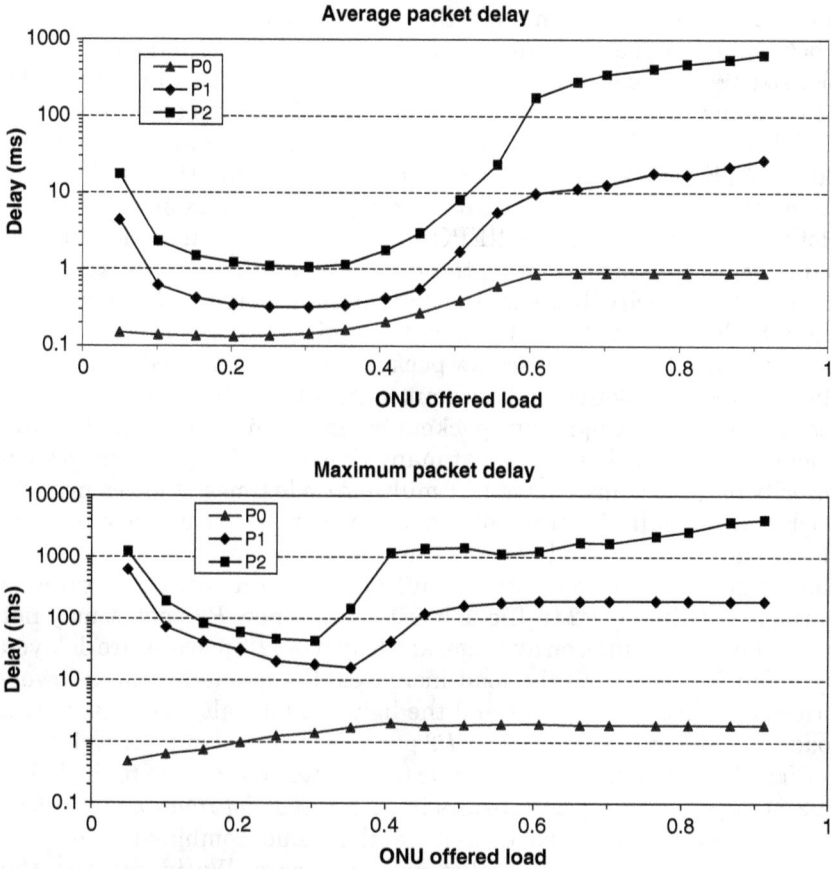

Figure 16.2   Packet delay for limited/priority service.

an EPON results in a very interesting phenomenon: as the load decreases from moderate (~0.25) to very light (~0.05), the average delay for the lower-priority classes *P1* and *P2* increases significantly. For example, for *P2* packets, the average packet delay at load 0.05 (or 5 Mbps) is 17.5 ms, almost 1600 percent higher than the 1.1-ms delay at a load 0.25 (25 Mbps). Similar behavior is observed for the maximum packet delay for *P1* and *P2* classes. We refer to this phenomenon as the *light-load penalty*.

16.3.1 Light-load penalty

A trace-level analysis of the polling scheme combined with priority queuing revealed the cause of light-load penalty. At the end of every timeslot, an ONU generates a new REPORT message containing the number of bytes that remain in the queue (*residual queue occupancy*). The residual queue occupancy is almost always less than the maximum slot size W^{max}, because the light-load penalty occurs only at light loads. This means that, for whatever slot size an ONU requested in the REPORT message, the OLT will grant the requested slot size through the next GATE message to that ONU. However, during the time lag between ONU's sending a REPORT and the arrival of its assigned timeslot (i.e., between sending a REPORT and transmission of the reported data), more packets arrive at the queue. Newly arrived packets may have higher priority than some packets already stored in the queue, and they will be transmitted in the next transmission slot before the lower-priority packets. Since these new packets were not reported to the OLT, the granted timeslot cannot accommodate all the stored packets. This leads to some lower-priority packets being preempted and left in the queue. This scenario may repeat many times, resulting in some lower-priority packets being delayed for multiple cycle times. A lower-priority packet will finally be transmitted when more lower-priority packets (bytes) accumulate (and reported to the OLT) behind a given packet than higher-priority packets cut in front of it. But since *P0* traffic is periodic (CBR) and *P2* traffic is bursty (i.e., a new *P2* packet may not arrive for a long time), on average, at a load 0.05, *P2* packets are delayed by about 70 cycles. As the load increases, the queue behind a lower-priority packet grows faster and the light-load penalty decreases. At a load 0.25, the average delay for *P2* packets is only about 4.2 cycles.

Since Ethernet packets cannot be fragmented (according to the IEEE 802.3), packet preemption results in an *unused slot remainder*, unless added higher-priority packets have the same combined size as a preempted lower-priority packet, which is rare. We investigate the properties of such remainders in Sec. 16.4.3.

It should be clear that the light-load penalty can never happen with the FCFS queuing discipline (with no priorities); later-arriving packets are appended to the end of the queue, and they cannot displace earlier packets from their places in the queue (i.e., arriving packets do not change the delineation of packets already stored in the queue and reported to the OLT).

Some higher-layer protocols (i.e., TCP Vegas [BMP94], or *cprobe* and *bprobe* [CC96]) rely on the packet delay as a measure of network congestion. The light-load penalty may have a detrimental effect on the operation of such protocols: a random fluctuation that reduces the load could increase the delay which, in turn, could be interpreted by a data source as increased congestion, and will force it to reduce its load (sending rate), thus increasing the delay even more. This chain reaction can lead to unstable behavior of the higher-layer protocol, or may prevent its proper operation altogether.

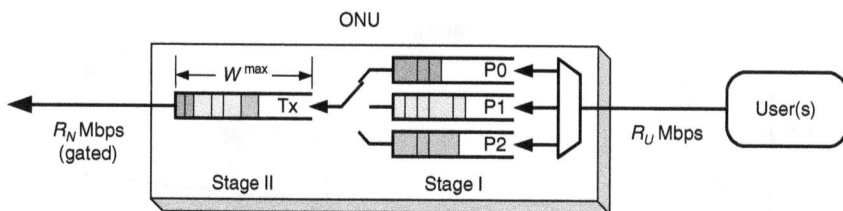

Figure 16.3   Tandem queue at an ONU.

Another reason EPON designers should eliminate or mitigate the light-load penalty is that it may encourage low-priority applications to artificially generate heavier-than-required load in order to get better performance from the network. While this may improve the performance for the lower-priority class, higher-priority classes will suffer.

16.4   Optimization Schemes

The light-load penalty is a result of combination of limited service at the OLT (inter-ONU scheduling) with strict priority queuing at an ONU (intra-ONU scheduling). Below, we consider two optimization schemes. The first scheme uses tandem queues at the ONU and is referred to as *limited/tandem* service. This scheme eliminates the light-load penalty through modifications to the intra-ONU scheduling; i.e., modifications are done to ONU.

The second scheme addresses the light-load penalty by changing the inter-ONU scheduling at the OLT. This scheme is called *CBR credit/priority*; in place of the limited service it uses a CBR credit service.

16.4.1   Tandem queuing

One way to eliminate the light-load penalty is to implement a tandem queue in an ONU (Fig. 16.3). In a tandem queue system, stage I consists of multiple priority queues and stage II consists of one *first-come, first served* (FCFS) queue. When a timeslot arrives, data packets from stage II are transmitted to the OLT, thereby vacating the queue; simultaneously, data packets from stage I are advanced into vacant spaces in the stage II queue. At the end of the current timeslot, the ONU reports to the OLT the occupancy of the stage II queue in order to get a corresponding slot size in the next cycle. The total size of the stage II buffer can be made exactly W^{max} bytes, so that the ONU never requests a slot greater than W^{max} bytes. This configuration will ensure that the given slot is always 100 percent utilized; i.e., the unused remainder is always zero.

Average packet delay

Maximum packet delay

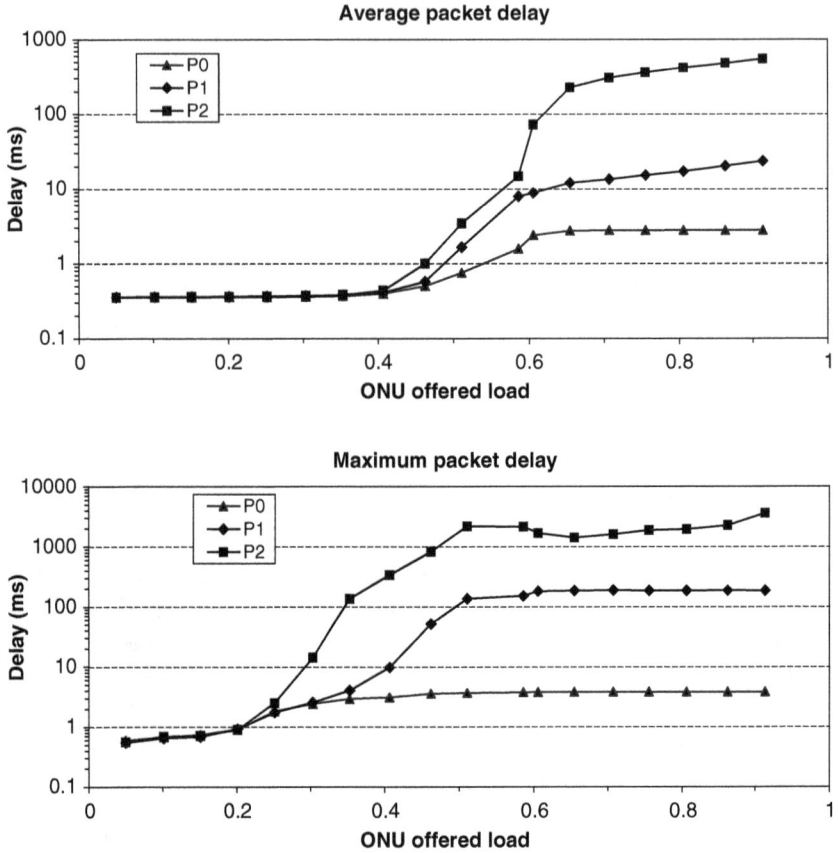

Figure 16.4   Packet delay for limited/tandem scheme.

Figure 16.4 presents the average packet delay in a limited/tandem configuration. We can observe that the light-load penalty has been eliminated. The drawback of this scheme, however, is the increased delay for the highest-priority class (*P0*). The average delay for *P0* has increased 3 times (under heavy-load conditions, the average delay in the two-stage scheme settles at 3 ms, while in the limited service, it was 1 ms). This can be intuitively explained by the fact that a packet, which arrived at a random time, will wait, on average, 0.5 cycle in the stage I queue and exactly 1 cycle in the stage II queue. If just one priority queue is implemented (without stage II), the average *P0* delay is 0.5 cycle time. The maximum delay for *P0* class has increased 2 times. The reason for this is also clear: a high-priority packet can spend at most 1 cycle time in each stage—2 cycle times total. In the limited/priority scheme, the maximum delay for *P0* packets was limited to 1 cycle time[1] (refer to Fig. 16.2).

The increased delay in a limited/tandem scheme could sometimes become a problem. All high-priority packets such as system alarms and failure indication may have to endure a longer delay. Furthermore, for voice traffic, ITU-T Recommendation G.114, *One-Way Transmission Time* [G.114], specifies 1.5-ms one-way propagation delay in an access network (digital local exchange). To keep the maximum packet delay within this bound, in the limited/priority scheme, the maximum cycle time T^{max} will have to be reduced to 1.5 ms, and in the limited/tandem scheme it should be reduced to 750 µs. Obviously, the shorter cycle time increases the guard time overhead ($overhead = NG/T^{max}$). With our default configuration parameters (see Table 14.1), the guard time overhead is equal to 1.07 and 2.13 percent for limited/priority and limited/tandem schemes, respectively. However, we note that, for the limited/tandem scheme, while the guard time overhead is higher, the total overhead is lower; this is so because the guard time overhead is compensated by a complete elimination of the unused slot remainder (frame delineation overhead). In Sec. 12.2.4, we derived the formula for the expected size of the unused slot remainder [Eq. (12.10)]. Using this equation, we can estimate the total overhead. For the packet size distribution from [SG01] and the limited/priority discipline with T^{max} = 1.5 ms, we have

$$overhead = \frac{N[G + E(R) \times 8 \text{ ns/byte}]}{T^{max}}$$

$$= \frac{16 \times (1 \mu s + 595 \text{ bytes} \times 8 \text{ ns/byte})}{1.5 \text{ ms}} \approx 6.14\,\%$$

For the limited/tandem scheme, the unused slot remainder is always zero; hence, for T^{max} = 750 µs, we get

$$overhead = \frac{NG}{T^{max}} = \frac{16 \times 1 \mu s}{750 \mu s} \approx 2.13\,\%$$

16.4.2 CBR credit

Another interesting solution to the light-load penalty (without increasing the delay of the highest-priority class beyond one cycle time as in the limited/tandem scheme) is to predict the amount of high-priority packets that are expected to arrive at the ONU and to adjust the

---

[1]This assumes that the queue always contained lower-priority packets, which could be preempted by a newly arrived higher-priority packet.

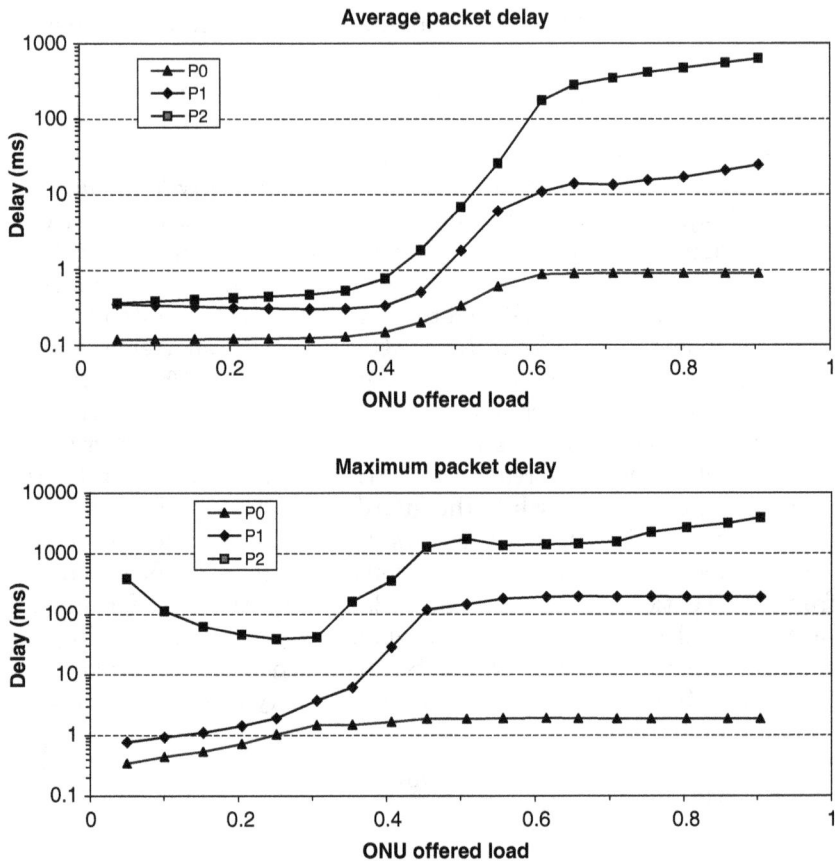

Figure 16.5   Packet delay for CBR credit/priority scheme.

granted timeslot size accordingly. Of course, to predict the traffic with any reasonable accuracy, we need to have some knowledge about the traffic profile. In our case, we have this knowledge about the *P0* traffic; namely, we know that this is a CBR flow with a given data rate. Therefore, when deciding on the size of the next timeslot for an ONU, the OLT can estimate the time of the next transmission and increase the timeslot size by the amount of CBR data it anticipates. Such an inter-ONU scheduler is called the *CBR credit* since the additional timeslot size increment (credit) is based on the known CBR arrival rate. The inter-ONU scheduler remains the default strict priority scheduler. This combination of intra- and inter-ONU scheduler is called CBR credit/priority.

Figure 16.5 shows the average and the maximum packet delay for the CBR credit/priority scheme. We can see that the light-load penalty is

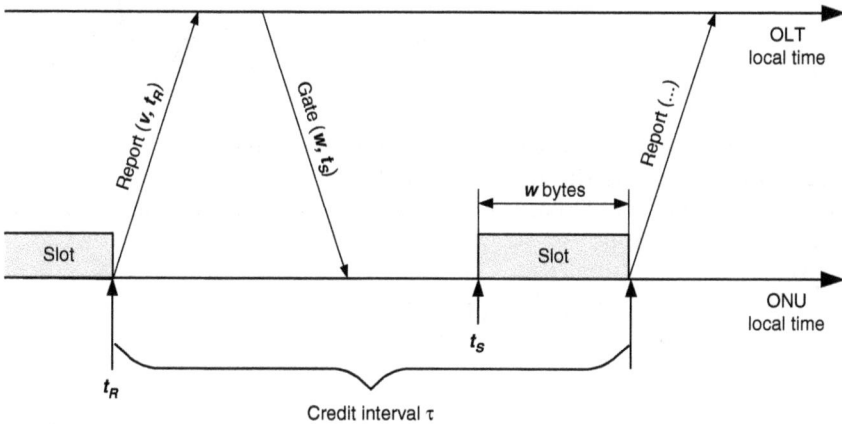

Figure 16.6 Calculation of the credit interval.

eliminated for average delay values. The penalty remains (in somewhat lesser degree) for the maximum delay values for the *P2* class only. This behavior is not unexpected, since the penalty for class *P2* is caused not only by *P0* frames (CBR) preempting *P2* frames, but by *P1* frames preempting *P2* as well. The CBR credit/priority scheme makes no attempt to predict the arrivals of *P1* traffic (which is a highly bursty, LRD traffic). Luckily, the probability of a *P1* packet displacing a *P2* packet at light load is not high, and the effect of the light-load penalty on the average delay of *P2* packets is not noticeable.

To compute the size of the credit, the OLT first determines the credit interval τ, shown in Fig. 16.6. The size of the next timeslot should be increased (credited) to accommodate CBR packets that are due to arrive during the credit interval. The credit interval can be calculated as follows. Given t_R = timestamp in a received REPORT message, t_S = start time of a granted timeslot, w = timeslot size, and R_N = EPON line rate, we have

$$\tau = t_S + \frac{w}{R_N} - t_R \tag{16.1}$$

Taking T^{CBR} = period of CBR packet arrivals (in seconds per packet), we can expect n^{CBR} CBR packets to arrive during the interval τ, where

$$n^{CBR} = \frac{\tau}{T^{CBR}} = \frac{t_S + w/R_N - t_R}{T^{CBR}} \tag{16.2}$$

But the granted timeslot size w itself depends on the number of additional CBR packets it should accommodate. Thus,

$$n^{\mathrm{CBR}} = \frac{1}{T^{\mathrm{CBR}}} \left(t_S + \frac{v + n^{\mathrm{CBR}} \times S^{\mathrm{CBR}}}{R_N} - t_R \right) = \frac{t_S + v/R_N - t_R}{T^{\mathrm{CBR}} - S^{\mathrm{CBR}}/R_N} \quad (16.3)$$

where S^{CBR} is the size of CBR packets (70 bytes in our experiment; see Sec. 16.2.1) and v is the number of bytes (requested timeslot size) reported by the ONU. Thus, the OLT assigns the timeslot size based on the following formula:

$$w_i = \min \begin{cases} v_i + \lceil n_i^{\mathrm{CBR}} \rceil \times \left(S^{\mathrm{CBR}} + \mathrm{IFG} \right) \\ W^{\max} \end{cases}$$

$$(16.4)$$

$$= \min \begin{cases} v_i + \left\lceil \dfrac{t_{S,i} + v_i/R_N - t_{R,i}}{T^{\mathrm{CBR}} - S^{\mathrm{CBR}}/R_N} \right\rceil \times \left(S^{\mathrm{CBR}} + \mathrm{IFG} \right) \\ W^{\max} \end{cases}$$

where w_i is the timeslot size assigned to ONU i; v_i is the requested timeslot size from ONU i; $t_{R,i}$ = timestamp of REPORT message received from ONU i; $t_{S,i}$ = start time of the timeslot assigned to ONU i; IFG = minimum interframe gap (includes 64-bit preamble as well as 96 bits of IFG); and $W^{\max}$ is the maximum limit on timeslot size.

The $t_{S,i}$ value is updated after each slot assignment as

$$t_{S,i+1} = t_{S,i} + \frac{w_i}{R_N} + G \quad (16.5)$$

i.e., the OLT expects the data (first bit) from ONU $i + 1$ to arrive exactly after the guard time G after the data (last bit) from ONU i.

The value $\lceil n^{\mathrm{CBR}} \rceil \times (S^{\mathrm{CBR}} + \mathrm{IFG})$ in Eq. (16.4) represents the CBR credit given to ONU i. The ceiling function is used to accommodate an integer number of packets. Obviously, in some instances, our prediction may be too generous, and the ceiling function will give an ONU more credit than it actually needs (i.e., only $\lfloor n^{\mathrm{CBR}} \rfloor$ packets may arrive at the ONU in interval τ). Below, we show that it is more efficient to give extra credit than not to give enough. If the OLT has credited $\lceil n^{\mathrm{CBR}} \rceil \times (S^{\mathrm{CBR}} + \mathrm{IFG})$ bytes, but only $\lfloor n^{\mathrm{CBR}} \rfloor$ CBR packets have arrived, the timeslot will have an unused remainder of size $S^{\mathrm{CBR}} + \mathrm{IFG}$ bytes exactly (that is, 70 + 20 = 90 bytes in our case). If, however, in an alternative case, the OLT conservatively has credited $\lfloor n^{\mathrm{CBR}} \rfloor \times (S^{\mathrm{CBR}} + \mathrm{IFG})$ bytes and $\lceil n^{\mathrm{CBR}} \rceil$ CBR packets have arrived, they will all be sent ahead of other lower-priority packets, displacing one or more lower-priority packets

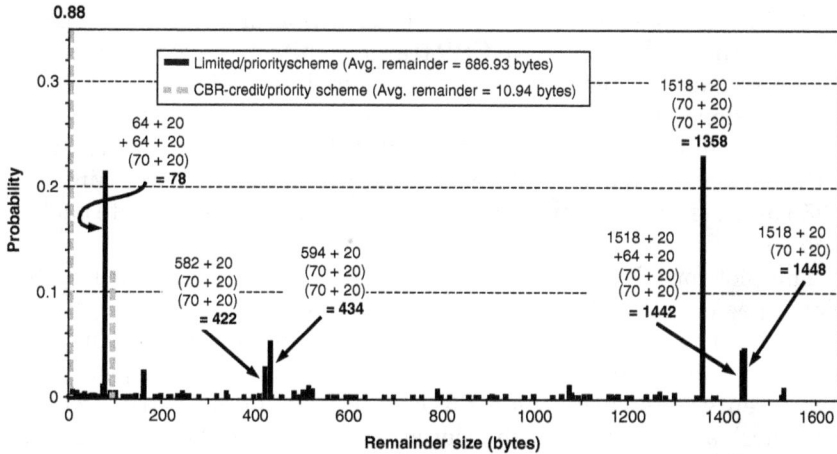

Figure 16.7   Histogram of unused slot remainder and most frequent preemption combinations.

from the timeslot. If this happens, the worst-case unused remainder is 1 less than the largest packet size (with associated IFG and preamble); that is, $1518 + 20 - 1 = 1537$ bytes of wasted timeslot space.

16.4.2.1 Dynamics of packet preemption. To illustrate the advantages of the CBR credit/priority scheme, we have built a distribution (histogram) for the unused timeslot remainder and compared it with this distribution for the limited/priority scheme (Fig. 16.7). These plots were obtained by simulating the transmission of 500 million packets through the EPON at an average ONU load of 0.05 (5 Mbps).

We measured and found the average unused slot remainder at a load of 0.05 to be approximately 687 bytes.[2] This is the average timeslot space which will remain unused with every credit misprediction. In the CBR credit/priority scheme, the unused remainder is reduced to only 11 bytes. Obviously, it is much cheaper (in terms of utilization) for the OLT to credit more than to credit less.

We also can observe an interesting periodicity in the limited/priority service remainder distribution. It can be explained by the very pronounced modality of the underlying packet size distribution. The trimodal packet size distribution has been observed in backbone [CMT98] and in access networks [SG01]. The three modes correspond to the most frequent packet sizes: 64 bytes—46 percent, 582/594 bytes—10 percent,

---

[2] The reader may notice that this remainder value does not agree with Eq. (12.10). Equation (12.10) requires the timeslot size to be independent of the packet sizes and significantly larger than the average packet size. Under a light load, this condition doesn't hold, as there are very few packets in a queue and timeslot size is based on sizes of these packets, i.e., is not independent.

and 1518 bytes—20 percent.[3] The inter-action of these three modes between themselves and with CBR traffic results in the periodic pattern of the remainder distribution. The plot in Fig. 16.7 illustrates what combinations of packets result in particular peaks of the remainder distribution. For example, the remainder of 1448 bytes is left when one CBR packet (70 bytes) displaces one 1518-byte packet. A remainder of 1358 bytes is left when two CBR packets displace one 1518-byte packet, and so on.

The plot for CBR credit/priority scheme shows two major distributional peaks: one at 0 bytes (88 percent) and the other at 90 bytes (12 percent). The peak at zero represents all timeslots where the OLT exactly predicted the number of CBR packet arrivals. The peak at 90 bytes represents cases when the ceiling function overestimated the CBR arrivals and granted the slot size for one extra CBR packet (i.e., when $\lfloor n^{CBR} \rfloor$ packets arrived instead of $\lceil n^{CBR} \rceil$ packets).

A limitation of the CBR credit/priority scheme is that external knowledge of the arrival process is necessary. Even though, for some time-critical applications, we may have such knowledge, it is by no means a universal case. This scheme can be applied for circuit emulation services with a CBR arrival process. A fairly straightforward modification of this scheme would allow it to be used with voice-over-packet traffic, even in *silence suppression* (SS) mode (i.e., when no packets are transmitted during silence intervals). Now, the OLT will start crediting an ONU when a talk spurt is detected and will stop crediting when silence is detected (the OLT can detect talk spurts and silence by the presence or absence of voice packets). The packet rate within a talk spurt is constant, and so the crediting scheme would work. The fact that the OLT will mispredict the beginning and end of a talk spurt should not introduce any significant overhead, since the average 3-s misprediction window (assuming average talk spurt is 1.65 s, and average silence interval is 1.35 s [DL86]) is much larger than the cycle time (2 ms maximum). Analysis of the prediction accuracy and efficiency of the CBR credit/priority scheme with the *P0* class consisting of voice-with-silence-suppression traffic remains a topic for future research.

16.4.3 Bandwidth utilization

Bandwidth utilization in EPON is determined by the cycle time, guard time, and average size of the unused timeslot remainder. Figure 16.8 presents the average timeslot remainder.

---

[3] Packet sizes include the Ethernet header (18 bytes). All payload sizes less than 46 bytes are padded to 46 bytes to comply with the 64-byte minimum IEEE 802.3 frame size.

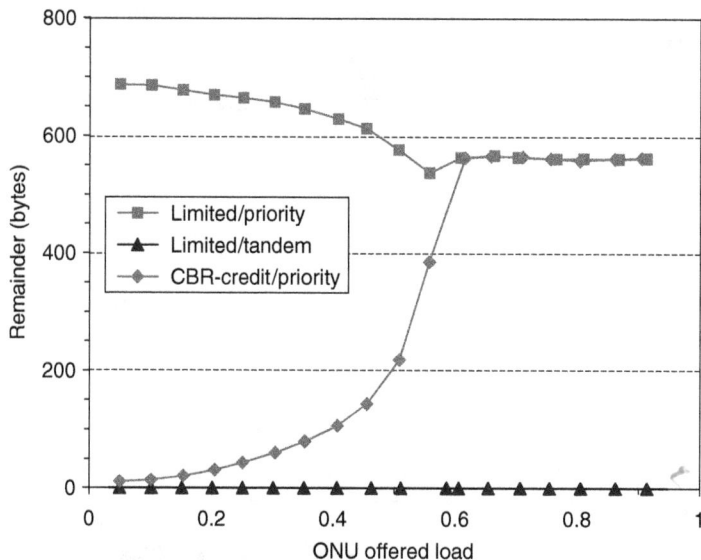

Figure 16.8   Average slot remainder.

At higher loads (0.6 and above), the average unused remainders in limited/priority service and CBR credit/priority schemes are the same and have an absolute value of approximately 560. This is expected since, at high loads, the ONUs request a window larger than W^{max} bytes. According to Eq. (16.4), the OLT will grant them a W^{max} -byte slot, ignoring the credit value. Therefore, the performance of the CBR credit/priority scheme at high loads is the same as in the limited/priority scheme.

Note that at high load the value of the remainder (560 bytes) is slightly lower than the 595 bytes predicted by Eq. (12.10). This equation assumes a single FCFS queue. In limited/priority service, if the $P1$ queue is blocked, the $P2$ queue will be tried, and if it has a small enough packet, this packet will be transmitted, thus further reducing the remainder.

However, Eq. (12.10), which assumes slot size independent of packet size and large enough to fit *many* packets, cannot be applied at light load. At light load, this independence assumption breaks down since ONUs request small timeslots, just enough to fit a few packets. Thus, the timeslot size distribution has strong correlation with packet size distribution. The increased remainder size in the limited/priority service plot at light loads is another manifestation of the light-load penalty. A large lower-priority packet, which is continuously being preempted by small higher-priority packets, will result in a larger remainder for many cycles.

(a) Average cycle

(b) Normalized cycle

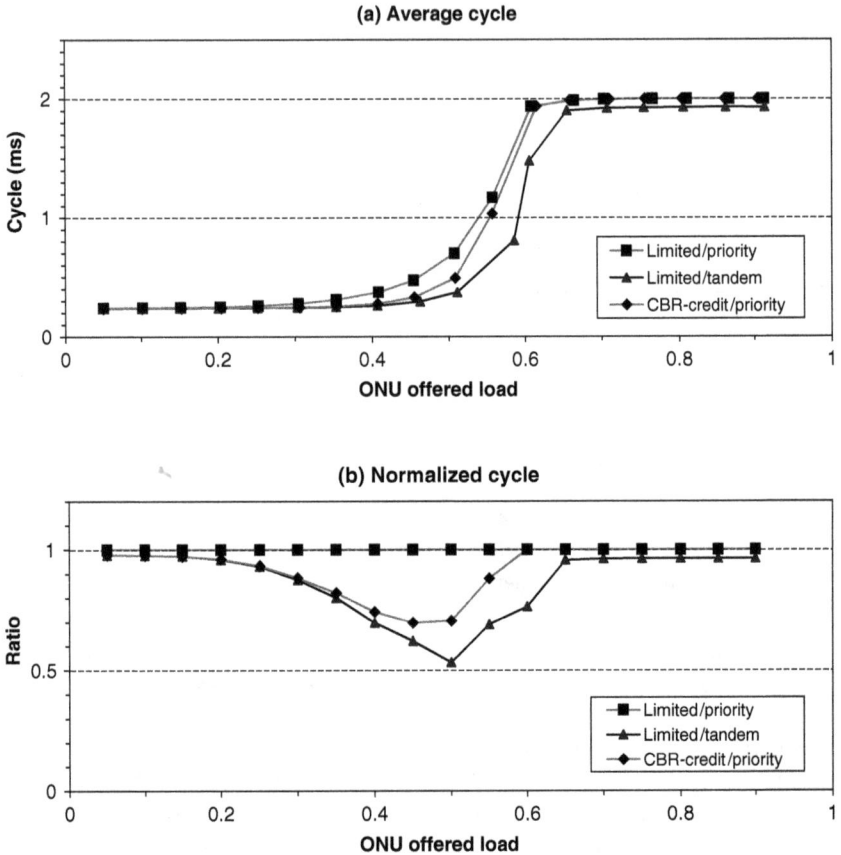

Figure 16.9 Average cycle time: (*a*) absolute value and (*b*) normalized to limited/priority service.

It is interesting to note that, in the CBR credit/priority scheme, despite granting a larger slot size to the ONUs, the cycle time is reduced. Figure 16.9 shows the average cycle time for all three schemes described above. At a load of 0.45, the average cycle time for the CBR credit scheme is 330 µs, a 30 percent reduction from 473 µs in the limited service scheme. This advantage is gained through the reduction in unused slot remainder, which results in increased network utilization. The limited/tandem system is found to have even larger cycle time reduction, because the unused remainder is always zero.

Both the cycle time and the slot remainder affect the bandwidth utilization in EPON. The following formula allows us to compute bandwidth utilization U:

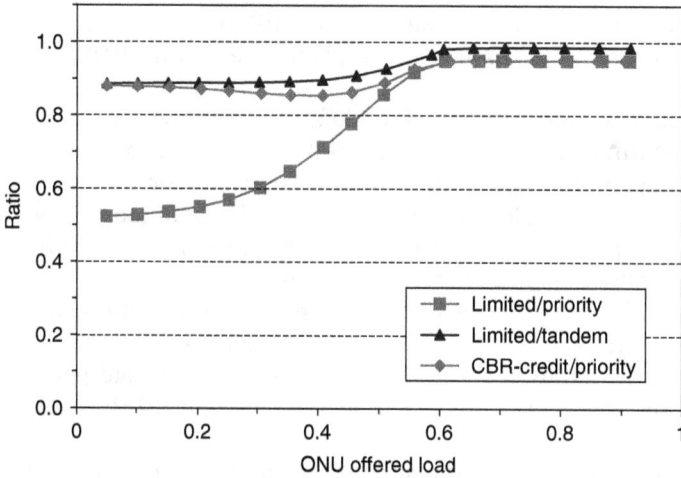

Figure 16.10   Bandwidth utilization.

$$U = 1 - \frac{N\left(GR_N + \overline{R} + W^{\text{REPORT}}\right)}{\overline{T}R_N} \qquad (16.6)$$

where

N	= number of ONUs
G	= guard time
$\overline{R}$	= average remainder
W^{REPORT}	= length of REPORT message
$\overline{T}$	= average cycle time
R_N	= EPON transmission rate (1 Gbps)

Based on Eq. (16.6), and the experimental results for $\overline{R}$ and $\overline{T}$, Fig. 16.10 presents the average bandwidth utilization values. We can see that both the limited/tandem scheme and the CBR credit/priority scheme result in considerable improvement in bandwidth utilization at light loads. The limited/tandem scheme also shows slightly better utilization at high loads due to its complete elimination of the unused remainder.

16.5   Summary

To successfully integrate Ethernet PONs into an access network environment, a whole host of technical issues need to be solved. This chapter focused on one important issue—support for differentiated classes of service. We limited our investigation to packet delay and bandwidth utilization characteristics.

We found that a combination of a default scheduling algorithm (priority scheduling) and MPCP (limited service) results in an interesting phenomenon where in some classes of traffic are treated unfairly when the network load is light. In fact, under light loads, ONUs with FCFS queue perform better than ONUs with priority scheduling. We called this phenomenon the light-load penalty.

To alleviate this penalty, we proposed and examined the characteristics of two optimization schemes with different tradeoffs. A limited/tandem (two-stage queuing) scheme eliminates the light-load penalty completely at the expense of increased packet delay for all the classes of traffic.

Another scheme (CBR credit/priority) attempts to predict high-priority packet arrivals. This scheme eliminates the light-load penalty for most packets (even though some low-priority packets are delayed excessively in a buffer, the number of such packets is small and does not affect the average packet delay). The limitation of this scheme is that some external knowledge of the traffic arrival process is needed.

Even though this chapter focused only on packet delay optimization and bandwidth utilization, they are by no means the only parameters affecting the network's performance. Other measures, such as guaranteed and best-effort bandwidth, jitter, and packet loss, are also important. The packet loss characteristic, while outside the scope of this chapter, is also an important performance parameter. The default packet discarding policy (drop-tail) shows that it is unfair to large packets; i.e., under heavy load, larger packets are more likely to not fit in the queue and be dropped (we call this the *large-packet penalty*). This behavior is understandable and even expected [this is one of the reasons for implementing the *random-early-discard* (RED) policy]. What is interesting in EPON settings is that the large-packet penalty affects higher-priority *P1* traffic more than it affects lower-priority *P2* traffic.

Being part of the IEEE 802 family of standards, EPON must be compliant with bridging defined in IEEE 802.1D, including compliance with CoS mechanisms in this standard. In this chapter, we focused only on strict priority scheduling because of its status as a default scheduling algorithm in IEEE 802.1D-compliant bridges and switches. Priority queuing is easy to implement. It provides low delay to high-priority traffic, but it may have some performance shortcomings such as better-than-needed performance for high-priority queues and starvation of low-priority queues.

There has been a large amount of research work done in developing scheduling algorithms with improved fairness in resource sharing (a family of *fair-queuing* protocols based on a concept of generalized processor sharing [PG93]). Integration of such schedulers in EPON is not a trivial task due to EPON's distributed nature and unique properties such as limited control-plane bandwidth, large propagation delay, and significant switching overhead. In Chap. 17, we investigate the feasibility of implementing a fair-queuing scheduling mechanism in EPON.

References

[802.1D]   *IEEE Standard for Information Technology - Telecommunications and information exchange between systems - Local and metropolitan area networks - Common specifications. Part 3: Media Access Control (MAC) Bridges*, ANSI/IEEE Standard 802.1D, 1998 edition. Available at http://standards.ieee.org/getieee802/download/802.1D-1998.pdf.

[BMP94]   L. Brakmo, S. O'Malley, and L. Peterson, "TCP Vegas: New techniques for congestion detection and avoidance," *Proceedings of the SIGCOMM '94 Symposium*, pp. 24–35, August 1994.

[CC96]   R. L. Carter and M. E. Crovella, *Measuring Bottleneck Link Speed in Packet-Switched Networks,* TR-96-006, Boston University Computer Science Department, March 15, 1996.

[CMT98]   K. Claffy, G. Miller, and K. Thompson, "The nature of the beast: Recent traffic measurements from an Internet backbone," in *Proceedings INET '98,* Geneva, Switzerland, July 1998. Available at http://www.isoc.org/inet98/proceedings/6g/6g_3.htm.

[DL86]   J. Diagle and J. Langford, "Models for analysis of packet voice communications systems," *IEEE Journal on Selected Areas of Communications,* vol. SAC-4, no. 6, pp. 847–855, September 1986.

[G.114]   ITU-T Recommendation G.114, *One-Way Transmission Time*, in Series G: Transmission Systems and Media, Digital Systems and Networks, Telecommunication Standartization Sector of ITU, May 2000.

[GW94]   M. W. Garrett and W. Willinger, "Analysis, modeling and generation of self-similar VBR video traffic," *Proceedings of ACM Sigcomm '94*, London, pp. 269–280, September 1994.

[PG93]   A. K. Parekh and R. G. Gallager, "A generalized processor sharing approach to flow control in integrated services networks: The single node case," *IEEE/ACM Transactions on Networking*, vol. 1, pp. 344–357, June 1993.

[SG01]   D. Sala and A. Gummalla, "PON functional requirements: Services and performance," presented at IEEE 802.3ah meeting in Portland, OR, July 2001. Available at http://www.ieee802.org/3/efm/public/jul01/presentations/sala_1_0701.pdf.

17

Objectives of EPON Scheduling Algorithm

In access network, an ONU may serve one or more subscribers and can have one or more queues assigned to each subscriber. Different queues belonging to one subscriber can be used, for example, to serve different classes of traffic (i.e., voice, video, and data) with different *quality-of-service* (QoS) guarantees. To satisfy the network requirements, a remote scheduler should meet the following objectives:

Scalability. The algorithm should support a large number of queues (several hundreds to several thousands). The algorithm should be efficient and scalable with the number of queues; i.e., overhead should not grow significantly with the number of queues served.

Guarantees. Unlike enterprise LANs, access networks serve noncooperative users; users pay for service and expect to receive their service regardless of the network state or the concurrent activities of other users. Therefore, the network operator must be able to guarantee a minimum bandwidth $B_i^{\min}$ to each queue i, assuming, of course, that the queue has enough data to send. Different services (queues) require different parameters. For example, voice service requires a delay bound of 1.5 ms [G.114], but needs only fixed and small bandwidth. Video traffic requires variable bandwidth, but can tolerate larger delay.

Fairness. To achieve multiplexing gain in EPON, excess bandwidth left by idle queues is redistributed among backlogged queues. The excess bandwidth B^{ex} should be allocated to backlogged queues in a fair and predictable manner, for example, in proportion to weights

φ assigned to each queue $(B_i^{ex} / \varphi_i = B_j^{ex} / \varphi_j)$. The fairness of bandwidth distribution should be preserved regardless of whether the queues are located in the same ONU or in different ONUs.

Isolation. A misbehaving user or application should not be able to disrupt the services of other users or applications. For example, if one subscriber generates a large volume of high-priority traffic, an EPON scheduler should be able to effectively isolate and limit this particular subscriber.

Robustness. A discrepancy may occur between OLT's knowledge of ONUs' states and actual ONUs' states, say, due to lost GATE or REPORT message. In case such a discrepancy occurs, the scheduling algorithm should not fatally fail. It should continue to function and return to its normal and efficient operation upon discrepancy removal.

In Chap. 16 we investigated EPON performance where intra-ONU scheduler implements priority queuing. It is generally known that in priority queuing, low-priority queues could starve. Previously, it was shown that with *preemptive* priority queuing, the lower-priority queues starve even under light load—a phenomenon called light-load penalty.

The problem of queue starvation at light loads can be mitigated by using a *nonpreemptive* priority queuing in ONUs. In nonpreemptive queuing, ONUs can transmit only previously reported packets, even if more higher-priority packets arrive after the last REPORT was sent. Nonpreemptive queuing can be implemented, for example, as a tandem queue (Sec. 16.4.1) or using pointers to the last reported packet in each queue. However, even in a nonpreemptive queuing system, under high load, the lower-priority queues would starve.

If an ONU, serving multiple users, implements a priority (CoS-based) queuing, then each of multiple users' data flows will be mapped to one of eight priority classes. A user whose traffic maps to a low-priority queue may get no service, while the user with high-priority traffic may get all the available bandwidth. Thus, no guarantees can be provided by CoS-based schedulers.

The fairness cannot be enforced either. Consider a simple example, where ONU k and ONU j both report same-priority queues with 10,000 bytes. In ONU k these bytes were generated by four users, each one contributing an equal share of 2500 bytes. In ONU j, on the other hand, the entire queue may be filled just by one user. If the OLT has available bandwidth for only 10,000 bytes, a fair scheduler would allocate 8000 bytes to ONU k and 2000 bytes to ONU j, so that each user gets a fair share of 2000 bytes. However, not knowing the exact composition of each queue (i.e., which user has contributed how many bytes), the scheduler is not able to make a correct decision. Allocating equal shares of 5000 bytes to both ONUs would give undeserved advantage to the user in ONU j.

It is clear that CoS-based schedulers cannot provide bandwidth guarantees, isolation, or fairness for different traffic flows. Using a priority queuing scheme, the scheduler *has no means to limit misbehaving high-priority queue without completely starving all lower-priority queues and cannot ensure fairness among same-priority traffic flows.*

To fix this problem and provide service protection (isolation) from misbehaving applications, ONU should implement some kind of ingress shaping on a per-flow basis. However, considering that available bandwidth depends on the state of *all ONUs* in EPON, such an ONU ingress shaping function should have a global view of the EPON. Without such global instantaneous feedback, ingress shapers have no choice but to trim incoming traffic to minimum rates guaranteed to each flow, even if excess bandwidth is available in the EPON. Not being able to utilize the excess bandwidth eliminates *multiplexing gain*—one of the main advantages of EPON over alternative subscriber access architectures. Furthermore, the absence of a standard protocol to dynamically control parameters of ingress shapers will necessitate proprietary solutions with detrimental effect on interoperability of EPON devices.

17.1 A Formal Definition of Fairness

As stated above, the objective of an EPON scheduler is to guarantee minimum service B^{min} to each queue and fairly share the excess service B^{ex}. Typically, bandwidth is distributed in timeslots; i.e., a timeslot (transmission window) of size W bytes is given to a queue once every T s (T is called the *cycle time*). Thus, it is convenient to define the minimum (guaranteed) timeslot size W_i^{min} that should be given to a queue i to guarantee its minimum bandwidth B_i^{min} ($W_i^{min} = B_i^{min}T$). We also define W^{cycle} as the total available service in one cycle time T (i.e., the number of bytes that can be transmitted in time T). Clearly, to guarantee minimum bandwidth, the sum of all W_i^{min} should not exceed W^{cycle}. The actual minimum slot size that a queue i gets in cycle k is $w_{i,k}^{min}$:

$$w_{i,k}^{min} = min \begin{cases} q_{i,k} \\ W_i^{min} \end{cases} \tag{17.1}$$

where $q_{i,k}$ is the length of queue i at the beginning of cycle k. Equation (17.1) states that a queue should never be given a slot larger than the amount of data the queue has accumulated. Total remaining transmission window size (excess bandwidth) w_k^{ex} left in cycle k after assigning all minimum slots to all the queues is equal to

$$w_k^{ex} = W^{cycle} - \sum_{i=1}^{N} w_{i,k}^{min} \tag{17.2}$$

We define a backlogged queue to be a queue that cannot or will not be served to exhaustion in one cycle (one queue transmission). The set of all queues backlogged in cycle k is denoted by Ω_k. Each queue i that remains backlogged after serving $w_{i,k}^{min}$ bytes (i.e., with $q_{i,k} > w_{i,k}^{min}$) should get a share of the excess bandwidth w_k^{ex} proportional to its weight φ_i, or

$$w_{i,k}^{ex} = w_k^{ex} \frac{\varphi_i}{\sum_{j \in \Omega_k} \varphi_j} \tag{17.3}$$

The subtle problem with the definition in Eq. (17.3) arises due to the fact that a queue shall not be given more slot size than it has data to transmit. Thus, if the guaranteed slot $w_{i,k}^{min}$ together with excess slot $w_{i,k}^{ex}$ exceeds the queue length $q_{i,k}$, the queue will be given a slot size equal to $q_{i,k}$. This means that queue i will be served to exhaustion, and thus it should not be considered a backlogged queue any more. Removing queue i from the set Ω_k of backlogged queues will affect the amount of remaining excess bandwidth as well as the share of each queue that remains backlogged. To capture this effect, we amend Eq. (17.3) as follows:

$$w_{i,k}^{ex} = \begin{cases} q_{i,k} - w_{i,k}^{min} & j \notin \Omega_k \\ \left(W^{cycle} - \sum_{j \in \Omega_k} W_i^{min} - \sum_{j \notin \Omega_k} q_{j,k} \right) \dfrac{\varphi_i}{\sum_{j \in \Omega_k} \varphi_j} & j \in \Omega_k \end{cases} \tag{17.4}$$

Equation (17.4) says that either the excess slot size given to a queue will be just enough to serve the queue to exhaustion (if queue i does not belong to a set of backlogged queues), or the queue will be served in proportion to its weight φ_i and the total number of bytes remaining available after serving to exhaustion all nonbacklogged queues and assigning the minimum guaranteed timeslots to all backlogged queues.

Whew! It is important to understand that this is a recursive definition, since the membership in set Ω_k is determined as $i \in \Omega_k$ iff $w_{i,k}^{min} + w_{i,k}^{ex} < q_{i,k}$.

Summing Eq. (17.1) and Eq. (17.4), we get the total timeslot size $w_{i,k}$ given to a queue i in cycle k:

$$
w_{i,k} = w_{i,k}^{min} + w_{i,k}^{ex}
$$

$$
= \begin{cases} q_{i,k} & i \notin \Omega_k \\[2em] W_i^{min} + \left(W^{cycle} - \displaystyle\sum_{j \in \Omega_k} W_j^{min} - \sum_{j \notin \Omega_k} q_{j,k} \right) \dfrac{\varphi_i}{\displaystyle\sum_{j \in \Omega_k} \varphi_j} & i \in \Omega_k \end{cases} \tag{17.5}
$$

Finally, the cumulative size of all slots assigned in one cycle cannot exceed the cycle capacity W^{cycle}. The cumulative slot size also cannot exceed the sum of all queue lengths (i.e., in the case when all the queues can be served to exhaustion in one cycle). This is reflected in Eq. (17.6):

$$
\sum_{i=1}^{N} w_{i,k} = \min \begin{cases} \displaystyle\sum_{i=1}^{N} q_{i,k} \\[1.5em] W^{cycle} \end{cases} \tag{17.6}
$$

It is easy to verify that Eq. (17.5) summed for all queues indeed complies with the requirement in Eq. (17.6).

A solution to a system of equations described by Eq. (17.5) constitutes a valid schedule compliant with the requirements for guaranteeing the minimum bandwidth and fairly sharing the excess bandwidth.

By specifying the minimum bandwidth B^{min} (or minimum timeslot W^{min}) and the weight φ per connection, the network operator can provision different types of services to subscribers (queues).

17.2 Fair Schedulers

Over the past 20 years, a lot of attention has been given to the problem of fair scheduling and fair resource allocation. One solution for this problem is *generalized processor sharing* (GPS) [PG93], an idealistic fluid model supporting fair resource sharing. Many practical algorithms were derived from the GPS model to support fair queuing in systems with atomic protocol data units (i.e., nondivisible cells or packets): *weighted fair queuing* (WFQ) [DKS90], *worst-case fair weighted fair queuing* (WF^{2}Q) [BZ96], *self-clocked fair queuing* (SCFQ) [Gol94], *start-time fair queuing* (STFQ) [GVC97], and many others. These algorithms were shown to distribute excess bandwidth among the queues almost fairly; i.e., at any moment of time, the amount of service a queue received would differ from an ideal fluid model by at most one maximum-size data unit (packet or cell).

However, among the multitude of existing fair scheduling algorithms, not many are suitable for EPON or for a remote scheduling system, in general. We define a *remote scheduling system* as a scheduling (resource-sharing) domain in which the queues (customers) and the scheduler (server) are located at a large distance from one another. EPON is just one example of a remote scheduling system; other examples include wireless (cellular) or cable TV networks. The properties of a typical remote scheduling system such as *significant queue switch-over overhead, large control-plane propagation delay,* and *limited control-plane bandwidth* do not allow easy adaptation of existing scheduling algorithms.

Significant queue switch-over overhead. When switching from one ONU to another, the receiver may need some additional time to readjust the gain since the power levels received from ONUs are different. In addition, the ONUs are required to keep lasers turned off between the transmissions (see Sec. 12.2.2). Turning a laser on and off is not an instantaneous process and will also contribute to the switch-over overhead. This leads to a significant overhead when switching from one ONU to another. For example, in EPON, the worst-case overhead is 2 µs.[1]

Large control-plane propagation delay. Control-plane delay is negligible for local schedulers (system-in-a-chip architectures or when queues and the scheduler are connected through a back plane). But in a remote scheduling system, the physical distances can be large, and delay can exceed by many times the packet transmission time. In addition, in systems like EPON, the control messages are in-band and can be transmitted only in a previously assigned timeslot. Thus, the control message delay increases even more, now due to waiting for the next timeslot to arrive. This results in the scheduler always operating with somewhat outdated information.

Limited control-plane bandwidth. Scheduling multiple clients (queues, ONUs) may require a separate control message to be sent

---

[1]The 2-µs time interval includes dead zone, laser on/off, automatic gain control (AGC), and clock-and-data recovery (CDR) times, and assumes no laser on/laser off overlap. Also see Sec. 12.2.2.

periodically from the scheduler located at the OLT to each client (GATE message) and from each client to the scheduler (REPORT message). Increasing the number of clients may give rise to scalability issues when a significant fraction of the total EPON bandwidth is consumed by the control messages.

Keeping in mind the above constraints and the objectives listed in the beginning of this chapter, we take a look at two configuration alternatives of the EPON scheduler: *direct* versus *hierarchical* scheduler.

17.2.1 Direct (single-level) schedulers

Applying a single-level algorithm to an EPON means that a scheduler located in the OLT would receive information from and individually schedule each consumer (queue) located in multiple ONUs (Fig. 17.1). Since the information about each individual queue is collected in one place, the centralized scheduler can easily ensure that the required service guarantees are preserved and that the excess bandwidth (if any) is fairly divided among backlogged queues.

The simplest approach to implement a direct scheduler is to allocate a separate logical link to each queue. This will eliminate any need for

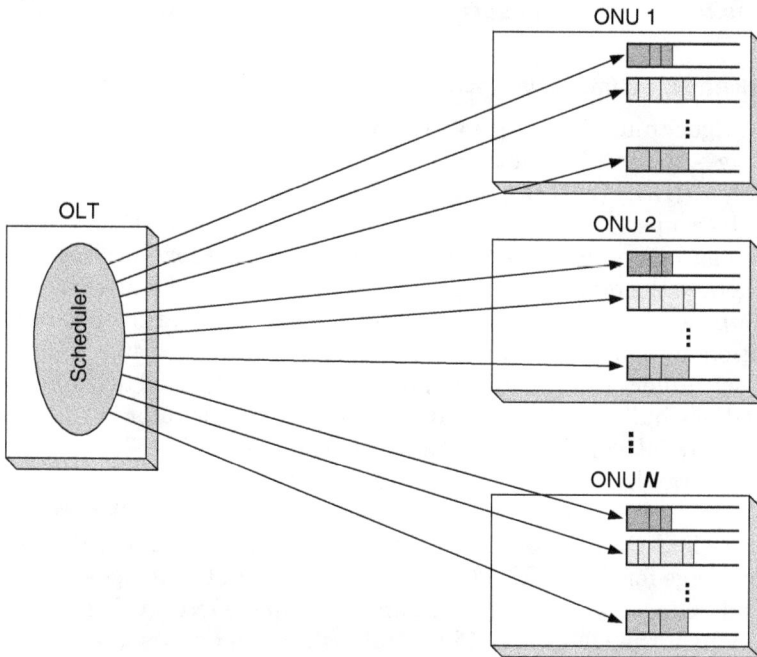

Figure 17.1   Direct (single-level) scheduling in EPON.

low-level scheduler or ingress shapers in the ONU and will concentrate all the intelligence in the OLT.

The OLT will receive a separate REPORT message from each individual LLID representing just one queue. Since the OLT issues a separate GATE message for each LLID, it can easily limit one queue while giving more excess bandwidth to another queue. The ONU in this case becomes very simple.

Such schemes were in fact implemented and work reasonably well if the number of queues per ONU is not very large. For systems with a large number of queues, the scheduling overhead becomes an issue, since a separate control message is required for each queue, instead of one message per ONU. Consider an EPON system with 32 ONUs, 64 subscribers per ONU, and 3 queues per subscriber, for a total of 6144 queues. This adds considerable overhead for control messages. For example, assuming that one-third of all queues are used for voice traffic with a delay bound of 1.5 ms [G.114], the OLT should be able to generate 2048 GATE messages within a 1.5-ms interval. But at a 1 Gbps EPON rate, it takes 1.38 ms to transmit this many GATE messages, which leaves almost no bandwidth for voice traffic itself. In addition, in the upstream channel, the number of guard bands may grow with the number of LLIDs, consuming even more bandwidth. Based on this observation, we conclude that, *in EPON, direct schedulers are non-scalable with the number of queues.*

17.2.2 Hierarchical (multilevel) schedulers

Several algorithms have been developed to support hierarchical scheduling—*hierarchical fair queuing* (HFQ) [BZ97], *hierarchical round-robin* (HRR) [KKK90], etc. In such schemes, all queues are divided into groups. The high-level scheduler schedules the groups (i.e., provides aggregated bandwidth per group) while the low-level schedulers schedule the queues within each group. The root scheduler treats each group as one consumer and has no information about the internal composition of each group. EPON can be naturally divided into a hierarchy of schedulers where the high-level scheduler is located in the OLT and schedules individual ONUs, and the low-level scheduler is located in each ONU and schedules queues within the ONU (Fig. 17.2).

In the hierarchical EPON scheduling scheme, the root scheduler (OLT) only schedules the intermediate nodes (ONUs). The OLT would receive one REPORT message from an ONU and would generate one GATE message for the ONU. The GATE and REPORT messages would grant and request an aggregated bandwidth per ONU (i.e., a large timeslot which the ONU would internally share among its queues). A hierarchical scheme solves the scalability issue due to elimination of

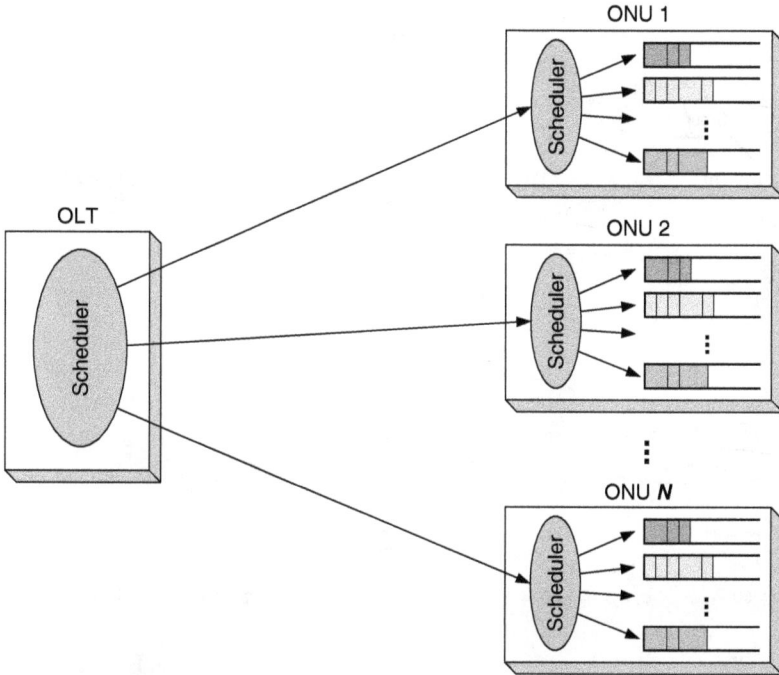

Figure 17.2 Hierarchical scheduling in EPON.

separate GATE and REPORT messages for each queue. It also solves the switch-over overhead issue due to the fact that all queues in one ONU are served consecutively with no guard times between their transmissions. (Guard times remain only when the OLT switches to serve the next ONU.)

17.2.2.1 Sibling-fair versus cousin-fair schedulers.

The challenging issue with a hierarchical scheme is to support fair resource distribution among queues in different groups (ONUs). Most hierarchical scheduling protocols allow fairness only among *siblings* (i.e., nodes having the same parent). We call such schedulers *sibling-fair* or *locally fair* schedulers. Figure 17.3 illustrates bandwidth distribution among 5 queues separated into two groups, A and B ($A = \{1, 2\}$ and $B = \{3, 4, 5\}$). Each queue i is characterized by its weight φ_i and its size (unfinished work) q_i. The amount of service each queue gets is denoted by w_i.

In Fig. 17.3a, the scheduler makes its bandwidth allocation decision based on the cumulative weight $\Sigma\varphi$ of each group. Queue 5 has less unfinished work and thus requires less service. Unused service left by queue 5 is distributed among its siblings 3 and 4, assuming schedulers at each level are work conserving. It can be easily observed that siblings

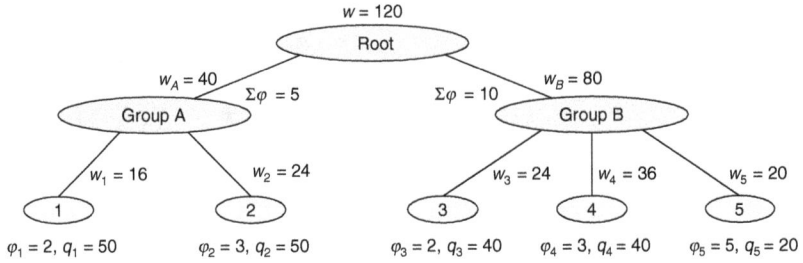

(a)

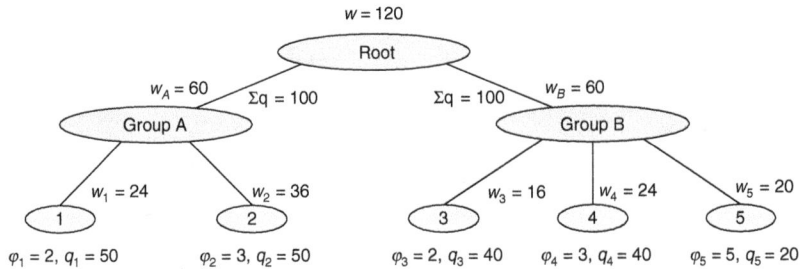

(b)

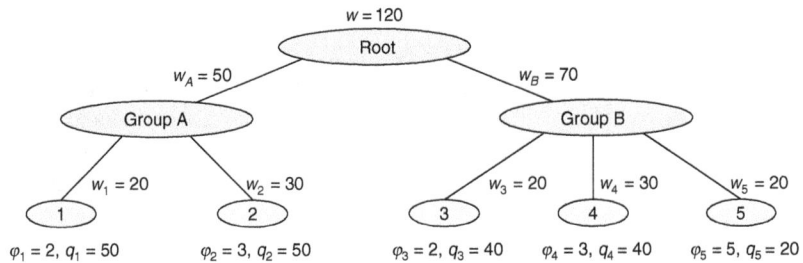

(c)

Figure 17.3 (*a*) Sibling-fair scheduling based on cumulative group weight $\Sigma\varphi$; (*b*) sibling-fair scheduling based on cumulative group work Σq; (*c*) cousin-fair scheduling.

at each level receive mutually fair service w; that is, for any queues i and j with sufficient amount of unfinished work q ($q_i \geq w_i$), the bandwidth is allocated in proportion to their weights $w_i / \varphi_i = w_j / \varphi_j$. However, the fairness does not extend across multiple groups. For example, we can see that $w_1 / \varphi_1 = w_2 / \varphi_2$ and $w_3 / \varphi_3 = w_4 / \varphi_4$, but $w_1 / \varphi_1 \neq w_3 / \varphi_3$ and $w_2 / \varphi_2 \neq w_4 / \varphi_4$. Similar outcome is observed if the service is distributed based on any other single value, e.g., the cumulative amount of unfinished work in a group, as shown in Fig. 17.3*b*.

Figure 17.3c illustrates the desired service distribution that achieves fairness among all leaves with sufficient unfinished work ($w_1/\varphi_1 = w_2/\varphi_2 = w_3/\varphi_3 = w_4/\varphi_4$). We call this scheme *cousin-fair* (or *globally fair*) scheduling, in contrast to the sibling-fair scheme described above. This scheduler does not provide fairness among intermediate nodes, but allows fairness among all leaves, no matter which group they belong to. Figure 17.3c shows that the bandwidth allocated to an intermediate node should dynamically change based on the state of all the leaves. While cousin fairness is illustrated here for a two-level system, it is easy to generalize the concept to a hierarchical scheduling system with an arbitrary number of levels.

We note that Rexford et al. [RGB96] reported a dynamic weight adjustment scheme that allows a hierarchical scheduler to be cousin-fair. In their algorithm, a root-level scheduler receives from each group a cumulative weight as a sum of weights of all nonempty queues in a group; i.e., it is the scheme shown in Fig. 17.3a. As soon as a busy queue becomes empty or an empty queue becomes busy, the root scheduler should learn the new weight. Thus, this algorithm is only suitable for systems with a small propagation delay and not for a remote scheduling system such as EPON.

17.3 Summary

On one hand, it seems obvious that the EPON scheduling algorithm should be hierarchical to be able to support a large number of independent subscribers, while remaining efficient in the presence of large delays, limited control-plane bandwidth, and large switch-over overhead. The main property of any hierarchical resource allocation algorithm is that the amount of information which each arbiter processes is proportional only to number of children of this node, and does not depend on the number of consumers at the bottom of the hierarchy.

On the other hand, to allow fairness among all the end consumers, the root scheduler should receive information from each consumer (especially if service consists of guaranteed and excess proportional-share parts, as we have stated in Sec. 17.1); i.e., it should be direct (nonhierarchical).

In Chap. 18, we will investigate a novel algorithm that successfully achieves both goals: it is hierarchical (each node knows only its immediate children) and it is cousin-fair, i.e., provides fairness among each end consumer.

References

[BZ96]   J. C. R. Bennett and H. Zhang, "WF^{2}Q: Worst-case fair weighted fair queuing," *Proceedings of INFOCOM `96*, pp. 120–128, San Francisco, March 1996.

[BZ97]   R. Bennett and H. Zhang, "Hierarchical packet fair queuing algorithms," *IEEE/ACM Transactions on Networking*, vol. 5, no. 5, pp. 675–689, October 1997.

[DKS90]   A. Demers, S. Keshav, and S. Shenker, "Analysis and simulation of a fair queuing algorithm," *Journal of Internetworking Research and Experience*, pp. 3–26, October 1990.

[G.114]   ITU-T Recommendation G.114, *One-Way Transmission Time*, in Series G: Transmission Systems and Media, Digital Systems and Networks, Telecommunication Standartization Sector of ITU, May 2000.

[Gol94]   S. J. Golestani, "A self-clocked fair queuing scheme for broadband application," *Proceedings of IEEE INFOCOM 94*, pp. 636–646, Toronto, Canada, June 1994.

[GVC97]   P. Goyal, H. M. Vin, and H. Cheng, "Start-time fair queuing: A scheduling algorithm for integrated services packet switching networks," *IEEE/ACM Transactions on Networking*, vol. 5, no. 5, pp. 690–704 October 1997.

[KKK90]   C. Kalmanek, H. Kanakia, and S. Keshav, "Rate controlled servers for very high-speed networks," *Proceedings of IEEE Globecom*, pp. 1264–1273, San Diego, December 1990.

[PG93]   A. K. Parekh and R. G. Gallager, "A generalized processor sharing approach to flow control in integrated services networks: The single node case," *IEEE/ACM Transactions on Networking*, vol. 1, pp. 344–357, June 1993.

[RGB96]   J. L. Rexford, A. G. Greenberg, and F. G. Bonomi, "Hardware-efficient fair queuing architectures for high-speed networks," *Proceedings of INFOCOM*, pp. 638–646, 1996.

18

Cousin-Fair Hierarchical Scheduling in EPON

In Chap. 17 we found that the EPON scheduler faces two conflicting requirements: on one hand, the algorithm should be hierarchical to be scalable; on the other hand, it should be direct to be globally fair (or cousin-fair) to all subscribers. In this chapter, we will investigate a new algorithm, called *fair queuing with service envelopes* (FQSE), which successfully achieves both goals: it is hierarchical (each node knows only its immediate children) and it is cousin-fair. FQSE can be generalized to various remote scheduling systems, e.g., wireless networks or coax-tree networks; however, our focus will be on its application to EPON.

18.1 Fair Queuing with Service Envelopes

FQSE is a hierarchical remote scheduling algorithm that distributes service in accordance with Eq. (17.5). The algorithm is based on a concept of a *service envelope* (SE). A service envelope represents the amount of service (timeslot size) given to a node as a function of some nonnegative value which we call the *satisfiability parameter* (SP). SP is a measure of how much the demand for bandwidth can be satisfied for a given node.

In a scheduling hierarchy, each node has its associated SE function. We distinguish the construction of a service envelope for a leaf (denoted E^*) from the construction of a service envelope for a nonleaf node (denoted E).

Envelope E^* is a piecewise linear function consisting of at most two segments (Fig.18.1, plots E_2^* and E_3^*). The first segment begins at a point with coordinates $\left(0,\ w_{i,k}^{min}\right)$ and ends at $\left((q_{i,k} - w_{i,k}^{min})/\varphi_i,\ q_{i,k}\right)$.[1] The ending SP value is chosen such that the slope of the first segment is exactly φ_i. The second segment has a slope equal to 0 and continues to infinity.

Intuitively, the meaning of the E^* function should be clear: as the satisfiability parameter changes, the E^* function determines the fair timeslot size for the given queue. In the worst case (i.e., when SP = 0), an exact $w_{i,k}^{min}$ -byte timeslot will be given to the queue (i.e., the queue will get its guaranteed minimum service). As the satisfiability parameter s increases, the queue will be given an additional timeslot (excess bandwidth) equal to $\varphi_i s$. When the timeslot size reaches $q_{i,k}$ (total queue length), it will not increase anymore, even if s increases. In case a queue has less data than its guaranteed slot size (that is, $q_{i,k} < W_i^{min}$), the E^* function will consist of only one segment with slope 0 (Fig. 18.1, plot E_1^*).

The service envelope E_i of a nonleaf node i is built as a sum of service envelopes of all the node's children:

$$E_{i,k} = \sum_{j \in D_i} E_{j,k} \tag{18.1}$$

where D_i is a set containing all children of node i. This is illustrated in Fig. 18.1, plot E_4.

We note two important properties of service envelopes. (The formal proofs of these properties can be found in [KMS+03].)

Property 1. The slope δ of any segment of a service envelope is always bounded by $0 \le \delta \le \Phi$, where Φ is the cumulative weight of all leaves $\Phi = \sum_{\text{all leaves}} \varphi$.

Property 2. Slopes of segments of any service envelope are decreasing; i.e., for any segments i and j, $\delta_i > \delta_j$ iff $i < j$.

The FQSE algorithm consists of alternating *requesting* and *granting* phases. The following are the steps of the algorithm.

18.1.1   Phase 1—Requesting service

At the end of transmission in a previously assigned timeslot, a node should generate a new service envelope and send it to its parent in a request message. After collecting service envelopes from all its children,

---

[1] Here, we use the same notation as in Sec. 17.1.

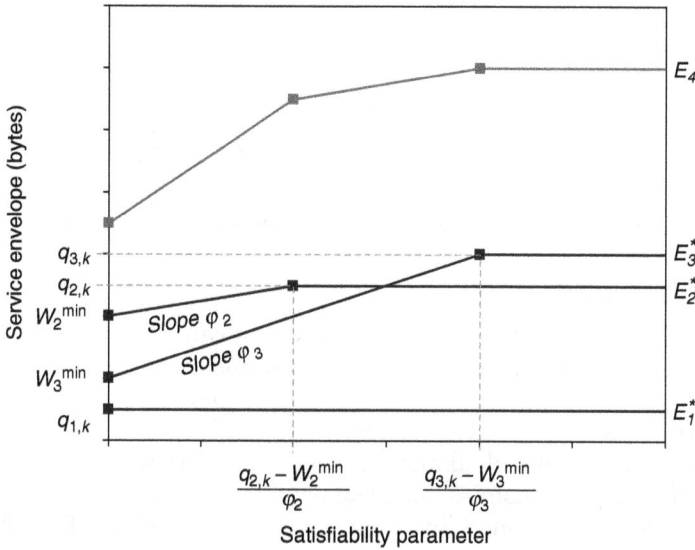

Figure 18.1   Construction of service envelopes.

an intermediate node would generate its own service envelope by summing all received envelopes. The intermediate node then sends this new service envelope to its parent.

A request message has a limited length and may contain at most K point coordinates representing knots of the piecewise linear service envelope. Since the number of children of any intermediate node i can be arbitrarily large, the service envelope of node i may contain an arbitrarily large number of points. In case the actual number of points $m_{i,k} > K$, node i will perform a piecewise linear approximation of the function $E_{i,k}$, such that the $m_{i,k}$-point function is described with only K points and can be transmitted in one request message. (The approximation procedure is described in Sec. 18.1.3.) The approximated function is denoted $\widetilde{E}_{i,k}$. Thus, Eq. (5.7) can be rewritten as

$$E_{i,k} = \sum_{j \in D_i} \widetilde{E}_{j,k} \tag{18.2}$$

where $\widetilde{E}_{j,k}$ is the approximated service envelope of the jth child of node i in cycle k.

The requesting phase ends when the root node receives service envelopes from all its children and calculates its own service envelope $E_{0,k} = \sum_{j \in D_0} \widetilde{E}_{j,k}$.

```
PROCESS_GRANT( t_{i,k}, s_k )

1       t = t_{i,k}
2       for each node j in D_i
        {
3           t_{j,k} = t
4           send grant ( t_{j,k}, s_k ) to node j
5           tx_time = time(Ẽ_{i,k}(s_k)) // transmission time needed to transmit Ẽ_{i,k}(s_k) bytes
6           t = t + tx_time
        }
```

Figure 18.2   PROCESS_GRANT procedure calculates start times for all children of node i.

18.1.2   Phase 2—Granting service

The root node knows the total number of bytes that can be transmitted in one cycle (W^{cycle}). When the root scheduler obtains the $E_{0,k}$ function in cycle k, it calculates the satisfiability parameter s_k as $E_{0,k}(s_k) = W^{\text{cycle}}$. Knowing the cycle start time and the satisfiability parameter s_k, the root node calculates the timeslot start time $t_{j,k}$ for each child $j \in D_0$ such that transmissions from each child will not overlap. This calculation is performed by a procedure PROCESS_GRANT ($t_{i,k}, s_k$) shown in Fig. 18.2. The timeslot start time $t_{j,k}$ and the satisfiability parameter s_k are then transmitted to each child j in a grant message.[2]

Upon receiving the grant message, each intermediate node invokes the same PROCESS_GRANT($t_{i,k}, s_k$) procedure to further subdivide the timeslot among its children. As an example, Fig. 18.3 illustrates the transmission start times and timeslot sizes calculated by the PROCESS_GRANT function for the scheduling hierarchy shown in Fig. 17.3.

The granting phase ends with each leaf node receiving the grant message. When leaf node i receives the grant message containing the timeslot start time $t_{i,k}$ and the satisfiability parameter s_k, it will calculate its own timeslot size $w_{i,k} = E^*_{i,k}(s_k)$. When the local clock in the leaf node reaches time $t_{i,k}$, the leaf node starts transmission and transmits $w_{i,k}$ bytes of data.[3]

18.1.3   Service envelope approximation schemes

In the requesting phase (phase 1), each intermediate node i will collect service envelopes from all its children and create service envelope $E_{i,k} = \sum_{j \in D_i} \widetilde{E}_{j,k}$, which it will send to its parent in a request control

---

[2] The timeslot start time in a grant message should be "precompensated" for the round-trip propagation delay as explained in Sec. 5.3.3.1.

[3] The actual transmission size may be less than $w_{i,k}$ bytes for the case of packet-based networks. The necessary adaptation mechanisms for variable-size, packet-based schedulers are discussed in Sec. 18.2.

Cycle start t_k

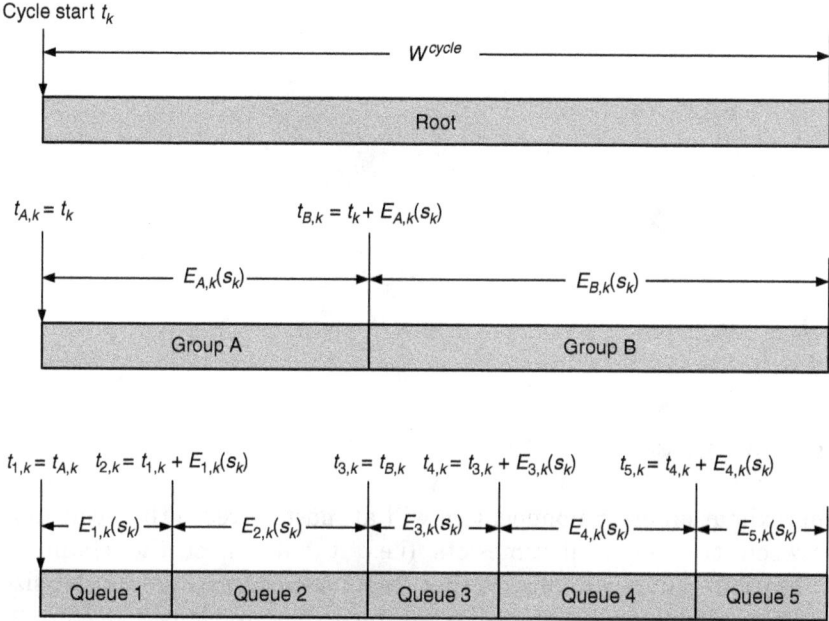

Figure 18.3 Granting phase for the scheduling hierarchy shown in Fig. 17.3c.

message. A request message can only accommodate a fixed number of points K. If node i has $|D_i|$ children, the $E_{i,k}$ function may have $m_{i,k}$ points: $1 \leq m_{i,k} \leq |D_i| \times (K-1) + 1$ (the first point always has SP = 0 and will coincide for all children). If the actual number of points $m_{i,k} > K$, node i will perform a piecewise linear approximation of the $E_{i,k}$ function, such that the $m_{i,k}$-point SE function is described with only K points. We denote the approximated function by $\widetilde{E}_{i,k}$.

The significance of the approximation step lies in the fact that it makes the FQSE a truly hierarchical algorithm: the amount of information that the scheduler at each level of the hierarchy has to process is proportional to only the number of children of this node and does not depend on the number of consumers at the bottom of the hierarchy. In EPON settings, this means that the OLT has to process one message per ONU, no matter how many queues this ONU has.

In performing the approximation, we set our objective at minimizing the maximum error [minimize $\max_{\forall s} |\widetilde{E}_{i,k}(s) - E_{i,k}(s)|$]. We also require that the error be nonnegative [$\widetilde{E}_{i,k}(s) \geq E_{i,k}(s), \forall s$]. It is easy to see that allowing a negative error would mean that a timeslot granted by a parent to a child [$\widetilde{E}_{i,k}(s)$] could be smaller than the timeslot assumed by the child [$E_{i,k}(s)$]. That may cause a collision with the data transmitted by some other child of parent node. Keeping the

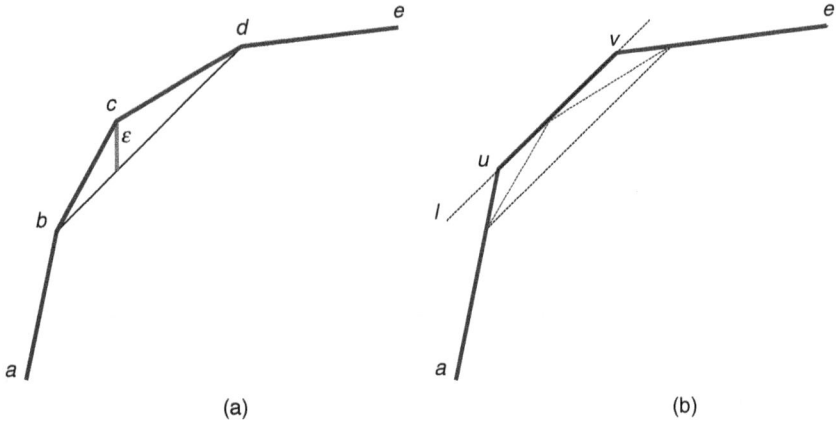

Figure 18.4   Approximation of SE function in Min-Error approach.

approximation error nonnegative will at most increase the dead zone between two adjacent timeslots (i.e., it will affect the channel's utilization), but it will ensure that each node can get its fair slot size and that no data collisions will occur due to slot overlaps.

We consider two alternative approximation schemes: *Min-Error* and *Min-Points*.

18.1.3.1 Min-Error approach.   In the Min-Error scheme, to perform the SE function approximation, we eliminate points one at a time, until only K points are left. The SE function is represented as an array of point coordinates $(s, E(s))$. Here are the steps of the approximation procedure:

Step 1.   By scanning through the original array find two adjacent segments (bc) and (cd) that have the smallest error ε, as illustrated in Fig. 18.4a.

Step 2.   Through point c, draw a line l parallel to segment bd.

Step 3.   Add points of intersection of line l with lines ab and de (points u and v) as illustrated in Fig. 18.4b.

Step 4.   Remove points b, c, and d.

Step 5.   Repeat steps 1 to 4 until no more than K points are left.

From Fig. 18.4, it is clear that the approximation error is always positive. However, this heuristic approach does not result in an optimal solution. Finding a pair of minimum-error segments does not count the error that a segment may have accumulated during previous steps. Estimation of the average error is not trivial, since the error is not uniformly distributed through the domain of the SE function.

To analyze the time complexity of the Min-Error approach, we observe that step 1 can be performed in $O(m_{i,k})$ time, where $m_{i,k}$ is the number of points in the original service envelope. The algorithm will iterate through each point on the curve while measuring the error ε, and will locate the point for which ε is minimum. Moreover, each time we walk through steps 1 to 4, the new approximated curve has 1 point less than the previous curve. Steps 1 through 4 are, therefore, carried out exactly $m_{i,k} - K$ times. Therefore, the running time of this scheme is

$$O\left(m_{i,k} + (m_{i,k} - 1) + (m_{i,k} - 2) + \cdots + \left[m_{i,k} - (m_{i,k} - K)\right]\right) = O\left(m_{i,k}^2 - K^2\right)$$

18.1.3.2 Min-Points approach. The Min-Points approach employs two functions MIN_POINT_APPROX( ) and CONSTRUCT_APPROX(E, ε). Given a fixed error ε and the original service envelope E, the CONSTRUCT_APPROX(E, ε) function tries to construct an approximated piecewise linear function $\widetilde{E}$ which has the minimum number of points, as explained in the following steps:

Step 1.   Construct a piecewise linear curve U (Fig. 18.5), all the points of which have an error ε from the original service envelope.

Step 2.   Draw a line L which coincides with the last segment of the original envelope E (segment gh in Fig. 18.5).

Step 3.   Extend the first segment ab until it intersects the curve U or the line L, whichever appears first. Call the point of intersection $\widetilde{a}$.

Step 4.   From $\widetilde{a}$, draw a tangent to the original service envelope E, so that it intersects E at point d. (It may happen that the tangent has the same slope as one of the segments, in which case the tangent intersects E at multiple points.)

Step 5.   Extend the tangent $\widetilde{a}d$ until it intersects the curve U or the line L again. Call the new intersection point $\widetilde{b}$.

Step 6.   Repeat steps 4 and 5 for each new point of intersection with curve U or line L until we find K such new points or we reach the last point of the original curve E (point h in the example on Fig. 18.5).

The procedure CONSTRUCT_APPROX(E, ε) returns SUCCESS if the approximation envelope can be constructed with K points (i.e., if the last point of E has been reached) and FAILURE, otherwise (Fig. 18.6). We considered two implementations of the CONSTRUCT_APPROX procedure that differ in their computational complexity: CONSTRUCT_APPROX_LINEAR(E, ε) and CONSTRUCT_APPROX_BINARY(E, ε).

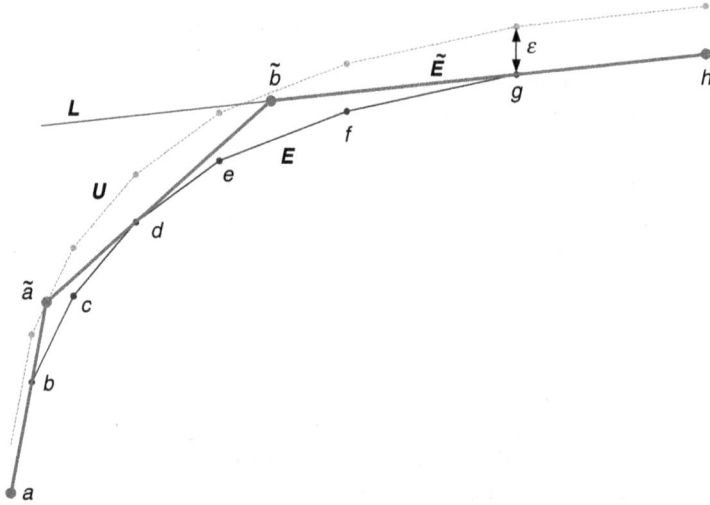

Figure 18.5   Approximating the service envelope E using the Min-Points approach.

The CONSTRUCT_APPROX_LINEAR(E, ε) procedure is shown in Fig. 18.6. It first linearly iterates through the points of the original service envelope E, searching for the point that lies on the tangent line, denoted `TangentLine`. Using function TEST_TANGENT(*Point*, *n*) (Fig. 18.7), every point E_n is tested on whether it belongs to `TangentLine` or not [i.e., whether the line *Point E_n* is tangential to E]. By relying on property 2 of service envelopes (Sec. 18.1), this test can be performed in $O(1)$ time. Indeed, since the slopes in the service envelope are strictly decreasing, the line *Point E_n* will be tangential to E if and only if the slope of $(E_{n-1}\ E_n)$ is larger than or equal to the slope of *Point E_n* and the slope of line $E_n\ E_{n+1}$ is smaller than or equal to the slope of *Point E_n*.

After the `TangentLine` is found, the procedure continues to iterate linearly through the remaining points of the original service envelope, now looking for an intersection of `TangentLine` with curve U [curve U is not created in advance, but rather is obtained by shifting original segments up by ε one at a time (the shifted segment is denoted `ShiftedSegment`) and checking if the `TangentLine` intersects the `ShiftedSegment`]. When such a point is found, it is added to the approximation curve $\widetilde{E}$ and a new search for tangent line begins.

Each point of the original service envelope E is passed only once, whether while searching for a `TangentLine` or searching for the intersection of `TangentLine` with `ShiftedSegment`. Therefore, CON-STRUCT_APPROX_LINEAR runs in $O(m)$ where m is the number of points in curve E as defined earlier.

```
      CONSTRUCT_APPROX_LINEAR(E, ε)
1         // first point in Ẽ corresponds to first point in E
2         Ẽ₁ = E₁
3         n = 2

4         for( k = 2 to m-1 )
5         {
6             // find tangent to E (linear search)
7             while(n < m-1 and TEST_TANGENT(Ẽₖ₋₁, n) > 0)
8                 n = n + 1
9             TangentLine = (Ẽₖ₋₁Eₙ)

10            // find next intersection point (linear search)
11            while( n < m and TangentLine( X(Eₙ) ) - Y(Eₙ) < ε )
12                n = n + 1

13            // find segment parallel to (Eₙ₋₁Eₙ) and shifted up by ε
14            // notation A ∩ B means point of intersection of lines A and B
15            ShiftedSegment = (Eₙ₋₁Eₙ) + ε
16            <x', y'> = TangentLine ∩ ShiftedSegment
17            <x'', y''> = TangentLine ∩ (Eₘ₋₁Eₘ)

18            // add kᵗʰ approximation point
19            if( x' < x'' )
20                Ẽₖ = <x', y'>
21            else
22            {
23                Ẽₖ = <x'', y''>
24                Ẽₖ₊₁ = Eₘ
25                return SUCCESS
26            }
27        }

28        return FAILURE
```

Figure 18.6   The CONSTRUCT_APPROX_LINEAR algorithm.

```
      TEST_TANGENT(Point, n)

1         if( SLOPE(Eₙ₋₁Eₙ) < SLOPE(PointEₙ) )
2             return -1
3         else if( SLOPE(EₙEₙ₊₁) > SLOPE(PointEₙ) )
4             return 1
5         else
6             return 0
```

Figure 18.7   TEST_TANGENT(point, n) function.

Procedure CONSTRUCT_APPROX_BINARY(E, ε) (Fig. 18.8) is similar to CONSTRUCT_APPROX_LINEAR(E, ε), but performs a binary search over the points of the original service envelope to locate the tangent line and its intersections. This must be done for at most K approximation points; therefore procedure CONSTRUCT_APP-ROX_BINARY(E, ε) runs in $O(K \log m)$ time.

```
     CONSTRUCT_APPROX_BINARY( E, ε )
1         // first point in Ẽ corresponds to first point in E
2         Ẽ₁ = E₁
3         n = 2

4         for( k = 2 to m-1 )
5         {
6             // find tangent to E (binary search)
7             low_n = n
8             high_n = m-1
9             while(high_n - low_n > 1 )
10            {
11                n = (high_n + low_n) / 2
12                t = TEST_TANGENT(Ẽₖ₋₁, n)
13                if( t < 0 )
14                    high_n = n
15                else if( t > 0 )
16                    low_n = n
17                else
18                    exit loop
19            }
20            TangentLine = (Ẽₖ₋₁Eₙ)

21            // find next intersection point (binary search)
22            low_n = n
23            high_n = m
24            while(high_n - low_n > 1 )
25            {
26                n = (high_n + low_n) / 2
27                if( TangentLine( X(Eₙ) ) - Y(Eₙ) > ε )
28                    high_n = n
29                else
30                    low_n = n
31            }

32            // find segment parallel to (Eₙ₋₁Eₙ) and shifted up by ε
33            // notation A > B means point of intersection of lines A and B
34            ShiftedSegment = (Eₙ₋₁Eₙ) + ε
35            <x', y'> = TangentLine ∩ ShiftedSegment
36            <x", y"> = TangentLine ∩ (Eₘ₋₁Eₘ)

37            // add kᵗʰ approximation point
38            if( x' < x" )
39                Ẽₖ = <x', y'>
40            else
41            {
42                Ẽₖ = <x", y">
43                Ẽₖ₊₁ = Eₘ
44                return SUCCESS
45            }
46        }

47        return FAILURE
```

Figure 18.8   The CONSTRUCT_APPROX_BINARY algorithm.

Since the values of K and m are known a priori, each node can choose between CONSTRUCT_APPROX_LINEAR and CONSTRUCT_APPROX_BINARY depending on whose running time is expected to be smaller. For example, if in cycle k, node i determines that $K \log m_{i,k} < m_{i,k}$, it will choose CONSTRUCT_APPROX_BINARY; otherwise it will choose CONSTRUCT_APPROX_LINEAR.

```
       MIN_POINT_APPROX( Δε )

1          lower_ε = 0
2          upper_ε = Δε

           // expanding search
3          while( CONSTRUCT_APPROX_xxx( E, upper_ε ) == FAILURE )
4          {
5              lower_ε = upper_ε
6              upper_ε = upper_ε × 2
7          }

           // contracting search
8          while(upper_ε - lower_ε > Δε )
9          {
10             ε = (lower_ε + upper_ε) / 2
11             if( CONSTRUCT_APPROX_xxx( E, ε ) == FAILURE )
12                 lower_ε = ε
13             else
14                 upper_ε = ε
15         }

16         CONSTRUCT_APPROX_xxx( E, upper_ε )
```

Figure 18.9   MIN_POINT_APPROX ( ) procedure.

The procedure MIN_POINT_APPROX($\Delta\varepsilon$) performs multiple invocations of either CONSTRUCT_APPROX_LINEAR(E, ε) or CONSTRUCT_APPROX_BINARY(E, ε) with different values of ε and finds the smallest value of ε for which a K-point (or less) approximation curve can be constructed. The final value of ε is found by performing a binary search on ε, as illustrated in Fig. 18.9.

The MIN_POINT_APPROX( ) procedure invokes CONSTRUCT _APPROX_LINEAR or CONSTRUCT_APPROX_BINARY at most O (log ε^{max}) times, where ε^{max} is the maximum error possible for any service envelope. Because only W^{cycle} bytes can be granted in 1 cycle, no service envelope needs to have any points with envelope value exceeding W^{cycle}.[4] Thus, no approximation error can exceed W^{cycle} (that is, $\varepsilon^{max} \leq W^{cycle}$).

To summarize, the running time of the Min-Points approximation scheme is either $O(K \log W^{cycle} \log m)$ or $O(m \log W^{cycle})$, depending on the choice between the CONSTRUCT_APPROX_LINEAR and CONSTRUCT_APPROX_BINARY functions.

---

[4] If, after summing all the envelopes received from the children, some points have envelope values above W^{cycle}, such points should be pruned from the resulting envelope.

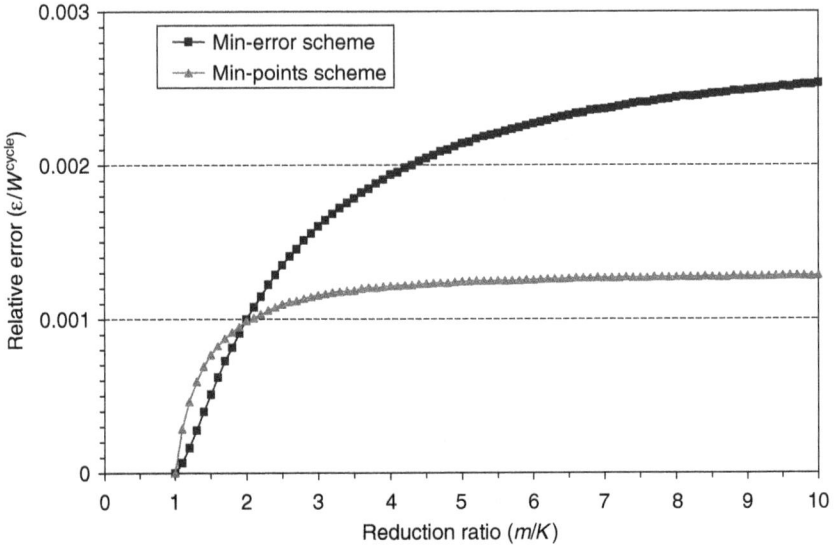

Figure 18.10   Approximation error for Min-Error and Min-Points schemes.

18.1.3.3 Approximation error. Both the Min-Error and Min-Points approximation schemes produce suboptimal solutions. We measured the performance of the Min-Error and Min-Points schemes with a large number of randomly generated service envelopes. Figure 18.10 presents the average relative approximation error (measured as $\varepsilon / W^{\mathrm{cycle}}$) for different reduction ratios $m_{i,k} / K$. Each point on a plot represents the average error measured over 10,000 service envelopes. It can be seen that the Min-Error approach performs better when the reduction ratio is less than 2 (i.e., the number of points in the original service envelope $m_{i,k} \leq 2K$), and the Min-Points scheme performs better if the reduction ratio is greater than 2.

The average approximation error (for Min-Points scheme) stabilizes at 0.00128; that is, we can expect each intermediate node to introduce 0.128 percent overhead. In EPON with 16 ONUs, we should expect $(16 + 1) \times 0.128$ percent ≈ 2.18 percent overhead due to approximation error.

18.1.4 FQSE complexity

We analyze the complexities of the requesting phase and the granting phase separately.

In the requesting phase, each node i should perform two operations: (1) obtain service envelope by summing service envelopes received from all its children and (2) perform an approximation, if necessary.

Each envelope received from a child may contain at most K points and is sorted by satisfiability parameter s. Thus, to calculate its service envelope, first, node i should merge points of received envelopes together while preserving the ordering, and then it should calculate the cumulative envelope values at each point. Performing pairwise merging, node i first merges $|D_i|/2$ pairs of K-point envelopes, resulting in $|D_i|/2$ $2K$-point envelopes. In the next iteration, these $|D_i|/2$ envelopes will be merged, resulting in $|D_i|/4$ $4K$-point envelopes. Node i will continue merging until, after $\log_2|D_i|$ steps, the last pair is merged into one $|D_i|K$-point envelope. Therefore, the complexity of this operation is

$$O\left(2K\,\frac{|D_i|}{2} + 4K\,\frac{|D_i|}{4} + \cdots + K|D_i|\right) = O\left(K|D_i|\log\left(|D_i|\right)\right)$$

In the following step, node i may need to perform an approximation procedure. As shown in Sec. 18.1.3.3, the complexity of this operation is bounded either by $O(K \log W^{\mathrm{cycle}} \log(K|D_i|))$ or by $O(K|D_i| \log W^{\mathrm{cycle}})$ (since $m_i = K|D_i|$). Thus, the overall complexity of the requesting phase at each node is bounded by $O(K|D_i|(\log|D_i| + \log W^{\mathrm{cycle}}))$ (using CONSTRUCT_APPROX_LIN-EAR) or by $O(K|D_i|\log|D_i| + K \log W^{\mathrm{cycle}} \log(K|D_i|))$ (using CONSTRUCT_APPROX_BINARY).

In the granting phase, each node invokes the PROCESS_GRANT( ) procedure (Fig. 18.2) to send a grant message to each child; therefore, the total work in the granting phase is $O(|D_i|)$.

18.1.5 Granting schemes

The FQSE algorithm requires the root scheduler to receive service envelopes from all its children before calculating the satisfiability parameter for the next cycle. Each intermediate node should also receive the envelopes from all the children before generating its own envelope. Figure 18.11 illustrates this granting scheme for the scheduling hierarchy shown in Fig. 17.3. The obvious drawback of this scheme is that each cycle will incur an overhead equal to one maximum round-trip delay plus message processing delay at each level in the hierarchy.

In an alternative approach (Fig. 18.12), the root node may segregate its children into two groups, A and B, and schedule each group independently. The root node would schedule nodes from group A while collecting requests from group B. Then, when all the requests from group B are collected, the root would schedule all nodes from group B, while receiving requests from group A. This scheme is free from the

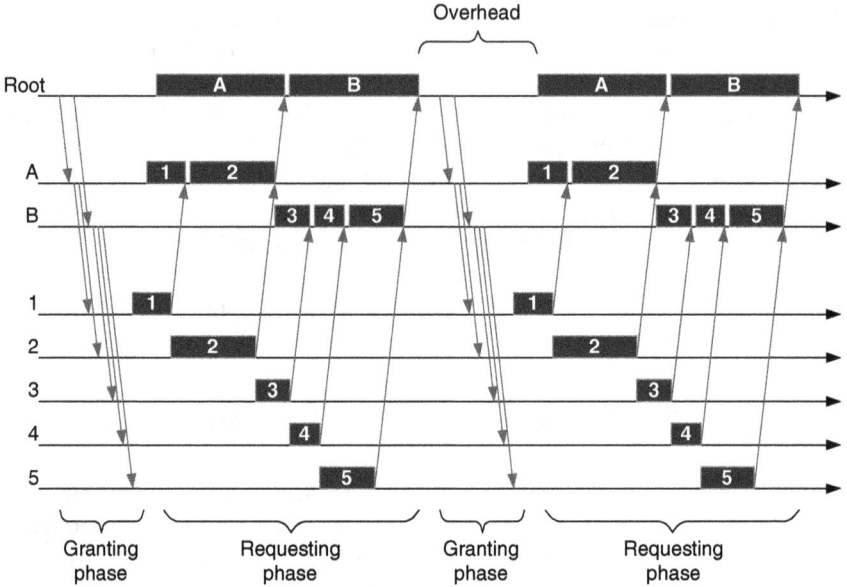

Figure 18.11   Collecting all requests before scheduling grants.

overhead shown in Fig. 18.11. But it provides fairness only among queues within each of the two groups.

The impact of the fact that fairness is only provided within each group can be lessened by carefully grouping the queues. For example, some SLAs may only require fixed guaranteed bandwidth and no excess bandwidth. Such queues do not participate in excess bandwidth sharing and therefore are good candidates for being grouped together. The rest of the queues could be placed in the other group, and they will get a fair share of the excess bandwidth. Another situation, when grouping may become even more beneficial, occurs when different types of customers are served by the same access network; e.g., business and residential subscribers may be separated into different groups.

18.2   FQSE Adaptation for EPON

So far, in our description, we assumed a fluid network model in which the transmission quanta can be infinitesimally small. The timeslot assignment guarantees fairness in terms of raw bandwidth, i.e., when an entire timeslot can be utilized if there are data available in the corresponding queue. This FQSE algorithm can be easily adapted for ATM traffic by simply measuring the timeslot size in units of ATM cells (53 bytes). In EPON, however, we are dealing with indivisible packets of

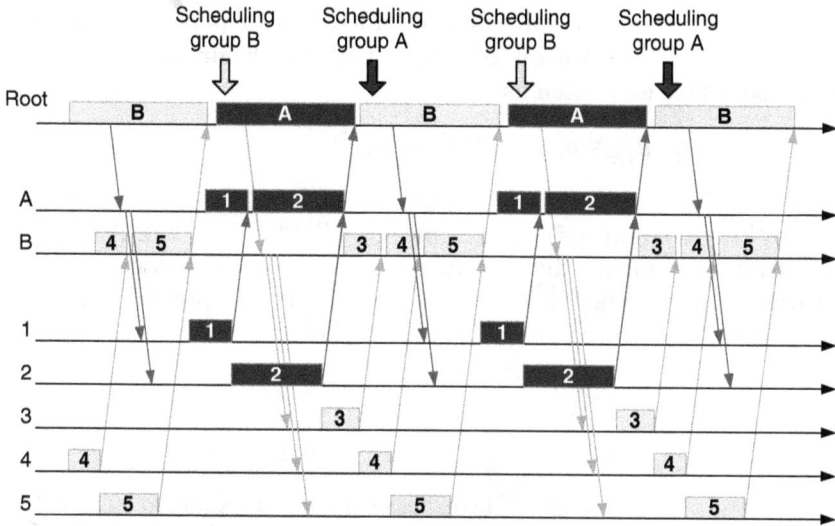

Figure 18.12   Scheduling two groups of nodes separately.

variable sizes. Ethernet packets cannot be fragmented; therefore, if the head-of-line (HOL) packet does not fit in the remaining timeslot, it will be deferred until the next timeslot, while the current timeslot will be left with an unused remainder. This creates two additional issues that the algorithm must address: *head-of-line blocking* and *bandwidth utilization*.

18.2.1   Head-of-line blocking

Head-of-line blocking is a result of coupling between bandwidth and latency in a time-sharing packet-based system. It is best explained by an example. Consider a case when a connection should be provisioned with a guaranteed bandwidth of 1 Mbps (and no excess bandwidth) and latency ≤ 1 ms (i.e., cycle time is 1 ms). This connection should be given a fixed timeslot of size 1 Mbps × 1 ms = 125 bytes. All Ethernet packets exceeding 125-byte size will be blocked in this case. Increasing the timeslot size would require a larger cycle time in order to keep the connection rate at a fixed value of 1 Mbps. However, larger cycle times will violate latency requirements.

To resolve the HOL blocking problem, FQSE allows a connection to occasionally request a minimum slot size larger than its guaranteed minimum slot size W^{min}. This approach may lead to a loss of *short-term fairness* when, in one cycle, a queue may be given a larger slot to accommodate a larger packet. To account for overused bandwidth (i.e., to maintain long-term fairness), we introduce a per-queue counter called *overdraft*. Overdraft of queue i at the beginning of cycle k (denoted $u_{i,k}$) is estimated as

$$u_{i,k} = u_{i,k-1} + W_{i,k-1}^{\min} - W_i^{\min} \tag{18.3}$$

where $W_{i,k}^{\min}$ = minimum timeslot size requested by queue i in cycle k and $W_i^{\min}$ = nominal minimum timeslot ($W_i^{\min} = B_i^{\min} T$). A positive overdraft value means that the queue consumed more service than it is entitled to. Denoting $S_{i,k}^{\text{HOL}}$ = size of HOL packet in queue i at the beginning of kth cycle, we calculate $W_{i,k}^{\min}$ as follows:

$$W_{i,k}^{\min} = \begin{cases} \max\left\{ S_{i,k}^{\text{HOL}}, W_i^{\min} \right\} & u_{i,k} \leq 0 \\ 0 & u_{i,k} > 0 \end{cases} \tag{18.4}$$

Equation (18.4) says that if no excess service was received by an ONU before the kth cycle, the ONU may request a larger timeslot in the kth cycle to accommodate a large HOL packet. This, of course, will be counted as a service overdraft ($u_{i,k} > 0$), and in the following few cycles the queue may be penalized by receiving less service, until overdraft becomes less than or equal to zero. At this point, if the next HOL packet exceeds $W_i^{\min}$, the queue will get excess service again.

With independent sources, it is reasonable to expect that, in each cycle, there will be approximately equal numbers of overdrafting and compensating sources. However, due to the stochastic nature of network traffic, an occasional fluctuation can occur when more sources overdraft on their requested bandwidth. Should this happen, the cycle time must increase to accommodate larger slots. Increased cycle time can affect the accuracy of bandwidth assignment and the SLA guaranteed to a subscriber. If a network operator decides to guarantee a fixed cycle time, the way to do it is by properly arranging or limiting the guaranteed bandwidth per queue. Clearly, the overdraft can happen only for queues with guaranteed minimum slot size $0 < W_i^{\min} \leq S^{\max}$, where $S^{\max}$ is the maximum packet size. To guarantee constant cycle time, the following equation should hold:

$$W^{\text{cycle}} \geq \sum_{i=1}^{N} \begin{cases} S^{\max} & 0 < W_i^{\min} < S^{\max} \\ W_i^{\min} & \text{otherwise} \end{cases} \tag{18.5}$$

Equation (18.5) states that even in the unlikely event that all the queues overdraft in the same cycle, their cumulative request size will not exceed W^{cycle}.

We analyze the effects of the loss of short-term fairness due to the HOL blocking avoidance mechanism in Sec. 18.3.3.

18.2.2 Bandwidth (timeslot) utilization

Ethernet traffic consists of nondivisible packets of variable sizes. In most cases, these packets cannot fill the slot completely (i.e., packet delineation in a buffer does not match slot size). This leads to an unused slot remainder and decreased bandwidth utilization. In Sec. 12.2.4 we derived a formula for the estimated size of the remainder for any packet size distribution. With the empirical trimodal packet size distribution reported in [SG01] and a single FIFO queue, the average size of the remainder is 595 bytes.

While each individual queue may be blocked on HOL packet, all the remainders together (assuming there are many queues in an ONU) constitute a considerable chunk of slot space, sufficient for sending several more complete packets. To utilize this bandwidth, we borrow the idea of per-queue *deficit counters* from the *deficit round-robin* (DRR) algorithm [SV96]. An unused remainder is added to the queue's deficit. When all queues have transmitted all their frames that fit in their granted slots $w_{i,k}$, the ONU performs a second pass and attempts to transmit the HOL packet from a queue with the highest value of its deficit counter. When a packet is transmitted, the value of the deficit counter is decremented by the size of the transmitted packet. The value of the deficit counter is retained between the cycles, and can accumulate if a queue does not get a chance to send additional data to compensate for previously unused remainders. This efficiently utilizes the bandwidth, with one remainder left per ONU, rather than one remainder per queue. Additionally, because an ONU can choose among the HOL packets in all the backlogged queues, the remainder left per ONU is considerably smaller than the one associated with a single FIFO queue. Figure 18.13 shows the average size of the unused ONU remainder as a function of the number of backlogged queues in the ONU.

It is interesting to notice the effects of multiplexing gain in slot utilization. When there is only one backlogged queue, the average remainder is 595 bytes, as predicted by Eq. (12.10). When the number of backlogged queues per ONU reaches 16, the average remainder drops to only 40 bytes. Further increase in the number of backlogged queues does not provide any significant improvement.

While using the deficit counter scheme, we should be aware of two problems: *starvation of an old queue* and *starvation of a new queue*. By *new queue* we mean a queue that became busy after a long idle interval; an *old queue* is a queue that remained busy for a long time.

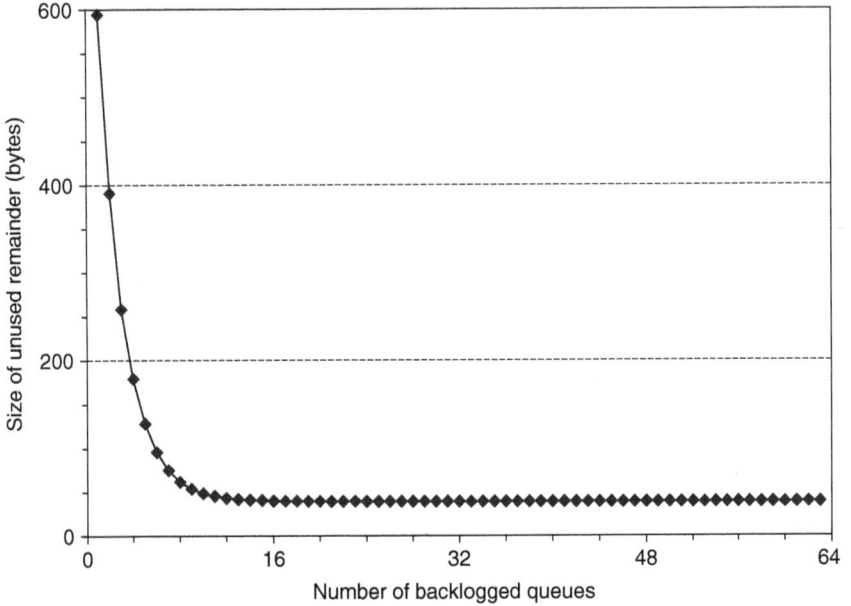

Figure 18.13   Size of unused remainder in an ONU.

As was explained in [SV96], idle queues should not accumulate the deficit. Allowing the deficit to accumulate during idle periods would permit a new queue to get an unfairly large bandwidth at the expense of old queues, which will lead to *starvation of old queues*. Thus, the deficit counter remains zero for all idle queues.

One important distinction with the DRR scheme is that the deficit cannot be completely satisfied; no matter how many queues the ONU has, the average unused slot remainder is not zero. In other words, after each transmission cycle, the cumulative unsatisfied deficit would increase by the size of the unused slot remainder (one remainder per ONU). For old queues, this deficit may accumulate for as long as the queues remain busy. When a previously idle queue becomes busy (i.e., a new queue appears), its deficit counter is much less than the counter of a queue that was busy for a long time (due to accumulation of unsatisfied deficit). This could lead to *starvation of a new queue*. To overcome this problem, after each transmission opportunity, all deficit counters will be decreased by the value of the smallest deficit counter among all the busy queues. In other words, we enforce a condition that the "most satisfied" old queue always has *deficit* = 0. Since deficit does not accumulate for idle queues, any new queue will have deficit 0, and thus would have the same chance for service as the most satisfied old queue.

18.3   FQSE Performance

In this study, we consider an EPON access network consisting of an OLT and N ONUs, each containing n queues. Most of the settings in our simulation experiments remain the same as in Chap. 14 (refer to Table 14.1). Table 18.1 lists additional or modified parameters used in this chapter.

We designate queues 1 to 4 in ONU A and queues 1 to 4 in ONU B as our test queues (see Fig. 18.14). The test queues are assigned the guaranteed bandwidth and weight as shown in Table 18.2.

The rest of the queues were used to generate background traffic (ambient throughput). Among the remaining queues, 18 queues in each ONU were assigned a guaranteed bandwidth of 1 Mbps and weight = 1, and the rest of the queues were best-effort queues (guaranteed bandwidth = 0 and weight = 1).

TABLE 18.1   Additional System Parameters

Parameter	Description	Value
n	Number of queues per ONU	64
Q	Buffer size for each queue	64 Kbytes
T	Cycle time	1 ms

TABLE 18.2   Examples of Queue Configurations

Queue	B^{min}	φ	Description
1	0	2	This is a best-effort service (no guaranteed bandwidth). Under very heavy network load this queue may get no service.
2	0	1	This is also a best-effort service. If the network load is not heavy, i.e., if excess bandwidth is available, this queue will get one-half the bandwidth that queue 1 gets (consider it a *good-effort* queue).
3	10 Mbps	0	This queue will never get any extra bandwidth, but it will always get its guaranteed bandwidth. No matter what the network load is, this queue will be able to transmit 10 Mbps of data. This configuration is good for circuit emulation services.
4	10 Mbps	1	This queue will always have its guaranteed bandwidth (10 Mbps) plus it will get an excess bandwidth if the network load is not high. When some excess bandwidth is available, this queue will be able to transmit exactly what queue 2 transmits plus 10 Mbps.

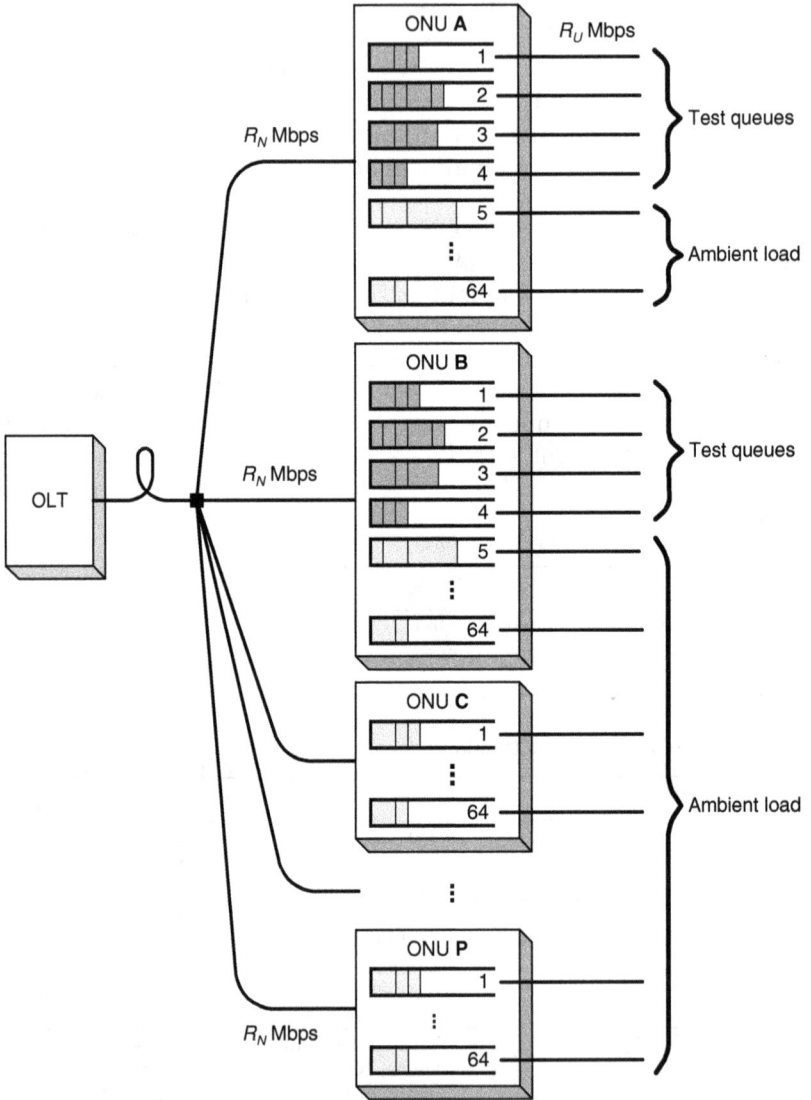

Figure 18.14 Experimental EPON model.

18.3.1 Fairness of FQSE

In this section, we analyze the fairness of FQSE by measuring the throughput of four queues (1 to 4) located in one ONU (ONU A). The throughput of each queue was measured over 1-s intervals at four different levels of ambient load (generated by all other queues in all the ONUs).

Each test queue was input a bursty traffic at an average load of 90 Mbps. Since the FQSE scheduler is work-conserving (i.e., it never grants to any queue a slot larger than the queue length), we expect that the burstiness of the input traffic would be reflected in the queues' throughput. To illustrate the effects of traffic burstiness, we analyzed queue throughput with two traffic types: *short-range dependent* (SRD) and *long-range dependent* (LRD). Both traffic types are bursty (consisting of alternating on and off periods) with burst sizes in SRD traffic having a negative exponential distribution and burst sizes in LRD having a heavy-tail distribution (e.g., Pareto distribution). The LRD traffic was generated using the method described in App. B and was verified to be self-similar with Hurst parameter 0.8.

First, we note that queue 3, which was configured to have 10 Mbps fixed bandwidth (weight $\varphi_3 = 0$), indeed has a constant throughput regardless of the ambient load (see Fig. 18.15).

The remaining queues were allowed to use the excess bandwidth, if available. In the first 25-s interval, the ambient load was kept relatively low, so each queue with nonzero weight (queues 1, 2, and 4) was able to send all arrived packets and never became backlogged. The average throughput of each queue was the same as the average load (90 Mbps). We can see that, in the case of LRD traffic (Fig. 18.15b), the throughput is bursty (even after averaging over 1-s intervals), reflecting the burstiness of the incoming data stream. In the case of SRD traffic (Fig. 18.15a), the averaging effects were much more pronounced, and the resulting plots are smoother.

When, at time $t = 25$, the ambient load increases to ~320 Mbps, queues 2 and 4 are not able to transmit all the incoming packets and become backlogged. From this moment on, they will maintain the fair relative throughput with queue 4 always being able to send 10 Mbps more than queue 2. Queue 1 is supposed to have twice the throughput of queue 2. This, however, would give queue 1 more than 90 Mbps bandwidth, so it only uses 90 Mbps and does not become backlogged until time $t = 50$. At $t = 50$, when the ambient load increases to ~500 Mbps, all four queues become backlogged and all are assigned fair bandwidth. Finally, at time $t = 75$, the ambient load increases even more (to ~660 Mbps), and the available excess bandwidth decreases to a very small amount. At this time, the throughput of each queue approaches its guaranteed bandwidth: 10 Mbps for queues 3 and 4 and 0 for queues 1 and 2.

18.3.2 Analysis of cousin fairness

Claim 1.   FQSE is a cousin-fair scheduling algorithm.

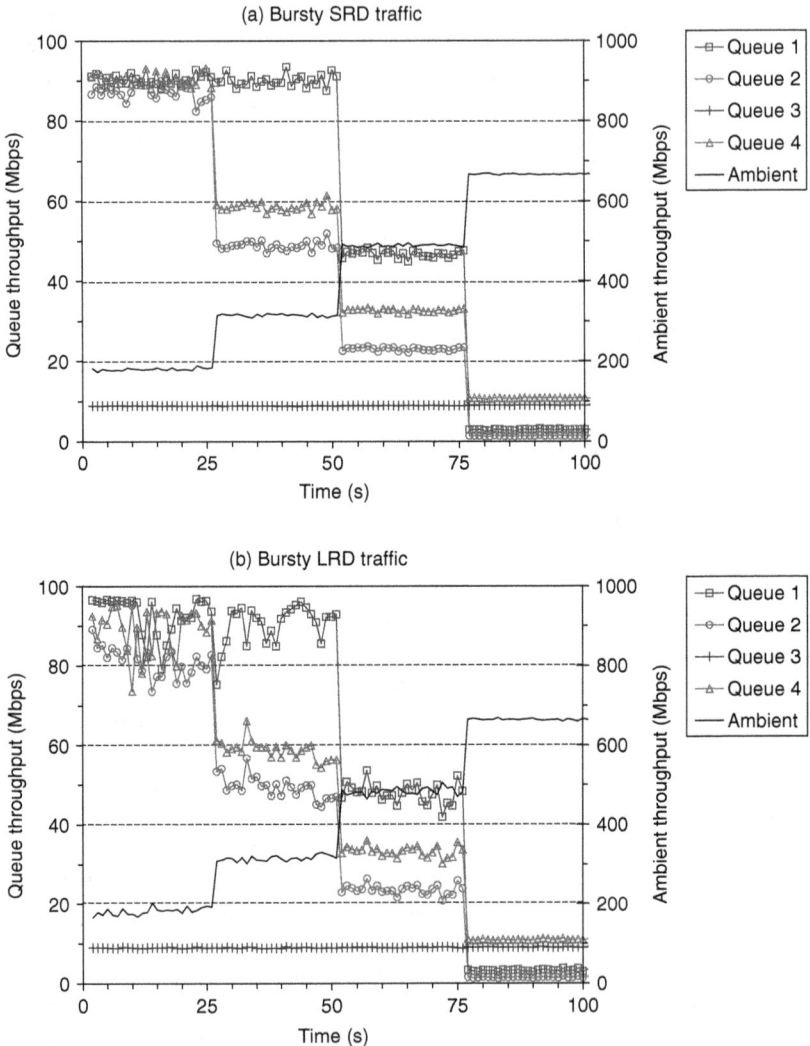

Figure 18.15   Throughput of test queues under different ambient loads.

Consider any two nonsibling leaves (queues) i and j. Let $E^*_{i,k}$ and $E^*_{j,k}$ be their respective service envelopes in cycle k, and let $w_{i,k}$ and $w_{j,k}$ be their timeslot sizes given satisfiability parameter $= s$ [$w_{i,k} = E^*_{i,k}(s)$ and $w_{j,k} = E^*_{j,k}(s)$]. We want to show that $w_{i,k}$ and $w_{j,k}$ are mutually fair timeslot sizes (i.e., excess bandwidth given to each queue is proportional to the queue's weight: $w^{ex}_{i,k} / \varphi_i = w^{ex}_{j,k} / \varphi_j$).

Proof. The claim is trivially true when the satisfiability parameter s completely satisfies one or both queues (i.e., when $w_{i,k} = q_{i,k}$ and/or

$w_{j,k} = q_{j,k}$). In this case, one or both queues will be served to exhaustion; the remaining (at most one) backlogged queue can take all the remaining bandwidth, and that will be fair.

Let us consider a case when both queues cannot be completely satisfied (i.e., in both functions $E^*_{i,k}$ and $E^*_{j,k}$, the coordinate s belongs to segments with slope $\neq 0$). Excess bandwidth given to queues i and j in this case is $w^{ex}_{i,k} = w_{i,k} - w^{min}_{i,k}$ and $w^{ex}_{j,k} = w_{j,k} - w^{min}_{j,k}$. We need to show that $w^{ex}_{i,k} / \varphi_i = w^{ex}_{j,k} / \varphi_j$. By construction of the E^* function, we have

$$\frac{w^{ex}_{i,k}}{\varphi_i} = \frac{w^{ex}_{j,k}}{\varphi_j}$$

$$\Rightarrow \quad \frac{w_{i,k} - w^{min}_{i,k}}{\varphi_i} = \frac{w_{j,k} - w^{min}_{j,k}}{\varphi_j}$$

$$\Rightarrow \quad \frac{w^{min}_{i,k} + s\varphi_i - w^{min}_{i,k}}{\varphi_i} = \frac{w^{min}_{j,k} + s\varphi_j - w^{min}_{j,k}}{\varphi_j}$$

Claim 1 holds no matter whether queues i and j have the same parent or not (i.e., this is a cousin-fair timeslot allocation).

To verify the property of cousin fairness experimentally, we compare the throughputs of four test queues located in ONU A with their counterparts (queues having the same guaranteed bandwidth and weight values) in ONU B. For each test queue i we plot the ratio of the throughput of queue i in ONU A to the throughput of queue i in ONU B (Fig. 18.16).

It can be seen that, for backlogged queues, this ratio approaches 1. For nonbacklogged queues, the ratio may deviate from 1, reflecting the burstiness of the input data stream. This behavior is expected for a work-conserving system.

18.3.3   Analysis of fairness bound

In Sec. 18.2.1, we explained that the HOL blocking avoidance mechanism could result in the short-term loss of fairness. In this section, we derive the bound for fairness error and compare it with experimental results. We start with the following claims:

Claim 2.   For any queue i with $W^{min}_i \leq S^{max}$, the running values of the overdraft counter $u_{i,k}$ are bounded as $1 - W^{min}_i \leq u_{i,k} \leq S^{max} - W^{min}_i$, if $W^{min}_i < S^{max}$; and $u_{i,k} = 0$, if $W^{min}_i \geq S^{max}$.

Proof.   Equations (18.3) and (18.4) can be combined as follows:

(a) Bursty SRD traffic

(b) Bursty LRD traffic

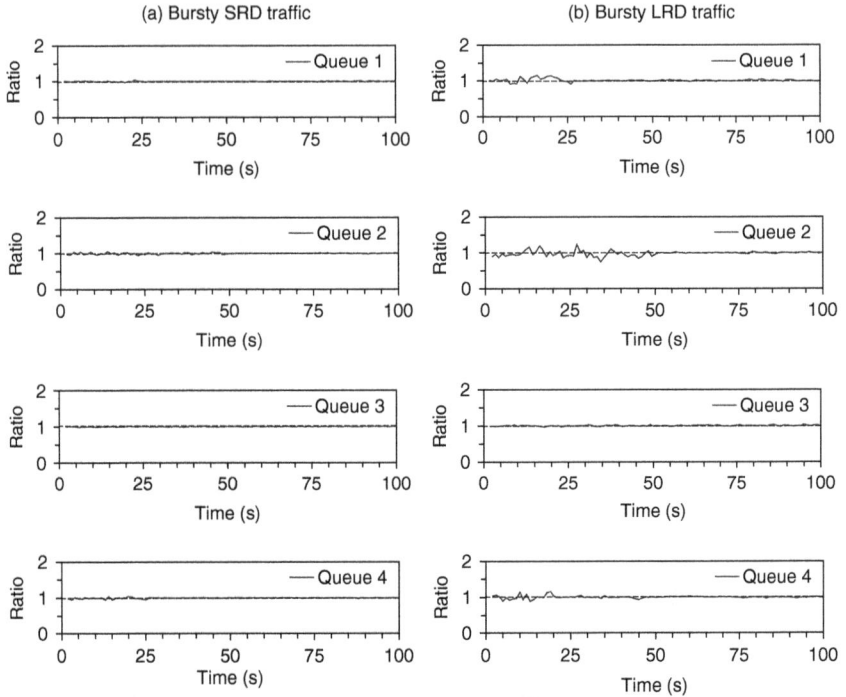

Figure 18.16   Ratio of throughputs for queues located in different ONUs.

$$u_{i,k} = \begin{cases} u_{i,k-1} & u_{i,k-1} \le 0 \text{ and } S_{i,k-1}^{\text{HOL}} \le W_i^{\min} & (18.6a) \\ u_{i,k-1} + S_{i,k-1}^{\text{HOL}} - W_i^{\min} & u_{i,k-1} \le 0 \text{ and } S_{i,k-1}^{\text{HOL}} > W_i^{\min} & (18.6b) \\ u_{i,k-1} - W_i^{\min} & u_{i,k-1} > 0 & (18.6c) \end{cases}$$

In Eq. (18.6a), the overdraft value in the next cycle remains the same as it was in the previous cycle. Equation (18.6b) results in an increased value of $u_{i,k}$. The value of $u_{i,k}$ will be the largest when $u_{i,k-1} = 0$ [by condition (18.6b), $u_{i,k}$ should be nonpositive] and $S_{i,k-1}^{\text{HOL}} = S^{\max}$. Eq. (18.6c) results in a decreased value of $u_{i,k}$. The value of $u_{i,k}$ will be the lowest when $u_{i,k-1} = 1$ [by condition (18.6c), $u_{i,k}$ should be positive].

Thus, we have $1 - W_i^{\min} \le u_{i,k} \le S^{\max} - W_i^{\min}$.

Also, from Eq. (18.6), we can derive bounds on the increment/decrement that the overdraft counter can have in one cycle.

Claim 3.   Increment that the overdraft counter may get in one cycle is bounded by $S^{\max} - W_i^{\min}$, and the decrement is bounded by $W_i^{\min}$; that is, for any k,

$$-W_i^{\min} \le u_{i,k} - u_{i,k-1} \le S^{\max} - W_i^{\min}$$

The proof follows directly from Eq. (18.6).

Definition. Let $w_i(k,n)$ be the cumulative optimal service that queue i should receive according to Eq. (17.5) during n cycles of the scheduler starting in cycle k:

$$w_i(k,n) = w_{i,k} + w_{i,k+1} + \cdots + w_{i,k+n-1}$$

and let $\widetilde{w}_i(k,n)$ be the actual cumulative service received by queue i during n cycles of the scheduler starting in cycle k:

$$\widetilde{w}_i(k,n) = \widetilde{w}_{i,k} + \widetilde{w}_{i,k+1} + \cdots + \widetilde{w}_{i,k+n-1}$$

($\widetilde{w}_{i,k}$ may be not equal to $w_{i,k}$ because of the HOL blocking avoidance mechanism; that is, $W_{i,k}^{\min} \ne W_i^{\min}$).

Theorem 1.   Service received by queue i in any interval of n cycles ($n = 1, 2, 3, \ldots$) during which the queue remains backlogged does not exceed the optimal service by more than $C_i^+(n)$, and does not fall short by more than $C_i^-(n)$. That is,

$$C_i^-(n) \le \widetilde{w}_i(k,n) - w_i(k,n) \le C_i^+(n)$$

where

$$C_i^-(n) = \max\left\{1 - S^{\max}, - nW_i^{\min}\right\}$$

$$C_i^+(n) = \min\left\{S^{\max} - 1, n\left(S^{\max} - W_i^{\min}\right)\right\}$$

Proof

$$\widetilde{w}_i(k,n) - w_i(k,n) = \sum_{k}^{k+n-1} \widetilde{w}_{i,k} - \sum_{k}^{k+n-1} w_{i,k}$$

$$= \sum_{k}^{k+n-1}\left(\widetilde{w}_{i,k}^{\min} + s_k\varphi_i\right) - \sum_{k}^{k+n-1}\left(w_{i,k}^{\min} + s_k\varphi_i\right) \quad (18.7)$$

$$= \sum_{k}^{k+n-1} \widetilde{w}_{i,k}^{\min} - \sum_{k}^{k+n-1} w_{i,k}^{\min}$$

where $w_{i,k}^{\min} = \min\{W_i^{\min}, q_{i,k}\}$ and $\widetilde{w}_{i,k}^{\min} = \min\{W_{i,k}^{\min}, q_{i,k}\}$.

Since queue i remains backlogged during the entire interval of n cycles, we can write $w_{i,k}^{\min} = W_i^{\min}$ and $\overline{w}_{i,k}^{\min} = W_{i,k}^{\min}$. Thus, Eq. (18.7) becomes

$$\widetilde{w}_i(k,n) - w_i(k,n) = \sum_{k}^{k+n-1} W_{i,k}^{\max} - \sum_{k}^{k+n-1} W_i^{\max}$$

$$= \sum_{k}^{k+n-1} W_{i,k}^{\max} - nW_i^{\max}$$

(18.8)

From Eq. (18.3), we have

$$u_{i,k+1} = u_{i,k} + W_{i,k}^{\max} - W_i^{\max}$$

$$u_{i,k+2} = u_{i,k+1} + W_{i,k+1}^{\max} - W_i^{\max}$$

$$\cdot$$
$$\cdot$$
$$\cdot$$

$$u_{i,k+n} = u_{i,k+n-1} + W_{i,k+n-1}^{\max} - W_i^{\max}$$

Alternatively,

$$u_{i,k+n} = u_{i,k} + \sum_{k}^{k+n-1} W_{i,k}^{\max} - nW_i^{\max}$$

(18.9)

Substituting Eq. (18.8) into Eq. (18.9), we get

$$\widetilde{w}_i(k,n) - w_i(k,n) = u_{i,k+n} - u_{i,k}$$

(18.10)

Equation (18.10) states that the difference between the actual and the optimal service during any interval of n cycles is equal to the difference between the values of the overdraft counter at the beginning and at the end of this interval.

Using the bounds on the overdraft counter from claim 2, we get

$$1 - S^{\max} \le u_{i,k+n} - u_{i,k} \le S^{\max} - 1$$

(18.11)

Alternatively, we can expand

$$u_{i,k+n} - u_{i,k} = \left(u_{i,k+n} - u_{i,k+n-1}\right) + \left(u_{i,k+n-1} - u_{i,k+n-2}\right) + \cdots + \left(u_{i,k+1} - u_{i,k}\right)$$

Using the bounds from claim 3, we get

$$-nW_i^{\min} \le u_{i,k+n} - u_{i,k} \le n\left(S^{\max} - W_i^{\min}\right)$$

(18.12)

Combining Eq. (18.11) and (18.12) and substituting the result into Eq. (18.10), we obtain upper and lower bounds on service (un)fairness as follows:

$$C_i^-(n) = \max\left\{1 - S^{\max}, - nW_i^{\min}\right\}$$

$$C_i^+(n) = \min\left\{S^{\max} - 1, n\left(S^{\max} - W_i^{\min}\right)\right\}$$

Corollary 1. Let queues i and j have the same guaranteed bandwidth ($W_i^{\min} = W_j^{\min}$) and the same weight ($\varphi_i = \varphi_j$). Then the amount of service that queues i and j get in any n-cycle interval ($n = 1, 2, 3, \ldots$) in which they remain backlogged would differ by no more than $C_i^+(n) - C_i^-(n)$.

Proof. From $W_i^{\min} = W_j^{\min}$ and $\varphi_i = \varphi_j$, it follows that. Thus,

$$\tilde{w}_i(k,n) - \tilde{w}_j(k,n) = \left[\tilde{w}_i(k,n) - w_i(k,n)\right] - \left[\tilde{w}_j(k,n) - w_j(k,n)\right]$$

Substituting $\overline{w}_i(k,n) - w_i(k,n)$ and $\overline{w}_j(k,n) - w_j(k,n)$ by their bounds from Theorem 1, we get $C_i^-(n) - C_j^+(n) \le \tilde{w}_i(k,n) - \tilde{w}_j(k,n) \le C_i^+(n) - C_j^-(n)$. Since $W_i^{\min} = W_j^{\min}$, we have $C_i^-(n) = C_j^-(n)$ and $C_i^+(n) = C_j^+(n)$. Thus,

$$\left|\tilde{w}_i(k,n) - \tilde{w}_j(k,n)\right| \le C_i^+(n) - C_i^-(n)$$

We illustrate Corollary 1 by an experiment in which we measure the difference in bandwidth allocated to two queues with the same guaranteed bandwidth and weight and located in different ONUs. For example, we choose queue 4 in ONU A and queue 4 in ONU B. Figure 18.17 presents two plots, the first being the maximum observed difference in bandwidths over the interval of 1000 s [measured as $|\overline{w}_i(k,n) - \overline{w}_j(k,n)| / nT$], and the second plot representing the maximum difference according to Corollary 1 (measured as $[C_i^+(n) - C_i^-(n)] / nT$).

We observe that the difference in the allocated bandwidths quickly declines as the sampling window size n increases. At 1-s sampling window ($n = 1000$), the maximum measured difference in allocated bandwidth was only ~18 kbps. Results in Fig. 18.17 illustrate the short-term nature of loss of fairness due to HOL blocking avoidance.

18.4 Summary

In this study, we have investigated FQSE, which is a novel algorithm that successfully combines hierarchical scheduling and cousin fairness.

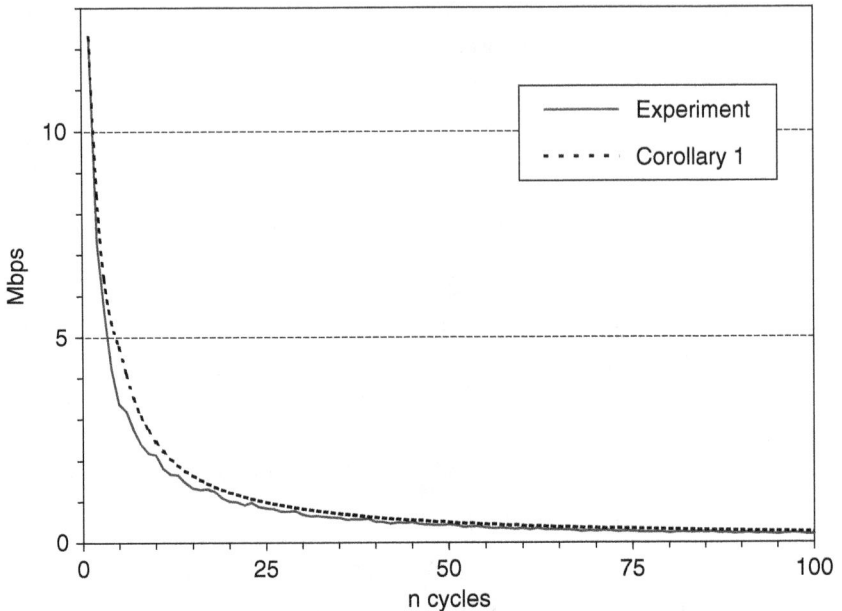

Figure 18.17   Maximum difference in bandwidth allocated to queue 4 in ONU A and B.

We have estimated FQSE's complexity and found that, at each node i, the scheduling work is proportional to the number of children $|D_i|$ of node i, and not to the total number of consumers, as it would be in a direct scheduling scheme.

We also proved the cousin fairness property of FQSE and derived bounds on its fairness index. We found that FQSE provides excellent fairness with a bound on fairness index of less than one maximum-size packet.

References

[KMS+03]   G. Kramer, B. Mukherjee, N. Singhal, A. Banerjee, S. Dixit, and Y. Ye, *Fair Queuing with Service Envelopes (FQSE): A Cousin-Fair Hierarchical Scheduler and Its Application in Ethernet PON*, Technical Report CSE-2003-6, Department of Computer Science, University of California, Davis, March 2003.

[SG01]   D. Sala and A. Gummalla, "PON functional requirements: Services and performance," presented at IEEE 802.3ah meeting in Portland, OR, July 2001. Available at http://www.ieee802.org/3/efm/public/jul01/presentations/sala_1_0701.pdf.

[SV96]   M. Shreedhar and G. Varghese, "Efficient fair queuing using deficit round-robin," *IEEE/ACM Transactions on Networking,* vol. 4, no. 3, pp. 375–385, June 1996.

19

Conclusion

Congratulations, dear reader! You are one of the most determined, one of the most persistent, one of those who finished the entire book. And it was not an easy reading (I know—I read it, too). But you weathered it all and have reached the end. Thank you for that!

I guess, now would be a good time to reveal that The EPON Book has not been written yet. Even though EPON took industry (and academia) by storm, advancing from a raw idea to an industrywide standard in 5 short years, its story is far from over.

EPON is the first public network architecture that traces its ancestry not to telephony or cable communications, but to private enterprise data networks. EPON is standardized by the IEEE 802.3 work group— a group that reigns in the LAN world, but has not ventured much beyond it. Ethernet in the First Mile was its first such attempt—a reconnaissance mission into yet unknown territory. It is not surprising that a great many things were done differently than they are traditionally done in the telecom world. But will the telecom world accept the EPON? Will network operators love it and mass-deploy it? Will EPON be successful. And, for that matter, what constitutes the success?

I am often asked if, in my opinion, EPON would be the ultimate access network architecture. And I always say, "I hope not." I hope, that EPON will break the grip of twisted pair on the first mile; I hope it will convince the most incumbent of incumbent network operators that all-optical access networks can be economical. And I hope it will allow new network applications to flourish, stimulating revenue growth and driving more traffic onto backbone routes, in turn generating renewed investment in metro and long-haul networks.

But most of all, I hope that when its time comes, EPON will gracefully yield to something even more advanced. Serving its purpose in the evolution of global networking infrastructure is the greatest success EPON can hope for. And that is why I hope EPON is not the ultimate architecture. But what will the next evolutionary step be and which upgrade path will take us there?

19.1 Upgrading EPON

By the amount of bandwidth it makes available to subscribers, EPON is a giant step forward compared to currently mass-deployed technologies, such as DSL or cable modem. With a line rate of 1 Gbps (current IEEE 802.3ah specification) and 16 to 64 subscribers per EPON, each user will get between 15 and 60 Mbps bandwidth. Still, unavoidably, as more bandwidth-intensive services become available to users, this capacity will get exhausted. It is, therefore, crucial for the success of EPON technology to provide a smooth path for future upgrades. It is hard to envision what upgrade scenario will be most favorable at the time when the EPON capacity will become a limiting factor. The best scenario will not require a forklift upgrade (i.e., will allow incremental expenses) and will capitalize on the most mature technology.

In a *wavelength-upgrade* scenario, some of the EPON ONUs will migrate to new wavelengths for both upstream and downstream traffic. While the data rate on each wavelength will remain the same, there will be fewer ONUs to share that bandwidth capacity. This procedure can be repeated again in the future and eventually will lead to a WDM PON system in which each subscriber is allocated its individual wavelengths. The major cost factor of such an upgrade is the necessity of tunable transmitters and receivers in ONUs. Also, the OLT must have multiple transceivers (tunable or fixed)—one for each wavelength. To allow an incremental upgrade, a new spectral region should be allocated for premium ONUs. This will allow nonpremium ONUs to continue using cheap 1310-nm lasers with high spectral width. Only the premium ONUs will have to be replaced by new ones operating at different wavelength or having tunable transceivers.

With the finalizing of the 10 Gbps Ethernet standard by the IEEE, *rate upgrade* appears as an attractive solution. To allow incremental upgrade cost, only a subset of ONUs may be upgraded to operate at higher rates. Thus, the rate-upgrade scenario will call for a mixed-rate EPON in which some ONUs operate at 1 Gbps and others at 10 Gbps. This upgrade would require high-speed electronics in the OLT which is able to operate at both rates. The downstream broadcasting traffic will have to remain on a slower channel, or be duplicated on both 1 Gbps and 10 Gbps channels.

In a *spatial-upgrade* scenario, a new trunk fiber is deployed from the CO to the splitter, and some branches are reattached to a new trunk fiber (and a new splitter). To avoid the cost of additional fiber deployment, this upgrade fiber can be predeployed at the time of the original deployment. Alternatively, some network operators consider EPON deployment with a splitter located in the CO. This EPON configuration will require as much fiber to be deployed as in the point-to-point configuration, but it will still require only one transceiver in the OLT. Such a topology will enable much higher-density equipment, which is very important in a limited CO space available to *competitive local exchange carriers* (CLECs) who have to rent CO space from *incumbent LECs* (ILECs). In a scenario where a splitter is located in the CO, to upgrade to higher bandwidth, some users will be reconnected to another EPON in the patch panel located in the same CO. Eventually, spatial upgrade may lead to a point-to-point architecture with an independent fiber running to each subscriber.

19.2   Open Access

Many access networks are now being built by neutral operators, such as municipalities, or utilities providers. These operators intend to attract multiple service providers, which would provide multiple and different services. Some service providers may offer only voice-based services, while others may specialize in video and conferencing. Such vendor-neutral access networks are known as open access networks. The objective is to bring down costs through free-market competition for subscriber access, and to avoid "lock-in" by incumbent broadband access service providers. It is expected that end users would subscribe to different service providers for different services. For example, in a typical home, while the parents may access video through a video content provider, the children may access some educational software provider by a different vendor, and the grandmother may be talking to someone using a video phone.

Since such an open access network is shared by both broadband users and service providers, an efficient and fair bandwidth allocation scheme is required. To attract service providers, technological mechanisms should be employed by the network operator to guarantee some performance levels to each service provider. Hence a scheduling algorithm that is fair to both service providers and end users—the two entities being located at opposite ends of the access channel, given different service-level agreement parameters for both these entities—is required.

These questions and many others remain the active and important research topics today.

A

Characteristics of Network Traffic

There is an extensive study showing that most network traffic flows [i.e., generated by http, ftp, variable-bit-rate (VBR) video applications, etc.] can be characterized by *self-similarity* and *long-range dependence* (LRD) (see [WTE96] for an extensive reference list).

Figure A.1 illustrates the scaling behavior of LRD traffic in comparison with that of a short-range dependent (SRD) traffic such as the one based on the Poisson process.

Consider a cumulative process $Y(t)$ with stationary increments, and let X_t be its incremental process:

$$X_t = Y(t) - Y(t - 1) \tag{A.1}$$

For example, $Y(t)$ can represent the number of bytes arriving up to time t, and X_t can represent the number of bytes arriving in 1 unit of time. The process $X_s^{(m)}$ is an aggregated process of X_t if

$$X_s^{(m)} = \frac{1}{m} \left(X_{sm-m+1} + X_{sm-m+2} + \ldots + X_{sm} \right) \tag{A.2}$$

Process X_t is said to be self-similar if X_t is indistinguishable from $X_s^{(m)}$. This is a very restrictive definition, especially considering the stochastic nature of the network traffic. Usually, *second-order* self-similarity is considered for the purposes of traffic description: *autocovariance* functions of the original and aggregated processes should have the same values.

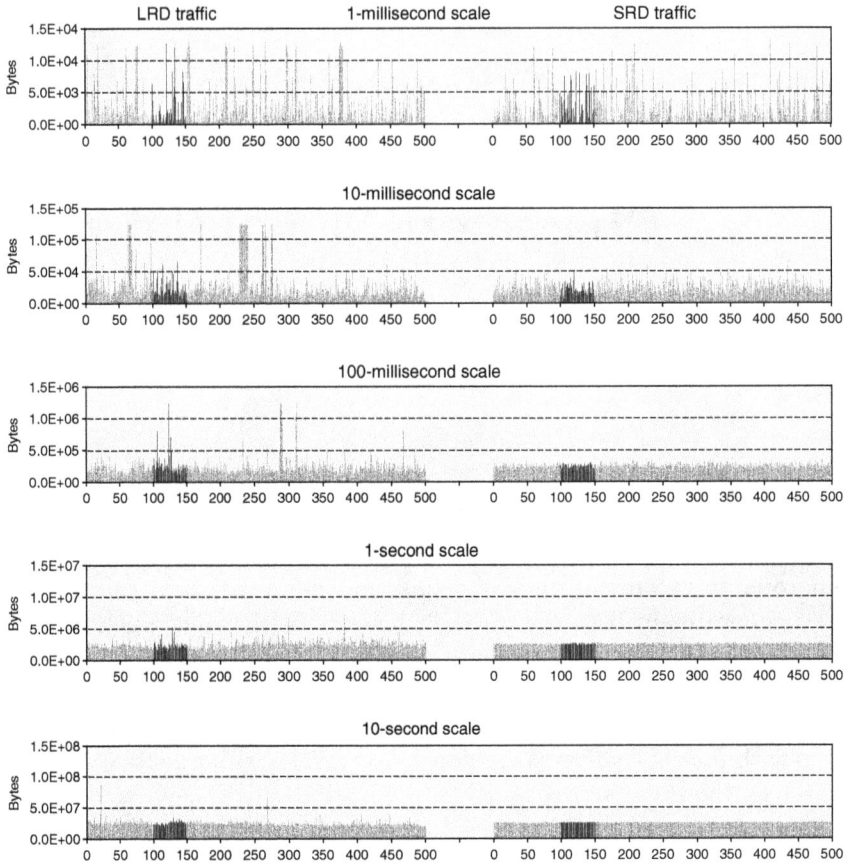

Figure A.1   Scaling behavior of LRD and SRD traffic flows.

Let

$$\gamma(k) = E[(X_t - \mu)(X_{t+k} - \mu)] \tag{A.3}$$

and

$$\gamma^{(m)}(k) = E[(X_s^{(m)} - \mu)(X_{s+k}^{(m)} - \mu)] \tag{A.4}$$

Then the process X_t is exactly second-order self-similar if

$$\gamma^{(m)}(k) = \gamma(k) \tag{A.5}$$

and asymptotically second-order self-similar if

$$\lim_{m \to \infty} \gamma^{(m)}(k) = \gamma(k) \tag{A.6}$$

A convenient measure of a process's distributional self-similarity is its Hurst parameter H. A process is self-similar with parameter H $(0 < H < 1)$ if

$$Y(t) \stackrel{\Delta}{=} k^{-H}Y(kt) \quad \text{for all } k > 0 \text{ and } t \geq 0 \tag{A.7}$$

i.e., the original and normalized aggregated processes should have the same distribution. Translating it to the corresponding stationary increment process, we get

$$X_s^{(m)} = \frac{1}{m}\left(X_{sm-m+1} + X_{sm-m+2} + \cdots + X_{sm}\right)$$

$$= \frac{1}{m}\left[Y(sm) - Y(sm - m)\right] \tag{A.8}$$

$$= \frac{1}{m}\left[Y(m) - Y(0)\right] \quad \text{(since increments are stationary)}$$

From Eq. (A.7), it follows that $Y(m) = m^H Y(1)$. Thus,

$$X^{(m)} = \frac{m^H}{m}\left[Y(1) - Y(0)\right] = m^{H-1}X \tag{A.9}$$

The self-similarity can be viewed as an ability of an aggregated process to "preserve" the burstiness of the original process, viz., the property of slowly decaying variance:

$$\text{var}(X^{(m)}) \sim m^{2H-2} \tag{A.10}$$

In the context of network traffic, this means that aggregating traffic over large time intervals reduces the burstiness very slowly (compared to non-self-similar traffic).

The property of long-range dependence refers to a nonsummable autocorrelation function $\rho(k)$:

$$\rho(k) = \frac{\gamma(k)}{\sigma^2} = \frac{E[(X_t - \mu)(X_{t+k} - \mu)]}{E[(X_t - \mu)^2]} \tag{A.11}$$

For $0 < H < 1, H \neq \frac{1}{2}$,

$$r(k) \sim H(2H - 1)k^{2H-2} \qquad k \to \infty \tag{A.12}$$

It is clear that if $\frac{1}{2} < H < 1$, then

$$\sum_{k=-\infty}^{\infty} \rho(k) = \infty \tag{A.13}$$

i.e., the process is long-range dependent.

The long-range dependence results from a heavy-tailed distribution of the corresponding stochastic process. Heavy-tailness refers to the rate of tail decay of the complementary distribution function. In a heavy-tailed distribution, the decay obeys the power law:

$$P[X > x] \sim cx^{-\alpha} \quad \text{as } x \to \infty \text{ and } 1 < \alpha < 2 \qquad (A.14)$$

As a result, the probability of an extremely large observation in the LRD process is nonnegligible. In the context of network traffic, this means that extremely large bursts of data (packet trains) and extremely long periods of silence (interarrival times) will occur from time to time. This is one of the reasons why analytic models employing traditional negative exponential distribution often provide overly optimistic estimates for the delays and queue sizes—the probability of an extreme event is negligible.

We refer the reader to [PW00] and [Ada97] for a more rigorous treatment of self-similarity and long-range dependence.

References

[Ada97]   A. Adas, "Traffic models in broadband networks," *IEEE Communications Magazine*, vol. 35, no. 7, pp. 82–89, July 1997.

[PW00]   K. Park and W. Willinger, "Self-similar network traffic: An overview," in K. Park and W. Willinger, eds., *Self-Similar Network Traffic and Performance Evaluation*, Wiley Interscience, 2000.

[WTE96]   W. Willinger, M. S. Taqqu, and A. Erramilli, "A bibliographical guide to self-similar traffic and performance modeling for modern high-speed networks," in *Stochastic Networks*, F. P. Kelly, S. Zachary, and I. Ziedins, eds., Oxford University Press, 1996, pp. 339–366.

B

Synthetic Traffic Generation

To obtain an accurate and realistic performance analysis, it is important to simulate the system behavior with appropriate traffic injected into the system. Most of the simulation experiments described in Part 4 of this book were performed with self-similar traffic. To generate self-similar traffic, we used the method described in [TWS97], in which the resulting traffic is an aggregation of multiple substreams, each consisting of alternating Pareto-distributed on/off periods.

Pareto distribution is a heavy-tailed distribution with the *probability density function* (pdf)

$$f(x) = \frac{\alpha b^{\alpha}}{x^{\alpha+1}} \quad x \geq b \tag{B.1}$$

where α is a shape parameter ($1 < \alpha < 2$) and b is a location parameter. Pareto distribution with $1 < \alpha < 2$ has a finite mean and an infinite variance.

In our implementation, each substream generates packets of constant size, although this size is different for different streams. To achieve the required packet size distribution (like trimodal distributions reported in [CMT98] or [SG01]), some substreams have higher relative load than the other substreams. Multiplexing (serializing) packets from different substreams produces self-similar traffic with the desired packet size distribution.

Each substream generates packets in groups (packet trains or bursts). The number of packets per burst (on period) follows the Pareto distribution with a minimum of 1 (i.e., the smallest burst consists of only 1 packet) and shape parameter $\alpha = 1.4$. The choice of α was prompted by measurements on actual Ethernet traffic performed by

Leland et al. [LT+94], who estimated the Hurst parameter to approximately be equal to 0.8 for moderate network load. The relationship between the Hurst parameter and the shape parameter α is $H = (3 - \alpha)/2$ (see [WT+95]). Thus, $\alpha = 1.4$ should result in $H = 0.8$.

Off periods (intervals between the packet trains) also follow the Pareto distribution, although with the shape parameter $\alpha = 1.2$. We used heavier tail for the distribution of the off periods because the off periods represent a stable state in a network; i.e., a network can be in the off state (no packet transmission) for an unlimited long time, while the durations of the on periods are ultimately limited by network resources and necessarily finite file sizes. The location parameter b for the off periods was chosen so as to obtain a desired load l_i from the given substream i:

$$l_i = \frac{E[\text{on}_i]}{E[\text{on}_i] + E[\text{off}_i]} \tag{B.2}$$

where $E[\text{on}_i]$ and $E[\text{off}_i]$ are expected lengths (durations) of on and off periods of source i. To generate Pareto-distributed values, we used the formula

$$X_{\text{PAR}} = \frac{b}{U^{1/\alpha}} \tag{B.3}$$

where U is a uniform random variable $(0 < U \leq 1)$. But since computers generate discrete values, we have to consider a truncated-tail Pareto distribution. Let us denote by $U^{\min}$ the smallest nonzero value of U. Then the largest Pareto-distributed value is $X^{\max} = b/(U^{\min})^{1/\alpha}$.

We can find the pdf $f_T(x)$ of a truncated-tail distribution as follows:

$$\int_b^{X^{\max}} f_T(x)\,dx = 1 \quad \Rightarrow \quad f_T(x) = \frac{f(x)}{\displaystyle\int_b^{X^{\max}} f(x)\,dx}$$

$$= \frac{\alpha b^\alpha}{x^{\alpha+1}} \times \frac{1}{1 - (b/X^{\max})^\alpha} \tag{B.4}$$

$$= \frac{\alpha b^\alpha}{x^{\alpha+1}\left(1 - U^{\min}\right)}$$

Then the expected value of a truncated-tail series is

$$E[X] = \int_{b}^{X^{max}} x f_T(x)\, dx$$

$$= \frac{\alpha b^\alpha}{1 - U^{min}} \times \frac{x^{1-\alpha}}{1-\alpha} \Bigg|_{b}^{X^{max}} = \frac{\alpha b^\alpha}{\alpha - 1} \times \frac{1 - \left(U^{min}\right)^{(\alpha-1)/\alpha}}{1 - U^{min}} \quad \text{(B.5)}$$

Now we can find the location parameter for off periods b_{off}. From Eq. (B.2) we get

$$E[\text{off}] = E[\text{on}] \times \left(\frac{1}{l} - 1\right) \quad \text{(B.6)}$$

After substituting Eq. (B.5) into Eq. (B.6), we have

$$\frac{\alpha_{off} b_{off}}{\alpha_{off} - 1} \times \frac{1 - \left(U^{min}\right)^{(\alpha_{off}-1)/\alpha_{off}}}{1 - U^{min}} = \frac{\alpha_{on} b_{on}}{\alpha_{on} - 1} \times \frac{1 - \left(U^{min}\right)^{(\alpha_{on}-1)/\alpha_{on}}}{1 - U^{min}}$$

$$\times \left(\frac{1}{l} - 1\right) \quad \text{(B.7)}$$

From Eq. (B.7), we find b_{off}:

$$b_{off} = b_{on} \times \frac{\alpha_{on}}{\alpha_{off}} \times \frac{\alpha_{off} - 1}{\alpha_{on} - 1} \times \frac{1 - \left(U^{min}\right)^{(\alpha_{on}-1)/\alpha_{on}}}{1 - \left(U^{min}\right)^{(\alpha_{off}-1)/\alpha_{off}}} \times \left(\frac{1}{l} - 1\right) \quad \text{(B.8)}$$

With our default values $b_{on} = 1$, $\alpha_{on} = 1.4$, $\alpha_{off} = 1.2$, and $U^{min} = 2^{-32}$, we get

$$b_{off} \approx 0.597 \times \left(\frac{1}{l} - 1\right) \quad \text{(B.9)}$$

Traffic in every ONU was generated by aggregating $n = 2910$ substreams (2 substreams for each packet size in the range from $S^{min} = 64$ bytes to $S^{max} = 1518$ bytes).

From Eq. (A.9), we get

$$\text{var}(X^{(m)}) = m^{2(H-1)} \text{var}(X) \quad \text{(B.10)}$$

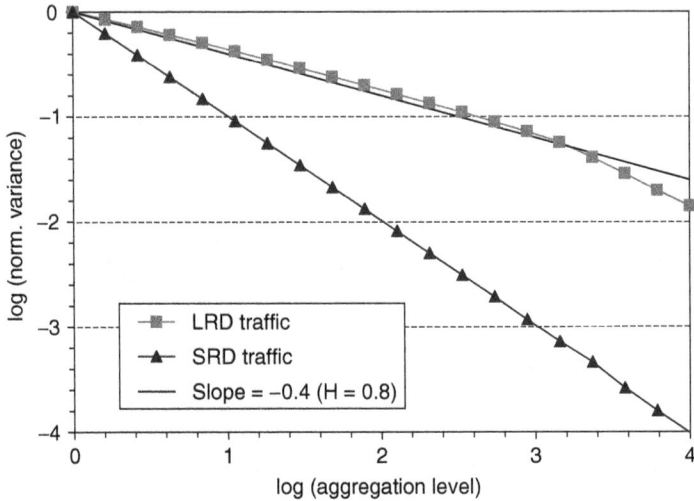

Figure B.1   Variance-time log-log plot.

or

$$\log \frac{\text{var}(X^{(m)})}{\text{var}(X)} = (2H - 2)\log m \qquad (B.11)$$

Equation (B.11) suggests that the log-log plot of variance versus aggregation level m should have a linear slope of value $2H - 2$. Figure B.1 shows a variance-time log-log plot we used to verify the correctness of our traffic generator. In a log-log plot, the x axis represents the logarithm of the aggregation parameter m, and the y axis represents the logarithm of the normalized variance. The "LRD traffic" plot shows a linear dependency (except in the tail region) with slope s value close to -0.4. From Eq. (B.11), we expect the log-log plot to have a slope $s = 2H - 2$. This results in $H = 1 + s/2 = 0.8$, as expected.

Note that, starting with log $m = 3.3$, the variance decay increases. There are two reasons for this. The first reason is the fact that synthetically generated traces (as well as those collected on real networks) have a truncated tail; i.e., after some threshold, the tail decays exponentially or faster. In our generator, the tail of the distribution function is truncated at the value $2^{32} - 1$; that is, the maximum burst size in our generator is $2^{32} - 1$ packets. However, the second reason is more important, i.e., the fact that simulations (as well as real network measurements) are performed with a finite set of observations. To build the variance-time plot in Fig. B.1, we generated 300 million packets. The maximum observed burst (length of the on period) was on the order of 100,000 packets long. Starting with the aggregation level of $10^{3.3}$ ms ≈ 2.4 s, the number of on periods that span multiple intervals of this size started to decrease rapidly. In other words, the number of on (and off) periods per interval started to increase proportionally to the increase of the interval size itself. This resulted (by the law of large numbers) in more pronounced averaging and, thus, in faster variance decay. Even though the faster variance decay is an artifact here, it does not affect the simulation results—the intervals (and, therefore, the delays) in the range of seconds are not of practical interest, as averaging of traffic on such a scale is beyond the buffering capacities or allowable delays.

The second plot, called *SRD traffic,* was obtained on the same generator, but with exponentially distributed on/off periods. This distribution possesses no long-range dependence, and its variance-time log-log plot is expected to have a slope of $-1 [\mathrm{var}(X^{(m)}) \sim m^{-1}]$. We plotted it just to verify that our variance normalizations and thus slopes are correct.

Figure B.2 illustrates the way the traffic was generated in an

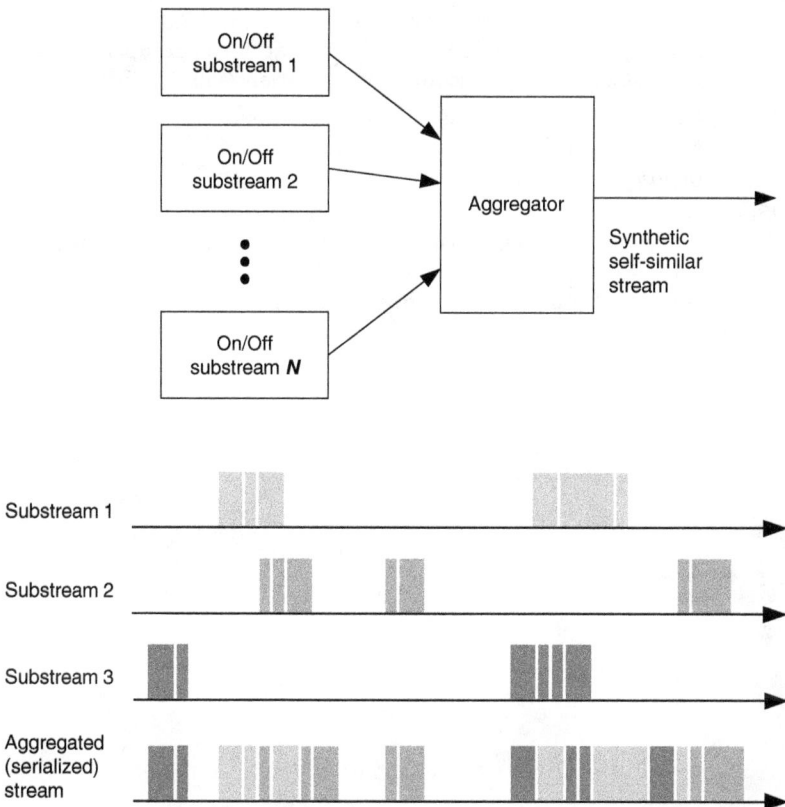

Figure B.2   Aggregation (multiplexing) of multiple substreams produces self-similar traffic.

individual packet source. Within the on period, every substream generates packets back to back. Packets generated by n substreams are aggregated (multiplexed) on a single line such that at least a 160-bit interval (96-bit interframe gap and 64-bit preamble) remains between any two adjacent packets. This multiplexed stream of packet represents a single traffic source.

References

[CMT98]   K. Claffy, G. Miller, and K. Thompson, "The nature of the beast: Recent traffic measurements from an Internet backbone," in *Proceedings INET `98,* Geneva, Switzerland, July 1998. Available at http://www.isoc.org/inet98/proceedings/6g/6g_3.htm.

[LT+94]   W. Leland, M. Taqqu, W. Willinger, and D. Wilson, "On the self-similar nature of Ethernet traffic (extended version)," *IEEE/ACM Transactions on Networking,* vol. 2, no. 1, pp. 1–15, February 1994.

[SG01]   D. Sala and A. Gummalla, "PON functional requirements: Services and performance," presented at IEEE 802.3ah meeting in Portland, OR, July 2001. Available at http://grouper.ieee.org/groups/802/3/efm/public/jul01/presentations/sala_1_0701.pdf.

[TWS97]   M. S. Taqqu, W. Willinger, and R. Sherman, "Proof of a fundamental result in self-similar traffic modeling," *ACM/SIGCOMM Computer Communication Review,* vol. 27, pp. 5–23, 1997.

[WT+95]   W. Willinger, M. Taqqu, R. Sherman, and D. Wilson, "Self-similarity through high-variability: Statistical analysis of Ethernet LAN traffic at the source level," in *Proc. ACM SIGCOMM '95*, pp. 100–113, Cambridge, MA, August 1995.

Bibliography

[802.1D]   IEEE Standard for Information Technology — Telecommunications and information exchange between systems — Local and metropolitan area networks — Common specifications. *Part 3: Media Access Control (MAC) Bridges*, ANSI/IEEE Standard 802.1D, 1998 edition. Available at http://standards.ieee.org/getieee802/download/802.1D-1998.pdf.

[802.3]   IEEE Standard for Information Technology — Telecommunications and information exchange between systems — Local and metropolitan area networks–Specific Requirements. *Part 3: Carrier Sense Multiple Access with Collision Detection (CSMA/CD) Access Method and Physical Layer Specification*, ANSI/IEEE Standard 802.3-2002, 2002 edition. Available at http://standards.ieee.org/getieee802/download/ 802.3-2002.pdf.

[Ada97]   A. Adas, "Traffic models in broadband networks," *IEEE Communications Magazine*, vol. 35, no. 7, pp. 82–89, July 1997.

[ANS01]   RHK Telecommunication Industry Analysis, *Access Network Systems: North America—Optical Access. DLC and PON Technology and Market Report,* Report RHK-RPT-0548, San Francisco, June 2001.

[BB01]   *Broadband 2001. A Comprehensive Analysis of Demand, Supply, Economics, and Industry Dynamics in the U.S. Broadband Market,* JP Morgan Securities, Inc., New York, April 2001.

[BMP94]   L. Brakmo, S. O'Malley, and L. Peterson, "TCP Vegas: New techniques for congestion detection and avoidance," *Proceedings of the SIGCOMM `94 Symposium*, pp. 24–35, August 1994.

[BZ96]   J. C. R. Bennett and H. Zhang, "WF^{2}Q: Worst-case fair weighted fair queueing," in *Proceedings of INFOCOM `96*, pp. 120–128, San Francisco, March 1996.

[BZ97]   R. Bennett and H. Zhang, "Hierarchical packet fair queuing algorithms," *IEEE/ACM Transactions on Networking*, vol. 5, no. 5, pp. 675–689, October 1997.

[CB96]   M. Crovella and A. Bestavros, "Self-similarity in World Wide Web traffic: Evidence and possible causes," in *Proceedings of ACM SIGMETRICS International Conference on Measurement and Modeling of Computer Systems,* May 1996.

[CC81]   G. C. Clark and J. B. Cain, *Error-Correction Coding for Digital Communications*, Kluwer Academic Press/Plenum Publishers, 1981.

[CC96]   R. L. Carter and M. E. Crovella, *Measuring Bottleneck Link Speed in Packet-Switched Networks,* TR-96-006, Boston University Computer Science Department, Mar. 15, 1996.

[CFL99]   W. Circiora, J. Farmer, and D. Large, *Modern Cable Television Technology: Video, Voice, and Data*, Morgan Kaufmann Publishers, 1999.

[Cho04]   Su-il Choi, "Multicasting in EPON," presented at IEEE 802.3 plenary meeting in Orlando, FL, March 2004. Presentation is available at http://www.ieee802.org/3/efm/public/comments/d3_1/pdfs/choi_p2mp_1_0304.pdf.

[Cla00]   S. Clavenna, "Metro optical Ethernet," *Lightreading* (www.lightreading.com), November 2000.

[CMT98]   K. Claffy, G. Miller, and K. Thompson, "The nature of the beast: Recent traffic measurements from an Internet backbone," in *Proceedings INET `98*, Geneva, Switzerland, July 1998. Available at http://www.isoc.org/inet98/proceedings/6g/6g_3.htm.

[CO01]   K. G. Coffman and A. M. Odlyzko, "Internet growth: Is there a "Moore's Law" for data traffic?" *Handbook of Massive Data Sets*, J. Abello, P. M. Pardalos, and M. G. C. Resende, eds., Kluwer, 2001.

[DKS90]   A. Demers, S. Keshav, and S. Shenker, "Analysis and simulation of a fair queueing algorithm," *Journal of Internetworking Research and Experience*, pp. 3–26, October 1990.

[DL86]   J. Diagle and J. Langford, "Models for analysis of packet voice communications systems," *IEEE Journal on Selected Areas of Communications*, vol. SAC-4, no. 6, pp. 847–855, September 1986.

[DSL04]   "2003 Global DSL subscriber chart (March 2, 2004)," *DSL Forum*, March 2004. Available at http://www.dslforum.org/PressRoom/2003_GlobalDSLChart_3.2.2004.pdf.

[Dwo01]   M. Dworkin, *Recommendation for Block Cipher Modes of Operation—Methods and Techniques,* National Institute of Standards and Technology, December 2001.

[EFM01]   IEEE 802.3 EFM Study Group, "Ethernet PON (EPON) and the PAR + 5 criteria," IEEE 802 interim meeting, St. Louis, MO, May 2001. Presentation is available at http://www.ieee802.org/3/efm/public/may01/pesavento_1_0501.pdf.

[FIPS197]   Federal Information Processing Standard 197, *Advanced Encryption Standard*, National Institute of Standards and Technology, U.S. Department of Commerce, Nov. 26, 2001.

[G.114]   ITU-T Recommendation G.114, *One-Way Transmission Time*, in Series G: Transmission Systems and Media, Digital Systems and Networks, Telecommunication Standardization Sector of ITU, May 2000.

[G.7041]   ITU-T Recommendation G.7041, *Generic Framing Procedure*, in Series G: Transmission Systems and Media, Digital Systems and Networks, Telecommunication Standardization Sector of ITU, October 2001.

[G.975]   ITU-T Recommendation G.975, *Forward Error Correction for Submarine Systems*, in Series G: Transmission Systems and Media, Digital Systems and Networks, Telecommunication Standardization Sector of ITU, October 2000.

[G984.3]   ITU-T Recommendation G.984.3, *Transmission Convergence Layer for Gigabit Passive Optical Networks*, in Series G: Transmission Systems and Media, Digital Systems and Networks, Telecommunication Standardization Sector of ITU, October 2003.

[Gol94]   S. J. Golestani, "A self-clocked fair queuing scheme for broadband application," *Proceedings of IEEE INFOCOM '94*, pp. 636–646, Toronto, Canada, June 1994.

[Gum+01]   A. Gummalla et al., "Multi-Point Control Protocol: Common framework," IEEE 802.3ah task force meeting, Austin, TX, November 2001. Available at http://www.ieee802.org/3/efm/public/nov01/kramer_1_1101.pdf.

[GVC97]   P. Goyal, H. M. Vin, and H. Cheng, "Start-time fair queuing: A scheduling algorithm for integrated services packet switching networks," *IEEE/ACM Transactions on Networking*, vol. 5, no. 5, pp. 690–704, October 1997.

[GW94]   M. W. Garrett and W. Willinger, "Analysis, modeling and generation of self-similar VBR video traffic," *Proceedings of ACM Sigcomm `94*, London, pp. 269–280, September 1994.

[Har00]   S. Hardy, "Verizon staffers find fiber-to-the-home cheaper than copper," *Lightwave*, vol. 17, no. 134, p. 1, December 2000.

[HOR87]   J. L. Hammond and P. J. P. O'Reilly, *Performance Analysis of Local Computer Networks*, Addison-Wesley, 1987.

[HPN02]   O-P. Hiironen, A. Pietiläinen, and A. Nylund, "Privacy in EPON," presented at IEEE 802.3ah meeting in Edinburgh, UK, May 2002. Available at http://www.ieee802.org/3/efm/public/may02/hiironen_1_0502.pdf.

[KKK90]   C. Kalmanek, H. Kanakia, and S. Keshav, "Rate controlled servers for very high-speed networks," *Proceedings of IEEE Globecom*, pp. 1264–1273, San Diego, December 1990.

[KM03]   G. Kramer and A. Maislos, "Laser control problem statement," IEEE 802.3ah task force meeting, Portonovo, Italy, September

2003. Available at http://www.ieee802.org/3/efm/public/sep03/
p2mp/kramer_1_0903.pdf.

[KMP02]   G. Kramer, B. Mukherjee, and G. Pesavento, "Interleaved
polling with adaptive cycle time (IPACT): A dynamic bandwidth
distribution scheme in an optical access network," *Photonic Network Communications*, vol. 4, no. 1, pp. 89–107, January 2002.

[KMS+03]   G. Kramer, B. Mukherjee, N. Singhal, A. Banerjee, S. Dixit, and
Y. Ye, *Fair Queuing with Service Envelopes (FQSE): A Cousin-
Fair Hierarchical Scheduler and Its Application in Ethernet
PON*, Technical Report CSE-2003-6, Department of Computer
Science, University of California, Davis, March 2003.

[LT+94]   W. Leland, M. Taqqu, W. Willinger, and D. Wilson, "On the self-
similar nature of Ethernet traffic (extended version)," *IEEE/
ACM Transactions on Networking*, vol. 2, no. 1, pp. 1–15, February 1994.

[Lun99]   B. Lung, "PON architecture 'futureproofs' FTTH," *Lightwave*,
vol. 16, no. 10, pp. 104–107, September 1999.

[Muk97]   B. Mukherjee, *Optical Communication Networks*, McGraw-Hill,
1997.

[Odl04]   A. Odlyzko, "Crisis and mythology in the telecom world,"
CommsDaySummit, Sydney, Australia, February 16, 2004.

[OSI94]   ISO/IEC 7498-1: 1994, *Information Technology—Open Systems
Interconnection—Basic Reference Model: The Basic Model*.

[PF95]   V. Paxson and S. Floyd, "Wide-area traffic: The failure of Poisson modeling," *IEEE/ACM Transactions on Networking*, vol. 3,
no. 3, pp. 226–244, June 1995.

[PG93]   A. K. Parekh and R. G. Gallager, "A generalized processor sharing approach to flow control in integrated services networks:
The single node case," *IEEE/ACM Transactions on Networking*, vol. 1, pp. 344–357, June 1993.

[PK99]   G. Pesavento and M. Kelsey, "PONs for the broadband local
loop," *Lightwave*, vol. 16, no. 10, pp. 68–74, September 1999.

[PW00]   K. Park and W. Willinger, "Self-similar network traffic: An
overview," in K. Park and W. Willinger, eds., *Self-Similar Network Traffic and Performance Evaluation*, Wiley Interscience,
2000.

[Ree03]   D. Reed, *"Copper Evolution,"* Federal Communications Commission, Technological Advisory Council III, Washington, April
2003. Available at http://www.fcc.gov/oet/tac/TAC_III_04_-
17_03/Copper_Evolution.ppt.

[RGB96]   J. L. Rexford, A. G. Greenberg, and F. G. Bonomi, "Hardware-
efficient fair queueing architectures for high-speed networks,"
Proceedings of INFOCOM, pp. 638–646, 1996.

[RS98]   R. Ramaswami and K. N. Sivarajan, *Optical Networks, A Practical Perspective*, Morgan Kaufmann, 1998.

[SG01]   D. Sala and A. Gummalla, "PON functional requirements: Services and Performance," presented at IEEE 802.3ah meeting in Portland, OR, July 2001. Available at http://www.ieee802.org/3/efm/public/jul01/presentations/sala_1_0701.pdf.

[SH+88]   J. R. Stern, C. E. Hoppitt, D. B. Payne, M. H. Reeve, and K. A. Oakley, "TPON—A passive optical network for telephony," *Fourteenth European Conference on Optical Communication (ECOC'88)*, vol. 1, pp. 203–206, Brighton, UK, September 1988.

[Ste94]   W. R. Stevens, *TCP/IP Illustrated*, vol. 1, Addison-Wesley, 1994.

[SV96]   M. Shreedhar and G. Varghese, "Efficient fair queuing using deficit round-robin," *IEEE/ACM Transactions on Networking*, vol. 4, no. 3, pp. 375–385, June 1996.

[TWS97]   M. S. Taqqu, W. Willinger, and R. Sherman, "Proof of a fundamental result in self-similar traffic modeling," *ACM/SIGCOMM Computer Communication Review*, vol. 27, pp. 5–23, 1997.

[Wic95]   S. B. Wicker, *Error Control Systems for Digital Communication and Storage*, Prentice-Hall, Inc., 1995.

[WT+95]   W. Willinger, M. Taqqu, R. Sherman, and D. Wilson, "Self-similarity through high-variability: Statistical analysis of Ethernet LAN traffic at the source level," in *Proc. ACM SIGCOMM '95*, pp. 100–113, Cambridge, MA, August 1995.

[WTE96]   W. Willinger, M. S. Taqqu, and A. Erramilli, "A bibliographical guide to self-similar traffic and performance modeling for modern high-speed networks," in *Stochastic Networks*, F. P. Kelly, S. Zachary, and I. Ziedins, eds., Oxford University Press, 1996, pp. 339–366.

Index

ABOUT THE AUTHOR

Glen Kramer, Ph.D., is a system architect for Teknovus, Inc., and an associate researcher at the University of California, Davis. He is a member of IEEE Standards Association and past editor of the EPON protocol clause in the IEEE 802.3ah "Ethernet in the First Mile" standard. One of the lead authors of the multi-point control protocol that became the baseline for the EPON architecture, Dr. Kramer has done extensive research on issues of EPON scheduling, fairness, and open access. He is the founder of EPON forum and teaches EPON tutorials and workshops at conferences around the world.

www.ingramcontent.com/pod-product-compliance
Lightning Source LLC
Chambersburg PA
CBHW050454190326
41458CB00005B/1280